**Springer Proceedings in Mathematics & Statistics**

Volume 305

**Springer Proceedings in Mathematics & Statistics**

This book series features volumes composed of selected contributions from workshops and conferences in all areas of current research in mathematics and statistics, including operation research and optimization. In addition to an overall evaluation of the interest, scientific quality, and timeliness of each proposal at the hands of the publisher, individual contributions are all refereed to the high quality standards of leading journals in the field. Thus, this series provides the research community with well-edited, authoritative reports on developments in the most exciting areas of mathematical and statistical research today.

More information about this series at http://www.springer.com/series/10533

Gareth A. Jones · Ilia Ponomarenko ·
Jozef Širáň
Editors

# Isomorphisms, Symmetry and Computations in Algebraic Graph Theory

Pilsen, Czech Republic, October 3–7, 2016

Springer

*Editors*
Gareth A. Jones
Department of Mathematical Sciences
University of Southampton
Southampton, UK

Ilia Ponomarenko
Steklov Institute of Mathematics
Russian Academy of Sciences
St. Petersburg, Russia

Jozef Širáň
Department of Mathematics and Statistics
The Open University
Milton Keynes, UK

ISSN 2194-1009 ISSN 2194-1017 (electronic)
Springer Proceedings in Mathematics & Statistics
ISBN 978-3-030-32807-8 ISBN 978-3-030-32808-5 (eBook)
https://doi.org/10.1007/978-3-030-32808-5

Mathematics Subject Classification (2010): 05E18, 05E30, 05C10, 05C50, 05C60, 05C85, 05A15, 05B03



This Springer imprint is published by the registered company Springer Nature Switzerland AG
The registered company address is: Gewerbestrasse 11, 6330 Cham, Switzerland

# Preface

The October 3–7, 2016 Workshop on Algebraic Graph Theory took place in the historic town of Pilsen, Czech Republic, organized jointly by the Faculty of Applied Sciences of the University of West Bohemia in Pilsen, the Pilsen Branch of the Center of Excellence—Institute for Theoretical Computer Science, and the Union of Czech Mathematicians and Physicists. The Program Committee was formed by Roman Nedela (University of West Bohemia, Pilsen, Czech Republic) and Mikhail Klin (Ben-Gurion University of the Negev, Beer-Sheva, Israel), and the Organizing Committee included Zdeněk Ryjáček, Přemysl Holub, Edita Rollová (University of West Bohemia, Pilsen, Czech Republic), and Ján Karabáš with Štefan Gyürki (Matej Bel University, Banská Bystrica, Slovak Republic).

The aim of the Workshop was to give the worldwide community of researchers in algebraic combinatorics the opportunity to gather together, exchange information, and present their advances and newest findings. The Workshop has not been part of any regular series and was brought about simply by rapid development and research progress in sub-disciplines such as coherent configurations and association schemes and the associated computational algebra, group actions on combinatorial structures, symmetric graph embeddings, and various other applications of algebra in the study of discrete objects.

The Workshop brought together a total of 33 researchers from all over the world, including a number of Ph.D. students. The list of invited plenary lecturers consisted of the following:

Sonia Balagopalan, Hebrew University, Jerusalem, Israel
Chris Godsil, University of Waterloo, Canada
Willem Haemers, Tilburg University, Netherlands
Leif K. Jørgensen, Aalborg University, Denmark
Gareth A. Jones, University of Southampton, UK
Aleksandar Jurišic, University of Ljubljana, Slovenia
Klavdija Kutnar, University of Primorska, Koper, Slovenia
Josef Lauri, University of Malta, Malta
Dragan Marušic, University of Primorska, Slovenia

Mikhail Muzychuk, Netanya Academic College, Israel
Christian Pech, Technical University of Dresden, Germany
Reinhard Pöschel, Technical University of Dresden, Germany
Ilia Ponomarenko, Russian Academy of Sciences, St. Petersburg, Russia
Sven Reichard, Technical University of Dresden, Germany
Gábor Somlai, Eötvös Loránd University, Budapest, Hungary
Jozef Širáň, Open University, Milton Keynes, UK
Qing Xiang, University of Delaware, USA
Matan Ziv-Av, Ben-Gurion University, Negev, Israel

The daily program consisted of 3–4 plenary lectures and 2–3 contributed paper presentations. Such a schedule left enough time and space for numerous informal discussions held in a friendly and collegial environment. The Wednesday conference trip took the participants to the world-famous Pilsen beer breweries.

This volume contains eight papers of varying length, all related to the plenary lectures given by some of the participants of the Workshop. Although representing only a small part of the wide-ranging program, the selection embodies the rich variety of mutually intertwined topics addressed in the presentations.

All participants are to be thanked for their contributions and for turning the Pilsen 2016 Workshop on Algebraic Graph Theory into a successful and memorable event.

Southampton, UK — Gareth A. Jones
St. Petersburg, Russia — Ilia Ponomarenko
Milton Keynes, UK — Jozef Širáň

# Acknowledgements

The organizers gratefully acknowledge support received from the Project LO1506 of the Czech Ministry of Education, Youth and Sports, allocated to the Pilsen Branch of the Center of Excellence—Institute for Theoretical Computer Science at the Faculty of Applied Sciences of the University of West Bohemia in Pilsen, which enabled to invite and cover a large number of plenary speakers.

Further, thanks for supporting the Workshop go to the Union of Czech Mathematicians and Physicists, the Faculty of Applied Sciences of the University of West Bohemia in Pilsen for additional technical and organizational aid, and Pilsen City Council for providing representative premises.

We also wish to thank all our anonymous referees for their cooperation, without which publication of this Volume would not be possible.

# Contents

# Orientably-Regular Maps on Twisted Linear Fractional Groups

**Grahame Erskine, Katarína Hriňáková and Jozef Širáň**

**Abstract** We present an enumeration of orientably-regular maps with automorphism group isomorphic to the twisted linear fractional group $M(q^2)$ for any odd prime power $q$.

**Keywords** Map · Orientably-regular map · Automorphism group · Twisted linear fractional group

## 1 Introduction

A (finite) orientably-regular map $\mathcal{M}$ is a cellular embedding of a connected graph in a compact, oriented surface, such that the group $\mathrm{Aut}^+(\mathcal{M})$ of all orientation-preserving automorphisms of the embedding is transitive, and hence regular, on arcs of the embedded graph. In a regular map, all vertices have the same valency, say $k$, and all faces are bounded by closed walks of the same length, say, $\ell$; the map is then said to be of type $(k, \ell)$. The group $\mathcal{A} = \mathrm{Aut}^+(\mathcal{M})$ is generated by two elements $x$ and $y$ of order $k$ and $\ell$ such that $x$ acts as a clockwise rotation of $\mathcal{M}$ about a vertex by $2\pi/k$ and $y$ acts as a clockwise rotation by $2\pi/\ell$ about the centre of a face incident with the vertex. The product $xy$ is then a rotation of $\mathcal{M}$ about the centre of an edge that is incident to both the vertex and the face.

Orientably-regular maps can be viewed as maps having the 'highest level' of orientation-preserving symmetry among general maps (i.e. cellular embeddings of graphs). Regularity of $\mathcal{A}$ on the arc set of the embedded graph enables one to identify the map $\mathcal{M}$ with the triple $(\mathcal{A}, x, y)$ in such a way that arcs, edges, vertices and faces

G. Erskine
Open University, Milton Keynes, UK

K. Hriňáková
Slovak University of Technology, Bratislava, Slovakia

J. Širáň (✉)
Open University, UK, Slovak University of Technology, Bratislava, Slovakia
e-mail: j.siran@open.ac.uk

 
G. A. Jones et al. (eds.), *Isomorphisms, Symmetry and Computations in Algebraic Graph Theory*, Springer Proceedings in Mathematics & Statistics 305,
https://doi.org/10.1007/978-3-030-32808-5_1

correspond to (say, left) cosets of the trivial group (i.e. to elements of $\mathcal{A}$) and of the subgroups $\langle xy\rangle$, $\langle x\rangle$ and $\langle y\rangle$ of $\mathcal{A}$. Incidence between arcs, edges, vertices and faces is given by non-empty intersection of the corresponding cosets, and the action of $\mathcal{A}$ on the cosets is simply given by left multiplication.

It follows that orientably-regular maps are, up to isomorphism, in a one-to-one correspondence with equivalence classes of triples $(G, x, y)$, where $G$ is a finite group admitting a presentation of the form $G = \langle x, y;\ x^k = y^\ell = (xy)^2 = \cdots = 1\rangle$, with two triples $(G_1, x_1, y_1)$ and $(G_2, x_2, y_2)$ being equivalent if there is a group isomorphism $G_1 \to G_2$ taking $x_1$ onto $x_2$ and $y_1$ onto $y_2$. This way, investigation of orientably-regular maps can be reduced to purely group-theoretic considerations. The corresponding algebraic theory has been developed in depth in the influential paper [12].

The study of orientably-regular maps has rich history and, save Platonic solids and Kepler's polyhedra, has roots in the late nineteenth century; for historical information and surveys, see e.g. [11, 12, 18, 20]. For an excellent introduction into fascinating connections between orientably-regular maps, Dyck's triangle groups, Riemann surfaces and Galois groups, we recommend [11]. In particular, classification of orientably-regular maps has implications in classification of Riemann surfaces, cf. [11, 12].

Since the concept of an orientably-regular map includes the underlying graph, the carrier surface and the supporting automorphism group, classification attempts for such maps in most cases follow one of these three directions. A number of influential results have been obtained in classification of orientably-regular maps in the first two directions; we refer to [20] for the most recent survey. In this paper, we focus on the third direction, that is, classification of orientably-regular maps by their automorphism groups, in which results are much less abundant.

Leaving the trivial case of Abelian groups aside, classification of orientably-regular maps with a given automorphism group has been completed for the following types of groups, ranked by complexity of their structure:

- groups of nilpotency class two [17];
- products of two cyclic groups one of which acts regularly on vertices of the map [5];
- groups with cyclic odd-order and dihedral even-order Sylow subgroups [4];
- $\mathrm{PSL}(2, q)$ and $\mathrm{PGL}(2, q)$, where $q$ is an arbitrary prime power [3, 16, 19];
- Ree groups ${}^2G_2(3^n)$ for odd $n > 1$, restricted to maps of type $(k, \ell)$ for $\ell = 3$ and $k = 7, 9$ and all prime $k \equiv 11 \bmod 12$ [9];
- Suzuki group $\mathrm{Sz}(2^n)$ for odd $n > 1$, restricted to maps of type $(5, 4)$ [10].

By this list, the only family of *simple* groups for which a classification of the corresponding orientably-regular maps is known are the groups $\mathrm{PSL}(2, q)$ for any prime power $q > 3$. Classification of orientably-regular maps with automorphism group $\mathrm{PGL}(2, q)$, the obvious degree-two extension of $\mathrm{PSL}(2, q)$, can be extracted from the corresponding classification for $\mathrm{PSL}(2, q^2)$ through the well-understood inclusion $\mathrm{PGL}(2, q) < \mathrm{PSL}(2, q^2)$, cf. [3, 19].

For odd $q$, however, the simple group $\mathrm{PSL}(2, q^2)$ admits another interesting extension of degree two, namely, the group $M(q^2)$, also known as a *twisted* linear fractional group. By a classical result of Zassenhaus [22], the groups $\mathrm{PGL}(2, q)$ and $M(q^2)$ are the only finite sharply 3-transitive groups (of degree $q+1$ and $q^2+1$, respectively). This motivates the question of classification of orientably-regular maps with automorphism group isomorphic to $M(q^2)$.

In this paper, we present a complete enumeration of (isomorphism classes of) orientably-regular maps with automorphism group isomorphic to $M(q^2)$. The results are strikingly different from those for the groups $\mathrm{PGL}(2, q)$ in many ways. To give three examples, we note that (a) all the orientably-regular maps for $\mathrm{PGL}(2, q)$ are reflexible, while this is not the case for $M(q^2)$; (b) the groups $\mathrm{PGL}(2, q)$ are also automorphism groups of non-orientable regular maps while the groups $M(q^2)$ are not; and (c) for any even $k, \ell \geq 4$ not both equal to 4, there are orientably-regular maps of type $(k, \ell)$ with automorphism group $\mathrm{PGL}(2, q)$ for infinitely many values of $q$, while for infinitely many such pairs $(k, \ell)$, there are no orientably-regular maps for $M(q^2)$ of that type for any $q$.

By our outline and the algebraic theory of [12], enumeration of orientably-regular maps with a given automorphism group $G$ reduces to enumeration of all triples $(G, x, y)$ with $G = \langle x, y;\ x^k = y^\ell = (xy)^2 = \cdots = 1\rangle$ up to conjugation by elements of $\mathrm{Aut}(G)$, that is, by considering triples $(G, x, y)$ and $(G, x', y')$ equivalent if there is an automorphism of $G$ taking $(x, y)$ onto $(x', y')$. We do this systematically for the twisted linear groups $G = M(q^2)$. In Sects. 2 and 3, we introduce the group $M(q^2)$ and study its subgroups. Sections 4, 5 and 6 deal with identifying 'canonical' forms of elements of $G$ and study their conjugacy in depth. In Sects. 7, 8 and 9, we develop arguments for counting 'canonical' pairs of elements of $G$. All the auxiliary facts are then processed in Sect. 10 to produce our main result:

**Theorem** *Let $q = p^f$ be an odd prime power, with $f = 2^\alpha o$ where $o$ is odd. The number of orientably-regular maps $\mathcal{M}$ with $\mathrm{Aut}^+(\mathcal{M}) \cong M(q^2)$ is, up to isomorphism, equal to*

$$\frac{1}{f}\sum_{d|o} \mu(o/d)h(2^\alpha d)\ ,$$

*where $h(x) = (p^{2x}-1)(p^{2x}-2)/8$ and $\mu$ is the Möbius function.*

We note that this result may be interpreted as counting generating pairs $(x, y)$ of $G = M(q^2)$ such that $(xy)^2 = 1$, up to conjugacy in $\mathrm{Aut}(G)$, which may be of independent interest to specialists in group theory. The final Sect. 12 contains related results and remarks.

## 2 The Twisted Linear Groups $M(q^2)$

For a finite field $F$, let $S(F)$ and $N(F)$ be the set of non-zero squares and non-squares of $F$. The general linear group $\mathrm{GL}(2, F)$ is the group of all non-singular $2 \times 2$ matrices with entries in $F$; restriction to matrices with determinant 1 gives the special linear group $\mathrm{SL}(2, F)$. The groups $\mathrm{PGL}(2, F)$ and $\mathrm{PSL}(2, F)$, the quotients of $\mathrm{GL}(2, F)$ and $\mathrm{SL}(2, F)$ by the corresponding centres, are known as the linear fractional groups. They can equivalently be described as groups of all transformations $z \mapsto (az + b)/(cz + d)$ of the set $F \cup \{\infty\}$ (with the obvious rules for calculations with $\infty$), with $ad - bc \neq 0$ and $ad - bc \in S(F)$ for $\mathrm{PGL}(2, F)$ and $\mathrm{PSL}(2, F)$, respectively. The group $\mathrm{PSL}(2, F)$ is an index 2 subgroup of $\mathrm{PGL}(2, F)$ unless $F$ has characteristic 2, in which case the two groups are the same.

Suppose now that $F$ admits an automorphism $\sigma$ of order 2, which happens if and only if $|F| = q^2$ for some prime power $q$, and $\sigma$ is then given by $x \mapsto x^q$ for every $x \in F$. If, in addition, $q$ is odd, then one may 'twist' the transformations described above by considering the permutations of $F \cup \{\infty\}$ defined by $z \mapsto (az + b)/(zc + d)$ if $ad - bc \in S(F)$ and $z \mapsto (az^\sigma + b)/(cz^\sigma + d)$ if $ad - bc \in N(F)$. These transformations form a group under composition, denoted $M(F)$ or $M(q^2)$, and called the *twisted fractional linear group*. Observe that $\mathrm{PSL}(2, F)$ is a subgroup of $M(F)$ of index two, again. By a well-known result due to Zassenhaus [22], the groups $\mathrm{PGL}(2, F)$ for an arbitrary finite field $F$, and $M(F)$ for fields of order $q^2$ for an odd prime power $q$, are precisely the finite, sharply 3-transitive permutation groups (on the set $F \cup \{\infty\}$ in both cases).

In this paper, we will focus on the twisted fractional linear groups, with the goal to classify the orientably-regular maps they support. For our purposes, however, it will be useful to work with a different representation of these groups. From this point on, let $F = \mathrm{GF}(q^2)$ for some odd prime power $q$ and let $F_0 \cong \mathrm{GF}(q)$ be its unique subfield of order $q$; let $F^*$ and $F_0^*$ be the corresponding multiplicative groups. Further, let $\sigma$ be the unique automorphism of $F$ of order 2; we have $x^\sigma = x^q$ for any $x \in F$, and $x^\sigma = x$ if and only if $x \in F_0$. If $A \in \mathrm{GL}(2, F)$, by $A^\sigma$ we denote the matrix in $\mathrm{GL}(2, F)$ obtained by applying $\sigma$ to every entry of $A$.

Let $J = \mathrm{GL}(2, F) \rtimes Z_2$, where multiplication in the semidirect product is defined by $(A, i)(B, j) = (AB^{\sigma^i}, i + j)$; equivalently, $J$ is an extension of $\mathrm{GL}(2, F)$ by the automorphism $\sigma$. To introduce a 'twisted' subgroup of $J$, for every $A \in \mathrm{GL}(2, F)$, we define $\iota_A \in Z_2 = \{0, 1\}$ by letting $\iota_A = 0$ if $\det(A) \in S(F)$ and $\iota_A = 1$ if $\det(A) \in N(F)$. We now let $K = \{(A, \iota_A);\ A \in \mathrm{GL}(2, F)\}$; multiplication in $K$ is, of course, given by $(A, \iota_A)(B, \iota_B) = (AB^{\sigma^{\iota_A}}, \iota_A + \iota_B)$ for any $A, B \in \mathrm{GL}(2, F)$. The group $K$ and its quotient groups will be of principal importance in what follows.

Let $K_0 = \{(A, 0);\ A \in \mathrm{GL}(2, F),\ \iota_A = 0\}$ be the subgroup of index 2 of $K$. The centre $L$ of $K_0$ consists of elements of the form $(D, 0)$, where $D \in \mathrm{GL}(F)$ is a scalar matrix; obviously $L$ is also a normal subgroup of both $K$ and $J$. It can be checked that the factor group $G = K/L$ is isomorphic to $M(q^2)$, and since $K$ has index 2 in $J$, the group $G = M(q^2)$ is (isomorphic to) a subgroup of index 2 of $\overline{G} = J/L$. The group $\overline{G}$ can alternatively be described as $G\langle\sigma\rangle$, the split extension of $G$ by $\langle\sigma\rangle \simeq Z_2$.

Observe also that the factor group $G_0 = K_0/L$ is isomorphic to $\mathrm{PSL}(2, F)$, and if $q$ is a prime, the group $J/L$ is isomorphic to $\mathrm{P\Gamma L}(2, q^2)$.

Elements $(A, i)L$, that is, cosets $\{(\delta A, i);\ \delta \in F^*\}$, of the factor groups $G = K/L$ and $\overline{G} = J/L$ will throughout be denoted $[A, i]$; they will be called *untwisted* if $i = 0$ and *twisted* if $i = 1$.

For our final enumeration, it will be necessary to determine the automorphism group of $M(q^2)$. While the result appears to be 'obvious', we provide a simple proof based on a fact which may be folklore to group-theorists.

**Lemma 1** *Let $U$ be a characteristic subgroup of a group $\tilde{U}$ of index 2. Suppose that the centre of $U$ is trivial and every automorphism of $U$ extends to an automorphism of $\tilde{U}$. Then* $\mathrm{Aut}(U) \cong Aut(\tilde{U})$.

***Proof*** The assumption of $U$ being characteristic in $\tilde{U}$ implies that every $h \in \mathrm{Aut}(\tilde{U})$ restricts to an $h_U \in \mathrm{Aut}(U)$. Since each automorphism of $U$ extends to an automorphism of $\tilde{U}$, the assignment $h \mapsto h_U$ is a group epimorphism $\vartheta : \mathrm{Aut}(\tilde{U}) \to \mathrm{Aut}(U)$. Suppose that $h \in \mathrm{Aut}(\tilde{U})$ is in the kernel of $\vartheta$, so that $h_U$ is the identity mapping on $U$. For every $x \in U$ and every $y \in \tilde{U}\backslash U$, we have $y^{-1}xy \in U$ and hence $y^{-1}xy = h(y^{-1}xy) = h(y)^{-1}xh(y)$, which implies that $h(y)y^{-1}$ commutes with $x$ for all $x \in U$. Observe that $h(y)y^{-1} \in U$, since $U$ was assumed to have index 2 in $\tilde{U}$. By triviality of the centre of $U$, we have $h(y)y^{-1} = 1$ and as this is valid for all $y \in \tilde{U}\backslash U$, we conclude that $h$ is the identity on $\tilde{U}$. It follows that the kernel of $\vartheta$ is trivial and so $\mathrm{Aut}(U) \cong Aut(\tilde{U})$. □

We now apply Lemma 1 to $U = G_0 = \mathrm{PSL}(2, q^2)$ and $\tilde{U} = G = M(q^2)$ for $q = p^f$, where $p$ is an odd prime and $f$ a positive integer. Being a simple subgroup of $M(q^2)$, the group $G_0$ is characteristic (and of index two) in $G$. It is well known (see e.g. [8]) that $\mathrm{Aut}(G_0) \cong \mathrm{P\Gamma L}(2, q^2) \simeq \mathrm{PGL}(2, q^2) \rtimes Z_{2f}$, with an element $(C, \varphi) \in \mathrm{PGL}(2, q^2) \rtimes Z_{2f}$ acting on $G_0$ by $X \mapsto (C^{-1}XC)^{\varphi}$. Now, any $(C, \varphi)$ is easily seen to extend to $G$ by $[X, \iota_X] \mapsto ([C, 0]^{-1}[X, \iota_X][C, 0])^{\varphi}$. By Lemma 1, we now obtain:

**Proposition 1** *The automorphism group of $M(q^2)$ is isomorphic to* $\mathrm{P\Gamma L}(2, q^2)$. □

## 3 Twisted Subgroups of $M(q^2)$

Let $q = p^f$ for an odd prime $p$ and a positive integer $f$; these will be fixed throughout. In this section, we will focus on the *twisted* subgroups of $M(p^{2f})$, that is, those isomorphic to $M(p^{2e})$ for suitable $e \le f$. From now on, we will use the notation $F_m = \mathrm{GF}(p^m)$ for a Galois field of order $p^m$ for $m \le f$ but keep letting $F = \mathrm{GF}(p^{2f})$. We begin by identifying the possible values of $e$.

**Lemma 2** *A group $M(p^{2e})$ is isomorphic to a subgroup of $M(p^{2f})$ if and only if $e$ is a divisor of $f$ such that $f/e$ is odd.*

***Proof*** If $e$ divides $f$, then $p^{2f} - 1 = (p^{2e} - 1)Q$ for $Q = p^{2e(d-1)} + p^{2e(d-2)} + \cdots + p^{2e} + 1$ and $d = f/e$. Assume that $d$ is odd. If $\xi$ is a primitive element of $F$, then $\xi^Q$ is a primitive element of $F_{2e} < F$. Further, the restriction of the assignment $x \mapsto x^{p^f}$ for $x \in F$ onto $F_{2e}$ coincides with the mapping $x \mapsto x^{p^e}$ for $x \in F_{2e}$. To see this, it is clearly sufficient to consider the effect of taking the powers of $x = \xi^Q$, for which we have $x^{p^e} = x^{p^f}$ if and only if $Qp^e \equiv Qp^f \mod (p^{2f} - 1)$. The congruence is further equivalent to $p^e \equiv p^f \mod (p^{2e} - 1)$ and also to $p^{2e} - 1$ dividing $p^{f-e} - 1$, which is true if and only if $2e$ is a divisor of $f - e = (d - 1)e$, and the last condition is satisfied because $d$ is odd. This shows that the restriction of the automorphism $\sigma : x \mapsto x^{p^f}$ of $F$ onto $F_{2e}$ coincides with the automorphism $\sigma' : x \mapsto x^{p^e}$ of $F_{2e}$. By construction of the twisted linear groups introduced in Sect. 2, it is now clear that $M(p^{2f})$ contains a copy of $M(p^{2e})$.

To prove the reverse implication, assume that $H \cong M(p^{2e})$ is a subgroup of $G = M(p^{2f})$. From the order of $H$ dividing the order of $G$, one sees that $e$ divides $f$ and our goal is to show that $f/e$ is odd. We prove this by induction on $f/e$. The case when $f/e = 1$ is obvious and so we assume that $f > e$. Let $\tilde{H}$ be a maximal subgroup of $G$ containing $H$. From Theorem 1.5 of [8], it follows that $\tilde{H}$ is isomorphic to the normaliser of $\mathrm{PSL}(2, p^{2g})$ in $G$ for some positive divisor $g$ of $f$ such that $f/g$ is an odd prime. As before, let $G_0$ be the (normal) subgroup of index 2 in $G$ isomorphic to $\mathrm{PSL}(2, p^{2f})$. The group $\tilde{H} \cap G_0$ is contained in a maximal subgroup of $G_0$ containing $\mathrm{PSL}(2, p^{2g})$. However, the group $\mathrm{PSL}(2, p^{2g})$ itself is a maximal subgroup of $G$ as $f/g$ is odd, cf. [8] again. It follows that $\tilde{H} \cap G_0$ must be isomorphic to $\mathrm{PSL}(2, p^{2g})$, and therefore $\tilde{H}$ must be isomorphic to $M(p^{2g})$. But $\tilde{H}$ contains also $M(p^{2e}) \cong H$ and therefore $e$ divides $g$. As $g/e < f/e$, we may apply our induction hypothesis by which $g/e$ is odd. Since $f/g$ was an odd prime, we conclude that $f/e$ is odd as well, completing the induction step. □

If $f/e$ is odd, a particularly important copy of $M(p^{2e})$ in $M(p^{2f})$ is formed by all the pairs $[X, \iota_X]$ with $X \in \mathrm{GL}(2, p^{2e})$ such that all entries of $X$ lie in the subfield $F_{2e}$ of $F$; this copy will be called *canonical*. The copy of $\mathrm{PSL}(2, p^{2e})$ in $M(p^{2f})$ formed by all the pairs $[X, 0]$ with $X \in \mathrm{SL}(2, F_{2e})$ will be called *canonical* as well. We now prove a useful auxiliary result on canonical subgroups.

**Proposition 2** *Let $f/e$ be an odd integer and let $H \cong M(p^{2e})$ be a subgroup of $G = M(p^{2f})$ such that $H$ contains the canonical copy of* $\mathrm{PSL}(2, p^{2e})$*. Then, $H$ is equal to the canonical copy of $M(p^{2e})$ in $G$.*

***Proof*** Let $H$ be a copy of $M(p^{2e})$ in $G$ such that $H_0 = H \cap \mathrm{PSL}(2, p^{2f})$ is equal to the canonical copy of $\mathrm{PSL}(2, p^{2e})$ in $G$. Obviously, $H_0$ is a normal subgroup of $H$ of index two. Let $[A, 1]$ be an element of $H \backslash H_0$, where $A$ is the $2 \times 2$ matrix with rows $(a, b)$ and $(c, d)$ for some $a, b, c, d \in F$ with $\delta = ad - bc \in N(F)$. We may assume that the entry $c$ in the lower left corner of $A$ is non-zero. Indeed, if $c = 0$ and $b \neq 0$, letting $D$ be an off-diagonal matrix with entries $-1$ and $1$, we may replace $[A, 1]$ with the product $[D, 0][A, 1] \in H \backslash H_0$, and if $A$ is a diagonal matrix, we may replace $[A, 1]$ with the product $[D', 0][A, 1] \in H \backslash H_0$ for a matrix $D'$ with rows $(1, 0)$ and $(1, 1)$. Then, since we are working with projective groups, we may assume that $c = 1$, so that $\delta = ad - b$.

By our assumption, the group $H_0$ also contains the element $[C, 0]$ with $C$ having rows $(1, 1)$ and $(0, 1)$. Normality of $H_0$ in $H$ implies that $[A, 1][C, 0][A, 1]^{-1} = [ACA^{-1}, 0] \in H_0$ and also $[A, 1]^{-1}[C, 0][A, 1] = [(A^\sigma)^{-1}CA^\sigma, 0] \in H_0$. Evaluating the products, we obtain

$$\varepsilon ACA^{-1} = \begin{pmatrix} a & b \\ 1 & d \end{pmatrix}\begin{pmatrix} 1 & 1 \\ 0 & 1 \end{pmatrix}\begin{pmatrix} d & -b \\ -1 & a \end{pmatrix} = \begin{pmatrix} \delta - a & a^2 \\ -1 & \delta + a \end{pmatrix}, \text{ and}$$

$$\varepsilon'(A^\sigma)^{-1}CA^\sigma = \begin{pmatrix} d^\sigma & -b^\sigma \\ -1 & a^\sigma \end{pmatrix}\begin{pmatrix} 1 & 1 \\ 0 & 1 \end{pmatrix}\begin{pmatrix} a^\sigma & b^\sigma \\ 1 & d^\sigma \end{pmatrix} = \begin{pmatrix} \delta^\sigma + d^\sigma & (d^\sigma)^2 \\ -1 & \delta^\sigma - d^\sigma \end{pmatrix}$$

for some $\varepsilon, \varepsilon' \in F^*$. Since $H_0$ is assumed to be equal to the canonical copy $\mathrm{PSL}(2, p^{2e})$ in $G$, all the remaining entries of the two matrices on the right-hand sides above must lie in $F_{2e}$. This readily implies that both $a, \delta, d^\sigma \in F_{2e}$, and since $F_{2e}$ is setwise preserved by $\sigma$, we also have $d \in F_{2e}$ and so $b = ad - \delta \in F_{2e}$ as well. We conclude that $A \in \mathrm{GL}(2, p^{2e})$ and hence the subgroup $H$ generated by $[A, 1]$ and $H_0$ is identical with the canonical copy of $M(p^{2e})$ in $G$. □

As a consequence, we prove that all twisted subgroups of $G$ are conjugate. Recall that $G_0$ denotes the (unique) subgroup of $G$ isomorphic to $\mathrm{PSL}(2, p^{2f})$.

**Proposition 3** *If $f/e$ is an odd integer, then all subgroups of $G = M(p^{2f})$ isomorphic to $M(p^{2e})$ are conjugate in $G_0$.*

***Proof*** Let $H \cong M(p^{2e})$ be a subgroup of $G$; letting $H_0 = H \cap G$ we obviously have $H_0 \cong \mathrm{PSL}(2, p^{2e})$. The classification of subgroups of $\mathrm{PSL}(2, p^{2f})$, nicely displayed in [15], tells us that in the case when $f/e$ is odd, all subgroups of $G_0$ isomorphic to $\mathrm{PSL}(2, p^{2e})$ are conjugate in $G_0$. It follows that there is an inner automorphism $\pi$ of $G_0$ such that $\pi(H_0)$ is equal to the canonical copy of $\mathrm{PSL}(2, p^{2e})$ in $G_0$. The group $\pi(H)$ is then isomorphic to $M(p^{2e})$ and contains the canonical copy of $\mathrm{PSL}(2, p^{2e})$. By Proposition 2, however, the group $\pi(H)$ must be *equal* to the canonical copy of $M(p^{2e})$ in $G$. In particular, all subgroups of $G$ isomorphic to $M(p^{2e})$ are conjugate in $G_0$. □

We conclude with a sufficient condition for a subgroup of $G$ to be twisted; this result will be of key importance later. In order to state it, we will say that a subgroup $H$ of $G$ *stabilises a point* if, in the natural action of $G$ on the set $F \cup \{\infty\}$ via linear fractional mappings from Sect. 2, there exists a point in $F \cup \{\infty\}$ fixed by all linear fractional mappings corresponding to elements of $H$. Also, for any positive divisor $g$ of $2f$, let $G_g$ be the canonical copy of $\mathrm{PSL}(2, p^g)$ in the group $G = M(p^{2f})$. Moreover, if $g$ is even, we let $G_g^*$ denote the copy of $\mathrm{PGL}(2, p^{g/2})$ in $G_g$ formed by (equivalence classes of) non-singular $2 \times 2$ matrices over $\mathrm{GF}(p^{g/2})$.

**Proposition 4** *Let $H$ be a subgroup of $G = M(p^{2f})$ not contained in the subgroup $G_0 = \mathrm{PSL}(2, p^{2f})$ and let $H_0 = H \cap G_0$. If $H_0$ does not stabilise a point and is* *neither dihedral nor isomorphic to $A_4$, $S_4$ or $A_5$, then $H$ is conjugate in $\overline{G}$ to a subgroup isomorphic to $M(p^{2e})$ for some positive divisor $e$ of $f$ such that $f/e$ is odd.*

***Proof*** By a handy summary [15] of Dickson's classification of subgroups of projective special linear groups over finite fields, subgroups of $G_0$ comprise point stabilisers, dihedral groups, $A_4$, $S_4$, $A_5$, and $\mathrm{PSL}(2, p^g)$ for divisors $g$ of $2f$ together with $\mathrm{PGL}(2, p^{g/2})$ for even divisors $g$ of $2f$. Our assumptions imply that $H_0$ must be isomorphic to one of the last two types of subgroups.

Let $e$ be the smallest positive integer such that $f/e$ is odd and $g$ divides $2e$. Invoking the classification of subgroups of $G_0$ again [15], and specifically the fact that any two copies of $\mathrm{PSL}(2, p^{2e})$ in $G$ are conjugate (by an element of $G$), we may assume that $H_0$ is a subgroup of the canonical copy $G_{2e}$ of $\mathrm{PSL}(2, p^{2e})$ in $G$. Now, by [15], the number of conjugacy classes of copies of $\mathrm{PSL}(2, p^g)$ in $G_{2e}$ is 1 or 2 according to whether $2e/g$ is odd or even. The two classes if $g$ divides $e$ are, however, by [1], fused under conjugacy in $\mathrm{PGL}(2, p^{2e})$ and hence also under conjugacy in $\overline{G}$; the same holds for copies of $\mathrm{PGL}(2, p^{g/2})$ if $g$ is even. We therefore may assume that $H_0$ is *equal* either to the canonical copy $G_g$ of $\mathrm{PSL}(2, p^g)$ in $G$ or to the canonical copy $G_g^*$ of $\mathrm{PGL}(2, p^{g/2})$ in $G$ if $g$ is even.

Let $[A, 1] \in H\backslash H_0$ and $[C, 0] \in H_0$ be the elements from the proof of Proposition 2. Considering the conjugation $[A, 1][C, 0][A, 1]^{-1} = [ACA^{-1}, 0] \in H_0$ and verbatim repeating the arguments from the proof of Proposition 2, except with $F_g$ in place of $F_{2e}$, we conclude that $\delta \in F_g$. However, a non-square element $\delta \in F = F_{2f}$ can live in a subfield $F_g < F_{2f}$ only if $2f/g$ is odd. As $g \mid 2e$ with $f/e$ odd and $e$ was the smallest positive integer with this property, oddness of $2f/g$ implies that $g = 2e$.

We have arrived at the conclusion that $H_0$ is *equal* either to the canonical copy $G_{2e}$ of $\mathrm{PSL}(2, p^{2e})$ or to the canonical copy $G_{2e}^*$ of $\mathrm{PGL}(2, p^e)$. But in the second case, by the argument in the preceding paragraph, we would have $\delta \in F_e$ while $2f/e$ is not odd, a contradiction. Therefore, $H_0$ must be equal to $G_{2e}$. The conclusion that $H \cong M(p^{2e})$ now follows from Proposition 2. □

## 4 Representatives of Twisted Elements

Recalling the notation introduced in Sect. 2, we begin by identifying elements in conjugacy classes of $K\backslash K_0$ that have a particularly simple form. To facilitate the description here and also in the sections that follow, we let $\mathrm{dia}(\alpha, \beta)$ and $\mathrm{off}(\alpha, \beta)$, respectively, denote the $2 \times 2$ matrix with diagonal entries $\alpha, \beta$ (from the top left corner) and zero off-diagonal entries, and the $2 \times 2$ matrix with off-diagonal entries $\alpha, \beta$ (from the top right corner) and zero diagonal entries.

For every element $(A, 1) \in K\backslash K_0$, we have $(A, 1)^2 = (AA^\sigma, 0)$. In the study of conjugacy in $K\backslash K_0$, it turns out to be important to understand the behaviour of the products $AA^\sigma$. Observe that if $\delta = \det(A) \in N(F)$, then $\det(AA^\sigma) = \delta\delta^\sigma \in N(F_0)$.

Let $(A, 1) \in K \backslash K_0$ and let $\{\lambda_1, \lambda_2\}$ be the spectrum of $AA^\sigma$ in the smallest extension $F'$ of $F$ of degree at most two in which $\sigma$ may still be assumed to be given by $x \mapsto x^q$. Since $A^\sigma A$ is both a conjugate and also a $\sigma$-image of $AA^\sigma$, we have $\{\lambda_1, \lambda_2\}^\sigma = \{\lambda_1, \lambda_2\}$. This means that either (1) $\lambda_i^\sigma = \lambda_i$ for $i = 1, 2$, or (2) $\lambda_1^\sigma = \lambda_2$ and $\lambda_2^\sigma = \lambda_1$. Note that (2) implies $\lambda_1^{q^2} = (\lambda_1^q)^q = \lambda_2^q = \lambda_1$ and, similarly, $\lambda_2^{q^2} = \lambda_2$. We conclude that $F' = F$ in both the situations (1) and (2) and so both $\lambda_1, \lambda_2$ are in $F$. Observe that $\lambda_1 \neq \lambda_2$, as otherwise, we would have $\lambda_1 = \lambda_1^\sigma \in F_0$ and $\det(AA^\sigma) = \lambda_1^2 \in S(F_0)$, a contradiction. Moreover, it follows that in the case (1) we have $\lambda_i \in F_0$ for $i = 1, 2$ with $\lambda_1\lambda_2 \in N(F_0)$, and in the case (2) $\lambda_i \in N(F) \subset F \backslash F_0$ since $\det(AA^\sigma) = \lambda_1\lambda_1^\sigma \in N(F_0)$.

We will now refine our considerations of $AA^\sigma$. As before, let $q = p^f$ for some odd prime $p$ and let $e$ be the *smallest* positive divisor of $f$ with $f/e$ odd such that $AA^\sigma = \varepsilon C$ for some $C \in \mathrm{SL}(2, p^{2e})$ and for some $\varepsilon \in F^*$. In other words, we look for the smallest subfield $F_{2e}$ of $F$, with $f/e$ odd, such that all entries of $C$ lie in $F_{2e}$; note that we may assume $C$ to have determinant 1 since the determinant of $AA^\sigma$ is a non-zero square of $F$. If $\{\mu, \mu^{-1}\}$ is the spectrum of $C$, we have, without loss of generality, $\lambda_1 = \varepsilon\mu$ and $\lambda_2 = \varepsilon\mu^{-1}$. Observe that since $\lambda_1, \lambda_2, \varepsilon \in F$, we have $\mu, \mu^{-1} \in F$. Now, $\mu, \mu^{-1}$ are roots of a quadratic polynomial over $F_{2e}$ and therefore, both belong to $F_{2e}$ or to a quadratic extension of $F_{2e}$. But as $f/e$ is odd, the field $F$ does not contain a quadratic extension of $F_{2e}$. We conclude that $\mu, \mu^{-1} \in F_{2e}$.

The facts in the previous paragraphs imply that if $(A, 1) \in K \backslash K_0$, then the matrix $AA^\sigma$ is diagonalisable over $F$ and $C$ is diagonalisable over $F_{2e}$. In particular, there exists a $P \in \mathrm{GL}(2, p^{2e})$ such that $P^{-1}CP = D'$ for $D' = \mathrm{dia}(\mu, \mu^{-1})$; multiplying by $\varepsilon$ then gives $P^{-1}AA^\sigma P = D$ for $D = \mathrm{dia}(\lambda_1, \lambda_2)$. Here, either $\lambda_1, \lambda_2 \in F_0$ with $\lambda_1\lambda_2 \in N(F_0)$, or $\lambda_1, \lambda_2 \in F \backslash F_0$ and $\lambda_1^\sigma = \lambda_2$. With $A$, $P$, $D$ and $D'$ as above, in $K$ we let $(B, 1) = (P, 0)^{-1}(A, 1)(P, 0) = (P^{-1}AP^\sigma, 1)$. Then,

$$(BB^\sigma, 0) = (P, 0)^{-1}(A, 1)(A, 1)(P, 0) = (P^{-1}AA^\sigma P, 0) = (D, 0) = (\varepsilon D', 0)$$

and it follows that $BB^\sigma = D = \varepsilon D'$.

We now derive more details about the matrix $B = P^{-1}AP^\sigma$; recall that $P \in \mathrm{GL}(2, p^{2e})$. Let $u_1, u_2$ be linearly independent (column) eigenvectors of $C$ and $AA^\sigma$ for the eigenvalues $\mu, \mu^{-1}$ and $\lambda_1, \lambda_2$, respectively; we have $Cu_1 = \mu u_1$, $Cu_2 = \mu^{-1}u_2$, and $AA^\sigma u_i = \lambda_i u_i$ for $i \in \{1, 2\}$. Taking the $\sigma$-image of the last equation and then multiplying by $A$ from the left, we obtain $AA^\sigma(Au_i^\sigma) = \lambda_i^\sigma(Au_i^\sigma)$ for $i = 1, 2$. This means that the column vectors $Au_i^\sigma$ are also eigenvectors of $AA^\sigma$ for the eigenvalues $\lambda_i^\sigma$, $i = 1, 2$. It follows that if $\lambda_i = \lambda_i^\sigma$ for $i = 1, 2$, then we must have $Au_i^\sigma = \varepsilon_i u_i$, and if $\lambda_i = \lambda_{3-i}^\sigma$, then $Au_i^\sigma = \varepsilon_{3-i}u_{3-i}$, in both cases for some $\varepsilon_1, \varepsilon_2 \in F$. The last bit we need is the fact that for the matrix $P$, we may take $P = (u_1, u_2)$, i.e. the matrix formed by the columns $u_1, u_2$, with entries in $F_{2e}$. Now, for $i = 1, 2$, in the case $\lambda_i = \lambda_i^\sigma$, we have $AP^\sigma = (Au_1^\sigma, Au_2^\sigma) = (\varepsilon_1 u_1, \varepsilon_2 u_2) = P\mathrm{dia}(\varepsilon_1, \varepsilon_2)$, and in the case $\lambda_i = \lambda_{3-i}^\sigma$ a similar calculation gives $AP^\sigma = (Au_1^\sigma, Au_2^\sigma) = (\varepsilon_2 u_2, \varepsilon_1 u_1) = P\mathrm{off}(\varepsilon_1, \varepsilon_2)$. This shows that our matrix $B = P^{-1}AP^\sigma$ is equal to $\mathrm{dia}(\varepsilon_1, \varepsilon_2)$ or to

$\text{off}(\varepsilon_1, \varepsilon_2)$ for suitable $\varepsilon_1, \varepsilon_2 \in F$, depending on whether $\lambda_1^\sigma$ is equal to $\lambda_1$ or $\lambda_2$. In both cases, of course, $\varepsilon_1\varepsilon_2 \in N(F)$.

Recalling our notation $[A, i]$ for the cosets $(A, i)L = \{(\delta A, i);\ \delta \in F^*\}$, the above calculations lead to the following result.

**Proposition 5** *Let $G = M(p^{2f})$ for some odd prime $p$. Then, every element of the form $[A, 1] \in G$ is conjugate in $\overline{G}$ to $[B, 1]$ with $B = \text{dia}(\lambda, 1)$ or $B = \text{off}(\lambda, 1)$ for some $\lambda \in N(F)$. If, in addition, $[AA^\sigma, 0] = [C, 0]$ for some $C \in \text{SL}(2, p^{2e})$ with $f/e$ odd, then $[B, 1] = [P, 0]^{-1}[A, 1][P, 0]$ for some $P \in \text{GL}(2, p^{2e})$, and $\lambda\lambda^\sigma \in F_{2e}$ or $\lambda/\lambda^\sigma \in F_{2e}$, depending on whether $B$ is equal to $\text{dia}(\lambda, 1)$ or to $\text{off}(\lambda, 1)$.*

***Proof*** We have proven everything except for the last assertion. We have seen that if $[AA^\sigma, 0] = [C, 0]$ for some $C \in \text{SL}(2, p^{2e})$ with $f/e$ odd, then $BB^\sigma = P^{-1}(AA^\sigma)P = \varepsilon\text{dia}(\mu, \mu^{-1})$ for some $\varepsilon \in F^*$, $\mu \in F_{2e}$ and some $P \in \text{GL}(2, p^{2e})$. If $B = \text{dia}(\lambda, 1)$, then we have $\text{dia}(\lambda\lambda^\sigma, 1) = BB^\sigma = \varepsilon C = \varepsilon\text{dia}(\mu, \mu^{-1})$, which implies that $\varepsilon = \mu$ and $\lambda\lambda^\sigma = \mu^2 \in F_{2e}$. In the case when $B = \text{off}(\lambda, 1)$, we have $\text{off}(\lambda, \lambda^\sigma) = BB^\sigma = \varepsilon C = \varepsilon\text{dia}(\mu, \mu^{-1})$, from which we obtain $\lambda/\lambda^\sigma = \mu^2 \in F_{2e}$. □

Let us have another look at conjugation in the group $\overline{G} = J/L$. Observe that if $(P, i) \in J$, then $(P, i)^{-1} = ((P^{\sigma^i})^{-1}, i)$. Conjugates of $(B, 1) \in K$ by $(P, i)$ have the form $(P, 0)^{-1}(B, 1)(P, 0) = (P^{-1}BP^\sigma, 1)$ if $i = 0$, and $(P, 1)^{-1}(B, 1)(P, 1) = ((P^\sigma)^{-1}B^\sigma P, 1)$ if $i = 1$. It follows that two elements $(B, 1)$ and $(B', 1)$ of $K$ are conjugate in $J$ if and only if $B' = P^{-1}BP^\sigma$ or $B' = (P^\sigma)^{-1}B^\sigma P$ for some $P \in \text{GL}(2, F)$. Taking the $\sigma$-image in the second case and passing onto $G = K/L$, we have.

**Proposition 6** *Two elements $[B, 1]$ and $[B', 1]$ of $G$ are conjugate in $\overline{G}$ if and only if $P^{-1}BP^\sigma = \varepsilon B'$ or $P^{-1}BP^\sigma = \varepsilon B'^\sigma$ for some $\varepsilon \in F^*$ and some $P \in \text{GL}(2, F)$.* □

We will write the two conditions of Proposition 6 in the unified form $P^{-1}BP^\sigma = \varepsilon B'^{(\sigma)}$, or, equivalently, $BP^\sigma = \varepsilon PB'^{(\sigma)}$ where $B'^{(\sigma)}$ is equal to $B'$ or $B'^\sigma$, depending on whether $i = 0$ or $i = 1$ when using the element $[P, i]$ for conjugation.

## 5 Conjugacy of Representatives of Twisted Elements

We continue with identification of elements of $\overline{G}$ that conjugate a diagonal (or an off-diagonal) element from Proposition 5 to another such element. As a by-product, we will be able to identify $\overline{G}$-stabilisers of our representatives of twisted elements in $G$.

We begin with the case when $B = \text{dia}(\lambda, 1)$ and $B' = \text{dia}(\lambda', 1)$; by Proposition 6, it is sufficient to find the non-singular matrices $P \in \text{GL}(2, F)$ and $\varepsilon \in F^*$ for which $BP^\sigma = \varepsilon PB'^{(\sigma)}$ in the sense of the notation introduced at the end of Sect. 4. Throughout the computation, we will use the symbols $\lambda^{(\sigma)}$ and $\lambda'^{(\sigma)}$ in an analogous way as explained for $B'^{(\sigma)}$. Assuming that $P$ has entries $\alpha, \beta, \gamma, \delta$, the above condition says that

$$\begin{pmatrix} \lambda & 0 \\ 0 & 1 \end{pmatrix} \begin{pmatrix} \alpha^\sigma & \beta^\sigma \\ \gamma^\sigma & \delta^\sigma \end{pmatrix} = \varepsilon \begin{pmatrix} \alpha & \beta \\ \gamma & \delta \end{pmatrix} \begin{pmatrix} \lambda'^{(\sigma)} & 0 \\ 0 & 1 \end{pmatrix} .$$

Evaluating the products, we obtain

$$\lambda\alpha^\sigma = \varepsilon\lambda'^{(\sigma)}\alpha \,, \quad \lambda\beta^\sigma = \varepsilon\beta \,, \quad \gamma^\sigma = \varepsilon\lambda'^{(\sigma)}\gamma \,, \quad \delta^\sigma = \varepsilon\delta \,.$$

By straightforward manipulation, this gives the following system of four equations:

$$\alpha(\lambda\lambda^\sigma - \varepsilon\varepsilon^\sigma\lambda'\lambda'^\sigma) = 0 \,, \quad \beta(\lambda\lambda^\sigma - \varepsilon\varepsilon^\sigma) = 0 \,, \quad \gamma(\varepsilon\varepsilon^\sigma\lambda'\lambda'^\sigma - 1) = 0 \,, \quad \delta(\varepsilon\varepsilon^\sigma - 1) = 0 \,.$$

From non-singularity of $P$, it follows that $\delta \neq 0$ or $\beta \neq 0$, that is, $\varepsilon\varepsilon^\sigma = 1$ or $\varepsilon\varepsilon^\sigma = \lambda\lambda^\sigma$.

Consider first the case $\varepsilon\varepsilon^\sigma = 1$, that is, $\varepsilon^{q+1} = 1$ for $q = p^f$. Since $\lambda, \lambda' \in N(F)$, we have $\lambda\lambda^\sigma \neq 1 \neq \lambda'\lambda'^\sigma$. Our equations together with $\varepsilon\varepsilon^\sigma = 1$ then imply that $\beta = \gamma = 0$. Hence, $\alpha, \delta \neq 0$, by non-singularity of $P$; in particular, $\lambda\lambda^\sigma = \lambda'\lambda'^\sigma$, or, equivalently, $(\lambda'/\lambda)^{q+1} = 1$. We are interested in conjugation in the group $\overline{G} = J/L$ and so we may assume that $\delta = 1$, which reduces the relations below our matrix equation to $\varepsilon = 1$ and $\alpha^{q-1} = \lambda'^{(\sigma)}/\lambda$. Since $(\lambda'/\lambda)^{q+1} = 1$, the equation $\eta^{q-1} = \lambda'/\lambda$ has $q-1$ solutions $\eta \in F^*$ (note that $|F^*| = q^2 - 1$). If $\lambda'^{(\sigma)} = \lambda'$, then all solutions of the equation $\alpha^{q-1} = \lambda'^{(\sigma)}/\lambda$ have the form $\alpha = \eta$, and if $\lambda'^{(\sigma)} = \lambda'^\sigma = \lambda'\lambda'^{q-1}$, then all solutions of this equation are $\alpha = \eta\lambda'$.

The second case to consider is $\varepsilon\varepsilon^\sigma = \lambda\lambda^\sigma$ $(\neq 1)$, which implies that $\alpha = \delta = 0$, and also $\lambda\lambda^\sigma\lambda'\lambda'^\sigma = 1$ since $\gamma, \beta$ now must be non-zero. By the same token as above, we may let $\gamma = 1$ without loss of generality. Then, our equations for $\gamma$ and $\beta$ in this case reduce to $\varepsilon\lambda'^{(\sigma)} = 1$ and $\lambda\beta^\sigma = \varepsilon\beta$, the latter now being equivalent to $\beta^{q-1} = 1/(\lambda\lambda'^{(\sigma)})$. Since now $(\lambda\lambda')^{q+1} = 1$, there are $q-1$ solutions $\zeta$ of the equation $\zeta^{q-1} = 1/(\lambda\lambda')$ in $F^*$. If $\lambda'^{(\sigma)} = \lambda'$, then we have $\beta = \zeta$, and if $\lambda'^{(\sigma)} = \lambda'^\sigma = \lambda'\lambda'^{q-1}$, we have $\beta = \zeta/\lambda'$. Summing up, we arrive at the following:

**Proposition 7** *Let* $B = \mathrm{dia}(\lambda, 1)$ *and* $B' = \mathrm{dia}(\lambda', 1)$ *for* $\lambda, \lambda' \in N(F)$. *If an element* $[P, i] \in \overline{G}$ *conjugates* $[B, 1]$ *to* $[B', 1]$, *then, without loss of generality,* $P = \mathrm{dia}(\omega, 1)$ *or* $P = \mathrm{off}(\omega, 1)$ *for suitable* $\omega \in F^*$. *Moreover:*
1. *If* $P = \mathrm{dia}(\omega, 1)$, *then* $\lambda\lambda^\sigma = \lambda'\lambda'^\sigma$, *and if this condition is satisfied, then* $[B, 1]$ *conjugates to* $[B', 1]$ *in* $\overline{G}$ *exactly by the* $q-1$ *elements* $[P, 0]$ *such that* $\omega = \eta$ *and the* $q-1$ *elements* $[P, 1]$ *with* $\omega = \eta\lambda'$, *where* $\eta \in F^*$ *is one of the* $q-1$ *solutions of the equation* $\eta^{q-1} = \lambda'/\lambda$.
2. *If* $P = \mathrm{off}(\omega, 1)$, *then* $\lambda\lambda^\sigma\lambda'\lambda'^\sigma = 1$, *and if this holds, then* $[B, 1]$ *conjugates to* $[B', 1]$ *in* $\overline{G}$ *exactly by the* $q-1$ *elements* $[P, 0]$ *with* $\omega = \zeta$ *and the* $q-1$ *elements* $[P, 1]$ *such that* $\omega = \zeta/\lambda'$, *where* $\zeta \in F^*$ *is one of the* $q-1$ *solutions of the equation* $\zeta^{q-1} = 1/(\lambda\lambda')$. □

We now repeat this process but now with matrices $B = \mathrm{off}(\lambda, 1)$ and $B' = \mathrm{off}(\lambda', 1)$. Conjugating by $[P, i]$ and assuming that $P$ has entries $\alpha, \beta, \gamma, \delta$, the unified form $BP^{\sigma} = \varepsilon PB'^{(\sigma)}$ of the condition of Proposition 6 now translates into the matrix equation

$$\begin{pmatrix} 0 & \lambda \\ 1 & 0 \end{pmatrix} \begin{pmatrix} \alpha^{\sigma} & \beta^{\sigma} \\ \gamma^{\sigma} & \delta^{\sigma} \end{pmatrix} = \varepsilon \begin{pmatrix} \alpha & \beta \\ \gamma & \delta \end{pmatrix} \begin{pmatrix} 0 & \lambda'^{\sigma^i} \\ 1 & 0 \end{pmatrix} .$$

It follows that

$$\alpha^{\sigma} = \varepsilon\delta\,, \quad \beta^{\sigma} = \varepsilon\lambda'^{\sigma^i}\gamma\,, \quad \lambda\gamma^{\sigma} = \varepsilon\beta\,, \quad \lambda\delta^{\sigma} = \varepsilon\lambda'^{\sigma^i}\alpha\,,$$

which, after some manipulation, yield the following two equations:

$$\gamma(\lambda^{\sigma} - \varepsilon\varepsilon^{\sigma}\lambda'^{\sigma^i}) = 0 \quad \text{and} \qquad \delta^{\sigma}(\lambda - \varepsilon\varepsilon^{\sigma}\lambda'^{\sigma^i}) = 0\,.$$

This all means that either (a) $\lambda = \varepsilon\varepsilon^{\sigma}\lambda'^{\sigma^i}$, and then $\beta = \gamma = 0$ and we may assume $\delta = 1$, or else (b) $\lambda^{\sigma} = \varepsilon\varepsilon^{\sigma}\lambda'^{\sigma^i}$, and then we have $\alpha = \delta = 0$ and, without loss of generality, $\gamma = 1$. Since $\varepsilon\varepsilon^{\sigma}$, $\lambda\lambda^{\sigma} \in F_0^*$, these conditions are equivalent to $\lambda/\lambda' \in F_0^*$ or $\lambda\lambda' \in F_0^*$, independently of the value of $i$, but in our analysis below, it is still useful to refer to $i$.

In the case (a), when $\lambda/\lambda'^{\sigma^i} = \varepsilon\varepsilon^{\sigma} \in F_0^*$, for every $i \in \{0, 1\}$ there are $q+1$ $(q+1)$th roots $\eta_{(i)}$ of $\lambda/\lambda'^{\sigma^i}$ in $F^*$. From $\delta = 1$, we have $a^{\sigma} = \varepsilon$ and $\lambda = \varepsilon\lambda'^{\sigma^i}\alpha$, that is, $\alpha^{q+1} = \lambda/\lambda'^{\sigma^i}$. This implies that $\alpha = \eta_{(i)}$ is one of the $(q+1)$th roots of $\lambda/\lambda'^{\sigma^i}$, giving $q+1$ conjugation elements $[P, i]$ such that $P = \mathrm{dia}(\eta_{(i)}, 1)$. In the case (b), $\lambda^{\sigma}/\lambda'^{\sigma^i} = \varepsilon\varepsilon^{\sigma} \in F_0^*$ and since also $\lambda\lambda^{\sigma} \in F_0^*$, we have $\lambda\lambda'^{\sigma^i} \in F_0^*$. It follows that for every $i \in \{0, 1\}$ there are $q+1$ $(q+1)$th roots $\zeta_{(i)}$ of $\lambda\lambda'^{\sigma^i}$ in $F^*$. From $\gamma = 1$, we obtain $\lambda = \varepsilon\beta$ and $\beta^{\sigma} = \varepsilon\lambda'^{\sigma^i}$, which means that $\beta^{q+1} = \lambda\lambda'^{\sigma^i}$. Consequently, $\beta = \zeta_{(i)}$ and we have in this second case, $q+1$ conjugation elements $[P, i]$ such that $P = \mathrm{off}(\zeta_{(i)}, 1)$. Realising that the condition (a) for $i = 0$ is equivalent to (b) for $i = 1$ (and equivalent to $\lambda/\lambda' \in F_0^*$) and, similarly, the condition (a) for $i = 1$ is equivalent to (b) for $i = 0$ (and equivalent to $\lambda\lambda' \in F_0^*$), we conclude that:

**Proposition 8** *Let* $B = \mathrm{off}(\lambda, 1)$ *and* $B' = \mathrm{off}(\lambda', 1)$ *for* $\lambda, \lambda' \in N(F)$. *Further, for* $i \in \{0, 1\}$, *let* $\eta_{(i)}, \zeta_{(i)} \in F^*$ *be any of the* $q+1$ *roots of the equation* $\eta_{(i)}^{q+1} = \lambda/\lambda'^{\sigma^i}$ *and* $\zeta_{(i)}^{q+1} = \lambda\lambda'^{\sigma^i}$, *respectively, Then, an element* $[P, i] \in \overline{G}$ *conjugates* $[B, 1]$ *to* $[B', 1]$ *if and only if* $\lambda/\lambda'^{\sigma^i}$ *or* $\lambda\lambda'^{\sigma^i}$ *are elements of* $F_0^*$. *In an equivalent form,* $[B, 1]$ *is conjugate to* $[B', 1]$ *if and only if either*

1. $\lambda/\lambda' \in F_0^*$, *in which case the conjugation is realised by exactly* $q+1$ *elements* $[P, 0]$ *with* $P = \mathrm{dia}(\eta_{(0)}, 1)$ *and exactly* $q+1$ *elements* $[P, 1]$ *with* $P = \mathrm{off}(\zeta_{(1)}, 1)$, *or*
2. $\lambda\lambda' \in F_0^*$, *by the conjugation realised by exactly* $q+1$ *elements* $[P, 0]$ *with* $P = \mathrm{dia}(\eta_{(1)}, 1)$ *and exactly* $q+1$ *elements* $[P, 1]$ *with* $P = \mathrm{off}(\zeta_{(0)}, 1)$. □

## 6 Conjugacy Classes of Twisted Elements

With the help of the calculations in the previous section, we can now prove a useful result about identification of suitable representatives of conjugacy classes (in the group $\overline{G}$) of elements of $G\backslash G_0$.

**Theorem 1** *Let $\xi$ be a primitive element of $F$ and let $[A, 1]$ be an element of $G$. Then, exactly one of the following two cases occur:*

1. *There exists exactly one odd $i \in \{1, 2, \ldots, (q-1)/2\}$ such that $[A, 1]$ is conjugate in $\overline{G}$ to $[B, 1]$ with $B = \mathrm{dia}(\xi^i, 1)$; the order of $[A, 1]$ in $G$ is then $2(q-1)/\gcd\{q-1, i\}$.*
2. *There exists exactly one odd $i \in \{1, 2, \ldots, (q+1)/2\}$ such that $[A, 1]$ is conjugate in $\overline{G}$ to $[B, 1]$ with $B = \mathrm{off}(\xi^i, 1)$, and the order of $[A, 1]$ in $G$ is $2(q+1)/\gcd\{q+1, i\}$.*

*Furthermore, we have*

3. *The stabiliser of $[B, 1]$ for $B = \mathrm{dia}(\lambda, 1)$, $\lambda \in N(F)$, in the group $\overline{G}$ is isomorphic to the cyclic group $Z_{2(q-1)}$ generated by (conjugation by) $[P, 1]$ for $P = \mathrm{dia}(\mu\lambda, 1)$ with a suitable $(q-1)$th root of unity $\mu$, except when $\lambda$ is a $(q+1)$th root of $-1$ and $q \equiv -1 \bmod 4$, in which case the stabiliser is isomorphic to $Z_{2(q-1)} \cdot Z_2$.*
4. *The stabiliser of $[B, 1]$ for $B = \mathrm{off}(\lambda, 1)$, $\lambda \in N(F)$, in the group $\overline{G}$ is isomorphic to the cyclic group $Z_{2(q+1)}$ generated by (conjugation by) $[P, 1]$ for $P = \mathrm{off}(\mu\lambda, 1)$, where $\mu$ is a suitable $(q+1)$th root of unity, except when $\lambda$ is a $(q-1)$th root of $-1$ and $q \equiv 1 \bmod 4$, when the stabiliser is isomorphic to $Z_{2(q+1)} \cdot Z_2$.*

***Proof*** By Proposition 5, each element $[A, 1] \in G$ is conjugate in $G$ to $[B, 1]$ with $B = \mathrm{dia}(\lambda, 1)$ or $B = \mathrm{off}(\lambda, 1)$ for some $\lambda \in N(F)$.

1. Suppose that $[A, 1]$ is conjugate in $G$ to $[B, 1]$, $B = \mathrm{dia}(\lambda, 1)$, $\lambda \in N(F)$. Letting $B' = \mathrm{dia}(\lambda', 1)$, Proposition 7 implies that the elements $[B, 1]$ and $[B', 1]$ are conjugate in $\overline{G}$ if and only if $\lambda\lambda^\sigma = \lambda'\lambda'^\sigma$ or $\lambda\lambda^\sigma\lambda'\lambda'^\sigma = 1$. If $\lambda = \xi^i$ and $\lambda' = \xi^{i'}$ for some odd $i$ and $i'$, the condition translates into $\xi^{(q+1)i} = \xi^{\pm(q+1)i'}$, which is equivalent to $(q+1)i \equiv \pm(q+1)i' \bmod (q^2-1)$ and which simplifies to $i \equiv \pm i'$ mod $(q-1)$. This proves Part 1 except for the order assertion. But, in $G$, for $B = \mathrm{dia}(\xi^i, 1)$, the order of $[B, 1]$ is twice the order of $[B, 1]^2 = [BB^\sigma, 0]$, which is equal to the order of $BB^\sigma = \mathrm{dia}(\xi^{(q+1)i}, 1)$ in $PSL(2, q^2)$, which is known to be equal to $(q^2-1)/\gcd\{q^2-1, (q+1)i\} = (q-1)/\gcd\{q-1, i\}$.

2. Now, let $[A, 1]$ be conjugate in $G$ to $[B, 1]$ for $B = \mathrm{off}(\lambda, 1)$ and $\lambda \in N(F)$. If $B' = \mathrm{off}(\lambda', 1)$, by Proposition 8, we see that $[B, 1]$ is conjugate to $[B', 1]$ in $\overline{G}$ if and only if $\lambda/\lambda'$ or $\lambda\lambda'$ are elements of $F_0^*$, or, equivalently, are powers of $\xi^{q+1}$. Letting $\lambda = \xi^i$ and $\lambda' = \xi^{i'}$ for odd $i, i'$, this condition translates into the congruence $i \equiv \pm i'$ mod $(q+1)$, proving Part 2 apart from the claim about the order. To fill in this last bit, the order of $[B, 1]$ in $G$ for $B = \mathrm{off}(\xi^i, 1)$ is two times the order of $[B, 1]^2 = [BB^\sigma, 0]$, which is equal to the order of $BB^\sigma = \mathrm{dia}(\xi^i, \xi^{qi})$

in $PSL(2, q^2)$. The latter is equal to the order of $\mathrm{dia}(1, \xi^{(q-1)i})$ in $PSL(2, q^2)$, i.e. to $(q^2-1)/\gcd\{q^2-1, (q-1)i\} = (q+1)/\gcd\{q+1, i\}$.

3. The values $\lambda = \lambda'$ automatically satisfy the first condition in Proposition 7, by which the stabiliser of $[B, 1]$ for $B = \mathrm{dia}(\lambda, 1)$ includes a subgroup $H$ formed by (conjugation by) the $(q-1)$ elements $[P, 0]$ for $P = \mathrm{dia}(\eta, 1)$ and the $q-1$ elements $[P, 1]$ with $P = \mathrm{dia}(\eta\lambda, 1)$, where $\eta$ ranges over the set of all $(q-1)$th roots of unity in $F$ (and this set is simply equal to $F_0^*$). We claim that $H$ is (isomorphic to) a cyclic group of order $2(q-1)$. Indeed, let $\lambda = \xi^i$ for odd $i$, $1 \le i \le (q-1)/2$. Then, letting $\eta = \xi^{(q+1)(1-i)/2}$, the group $H$ is generated by $[P, 1]$, where $P = \mathrm{dia}(\eta\lambda, 1)$. To see this, observe that $[P, 1]^2 = [PP^\sigma, 0] = [Q, 0]$, where $Q = \mathrm{dia}(\eta^2\lambda\lambda^\sigma, 1)$. With $\lambda$ and $\eta$ as above, we have $Q = \mathrm{dia}(\xi^{q+1}, 1)$, which means that the order of $[Q, 0]$ is $q-1$ and so $[P, 1]$ generates $H$, as claimed.

Observe that the values $\lambda = \lambda'$ may also satisfy our second condition $\lambda\lambda^\sigma\lambda'\lambda'^\sigma = 1$ from Proposition 7. This happens if and only if $\lambda$ is a $(q+1)$th root of $-1$ in $F$ and $q \equiv -1 \bmod 4$, which is what we now assume; note that now $\lambda^\sigma = -\lambda^{-1}$. Then, if $\zeta$ and $\zeta'$ are $(q-1)$th roots of $(\lambda\lambda')^{-1} = \lambda^{-2}$ as in the calculations immediately preceding Proposition 7, we have $\zeta'/\zeta \in F_0^*$, $\lambda^{q-1} = -\zeta^{q-1}$, and $[\mathrm{off}(\zeta'/\lambda, 1), i][\mathrm{off}(\zeta/\lambda, 1), i] = [\mathrm{dia}((-1)^i\zeta'/\zeta, 1), 0]$ for $i \in \{0, 1\}$. In particular, letting $Q = \mathrm{off}(\zeta/\lambda, 1)$, the element $[Q, 1]^2 = [\mathrm{dia}(-1, 1), 0]$ lies in the cyclic group $H$; hence $[Q, 1]$ has order 4. Moreover, one may check that for every $\eta \in F_0^*$, we have $[Q, 1]^{-1}[P, 0][Q, 1] = [P, 0]^{-1}$ if $P = \mathrm{dia}(\eta, 1)$ and $[Q, 1]^{-1}[P, 1][Q, 1] = [P, 1]^{-1}[Q, 1]^2$ if $P = \mathrm{dia}(\eta\lambda, 1)$.

Summing up, the stabiliser of an element $[B, 1]$ with $B = \mathrm{dia}(\lambda, 1)$ in $\overline{G}$ for $\lambda \in N(F)$ is isomorphic to $H \cong Z_{2(q-1)}$ except for the case when $q \equiv -1 \bmod 4$ and $\lambda$ is a $(q+1)$th root of $-1$ and then the stabiliser of $[B, 1]$ is isomorphic to an extension $H^*$ of $H$ by $[Q, 1]$ with $[Q, 1]^2 \in H$, so that $H^* \cong Z_{2(q-1)} \cdot Z_2$, as claimed.

4. The values $\lambda = \lambda'$ automatically satisfy the condition (1) from Proposition 8. By the calculations preceding this Proposition, the stabiliser of $[B, 1]$ for $B = \mathrm{off}(\lambda, 1)$ contains a subgroup formed by the $(q+1)$ elements $[P, 0]$ for $P = \mathrm{dia}(\eta, 1)$ and the $q-1$ elements $[P, 1]$ with $P = \mathrm{off}(\eta\lambda, 1)$, where this time $\eta$ ranges over the set of all $(q+1)$th roots of unity in $F$. The collection of these $2(q+1)$ elements is closed under multiplication and hence forms a group $H$ of order $2(q+1)$; we claim that $H$ is cyclic. To demonstrate this, let $\lambda = \xi^i$ for some odd $i$ such that $1 \le i \le (q+1)/2$. Then, using $\eta = \xi^{(q-1)(1+i)/2}$, the element $[P, 1]$ with $P = \mathrm{off}(\eta\lambda, 1)$ is a generator of $H$. Indeed, a calculation shows that $[P, 1]^2 = [PP^\sigma, 0] = [Q, 0]$, where $Q = \mathrm{dia}(\eta^2\lambda^{1-q}, 1)$, and with the given $\lambda$ and our choice of $\eta$, we have $Q = \mathrm{dia}(\xi^{q-1}, 1)$. It follows that the order of $[Q, 0]$ is $q+1$ and so $[P, 1]$ generates $H$.

Here, the values $\lambda = \lambda'$ may also fulfil the condition (2) from Proposition 8, that is, $\lambda^2 \in F_0^*$. Since we know that $\lambda \in N(F)$, it can be checked that $\lambda^2 \in F_0^*$ is equivalent to $\lambda^{q-1} = -1$ and implies $q \equiv 1 \bmod 4$. Again, the calculations made immediately before Proposition 8 imply the following: Letting $\zeta$ be any of the $q+1$ $(q+1)^{\mathrm{st}}$ roots of $\lambda^2$ in $F$, in the stabiliser of $[B, 1]$ we have additional $q+1$ elements $[P, 1]$ such that $P = \mathrm{dia}(\zeta/\lambda, 1)$, as well as $q+1$ elements $[P, 0]$ such that $P = \mathrm{off}(\zeta, 1)$. Further calculations show that if $\zeta$ and $\zeta'$ are $(q+1)$th roots of $\lambda^2$, $P =$

$\mathrm{off}(\zeta, 1)$ and $P' = \mathrm{off}(\zeta', 1)$, then we have $[P', 0][P, 0] = [\mathrm{dia}(\zeta'/\zeta, 1), 0] \in H$ since it can be checked that $\zeta'/\zeta \in F_0^*$. Similarly, if $Q = \mathrm{dia}(\zeta/\lambda, 1)$ and $Q' = \mathrm{dia}(\zeta'/\lambda, 1)$, then $[Q', 1][Q, 1] = [Q'Q^\sigma, 0] = [\mathrm{dia}((\zeta'/\zeta)(\zeta/\lambda)^{q+1}, 1), 0] \in H$. In particular letting $\zeta' = \zeta$ we have $(\zeta/\lambda)^{q+1} = \lambda^2/(\lambda^{q-1}\lambda^2) = -1$ and so $[Q, 1]^2 = [\mathrm{dia}(-1, 1), 0] \in H$. We thus have a situation analogous to the Case 3 of this proof: $[Q, 1]$ has order 4, conjugation of an element $[P, 0] \in H$ with $P = \mathrm{dia}(\eta, 1)$ by $[Q, 1]$ inverts $[P, 0]$, while conjugation of a $[P, 1] \in H$ such that $P = \mathrm{off}(\eta\lambda, 1)$ by $[Q, 1]$ gives $[P, 1]^{-1}[Q, 1]^2$.

In summary, the stabiliser of an element $[B, 1]$ for $B = \mathrm{off}(\lambda, 1)$ is isomorphic to $H \cong Z_{2(q+1)}$ if $\lambda$ is not a $(q-1)$th root of $-1$, and isomorphic to an extension $H^*$ of $H$ by $[Q, 1]$ (where $[Q, 1]^2 \in H$), implying that $H^* \cong Z_{2(q-1)} \cdot Z_2$. This completes the proof. □

Let us remark that the exceptional cases in the items 3 and 4 above correspond precisely to elements $[B, 1]$ of order 4 in $G$. By inspecting possible orders of $[B, 1]$, we also have

**Corollary 1** *Every element of $G\backslash G_0$ has order divisible by* 4. □

## 7 Non-singular Pairs and Twisted Subgroups

Our aim in this and the following two sections is to determine representatives of selected conjugacy classes $\{(x, y)^g;\ g \in \overline{G} = G\langle\sigma\rangle\}$ of elements $x, y \in G$ satisfying $(xy)^2 = 1$, and make important conclusions about subgroups the corresponding pairs $(x, y)$ generate. We note that the action of $\sigma$ need not be considered separately, because $[I, 1][A, i][I, 1] = [A^\sigma, i]$ for $i \in \{0, 1\}$, which means that the action of $\sigma$ is equivalent to conjugation by the element $[I, 1] \in \overline{G}$.

Since we want $xy$ to have order 2, both $x$ and $y$ as above must lie in $G\backslash G_0$ because, by Corollary 1, there are no involutions in $G\backslash G_0$. By Theorem 1, we may assume that $y = [B, 1]$ for $B = \mathrm{dia}(\lambda, 1)$ or $B = \mathrm{off}(\lambda, 1)$ for a suitable $\lambda \in N(F)$. Letting $x = [A, 1]$, the pair $x, y$ may in general generate a proper subgroup of $G = M(p^{2f})$; such cases will still be of interest for our intended classification of orientably-regular maps as long as the subgroup $\langle x, y\rangle$ is twisted, that is, isomorphic to $M(p^{2e})$ for a suitable divisor $e$ of $f$.

We now identify conditions on $A$ implied by the requirement that $([A, 1][B, 1])^2$ be the identity in $G$ and begin with the case when $B = \mathrm{dia}(\lambda, 1)$. Let $A \in \mathrm{GL}(2, F)$ have rows $(a, b)$ and $(c, d)$, with determinant $ad - bc \in N(F)$. Then, $[A, 1][B, 1] = [AB^\sigma, 0]$, where

$$AB^\sigma = \begin{pmatrix} a\lambda^\sigma & b \\ c\lambda^\sigma & d \end{pmatrix}.$$

Since $AB^\sigma$ lies in $\mathrm{PSL}(2, F)$, it has order 2 if and only if its trace $a\lambda^\sigma + d$ is equal to zero. If one of $a, d$ was equal to zero, both would have to be zero and then $[A, 1]$ and $[B, 1]$ would clearly not generate a twisted subgroup of $G$. Therefore both $a, d$

are non-zero and we may assume without loss of generality that $a = -1$ and $d = \lambda^\sigma$. We will thus consider only elements $[A, 1] \in G$ of the form

$$A = \begin{pmatrix} -1 & b \\ c & \lambda^\sigma \end{pmatrix} , \quad u = bc + \lambda^\sigma \in N(F) . \tag{1}$$

Next, consider the case when $B = \mathrm{off}(\lambda, 1)$. For a matrix $A \in \mathrm{GL}(2, F)$ with rows $(a, b)$ and $(c, d)$ such that $ad - bc \in N(F)$, we now have $[A, 1][B, 1] = [AB^\sigma, 0]$, where

$$AB^\sigma = \begin{pmatrix} b & a\lambda^\sigma \\ d & c\lambda^\sigma \end{pmatrix} .$$

Again, $AB^\sigma \in \mathrm{PSL}(2, F)$ has order 2 if and only if its trace $b + c\lambda^\sigma$ is equal to zero. If one of $b, c$ was equal to zero, we would have $b = c = 0$, but then $[A, 1]$ and $[B, 1]$ would again not generate a twisted subgroup of $G$. Therefore both $b$ and $c$ are non-zero and we may assume that $c = -1$ and $b = \lambda^\sigma$. It follows that, without loss of generality, we only need to consider elements $[A, 1] \in G$ such that

$$A = \begin{pmatrix} a & \lambda^\sigma \\ -1 & d \end{pmatrix} , \quad u = ad + \lambda^\sigma \in N(F) . \tag{2}$$

With $A$ and $B$ as above, we can still identify obvious instances when $[A, 1]$ and $[B, 1]$ do not generate a twisted subgroup of $G$. This is certainly the case if

(i) both $[A, 1]$ and $[B, 1]$ have order 4, as then the two elements generate a solvable group, cf. [6], or
(ii) $B = \mathrm{dia}(\lambda, 1)$ and $A$ is an upper- or a lower-triangular matrix, as then $[A, 1]$ and $[B, 1]$) generate a triangular subgroup of $G$, or else
(iii) $B = \mathrm{off}(\lambda, 1)$ and $A$ is an off-diagonal matrix, as then $[A, 1]$ and $[B, 1]$ clearly do not generate a twisted subgroup of $G$.

For $B = \mathrm{dia}(\lambda, 1)$ and $A$ given by (1) and for $B = \mathrm{off}(\lambda, 1)$ and $A$ given by (2), an ordered pair $([A, 1], [B, 1])$ not satisfying any of (i), (ii) and (iii) will be called *non-singular*.

We are now in a position to classify the subgroups of $G = M(p^{2f})$ generated by non-singular pairs. To do so, we will again use knowledge of the situation in the subgroup $G_0 \simeq \mathrm{PSL}(2, p^{2f})$ of $G$. Recall that a subgroup $H$ of $G = M(p^{2f})$ was said to stabilise a point if there exists an element in $F \cup \{\infty\}$ fixed by all linear fractional mappings corresponding to elements of $H$; also, $G_0$ denotes the (unique) copy of $\mathrm{PSL}(2, p^{2f})$ in $G$.

**Proposition 9** *Let $H$ be a subgroup of $G$ generated by a non-singular pair* $([A, 1], [B, 1])$. *Then, $H$ is isomorphic to $M(p^{2e})$ for some positive divisor $e$ of $f$ such that $f/e$ is odd.*

***Proof*** Let $H_0 = H \cap G_0$. The classification of [15] tells us that subgroups of $G_0$ fall into the following categories: point stabilisers, dihedral groups, $A_4$, $S_4$, $A_5$, and

$\mathrm{PSL}(2, p^g)$ for divisors $g$ of $2f$ together with $\mathrm{PGL}(2, p^{g/2})$ for even divisors $g$ of $2f$. For our subgroup $H_0$, we subsequently rule out all but the penultimate case.

We first show that $H_0$ neither stabilises a point nor is dihedral. The first coordinate of every element of the group $H_0$ can be written as a product of the form $X_1Y_1^\sigma \ldots X_mY_m^\sigma$, where $X_i, Y_i \in \{A, B\}$, $1 \le i \le m$. It follows that $H_0$ is generated by the three elements $[BB^\sigma, 0]$, $[BA^\sigma, 0]$ and $[AB^\sigma, 0]$; note that $AA^\sigma = AB^\sigma(BB^\sigma)^{-1}BA^\sigma$. A straightforward calculation (details of which we leave to the reader) shows that for our generators $[A, 1]$ and $[B, 1]$ of $H$, there is no point in $F \cup \{\infty\}$ fixed by all three of the linear fractional mappings corresponding to $BB^\sigma$, $BA^\sigma$ and $AB^\sigma$. It follows that the subgroup $H_0 = \langle [BB^\sigma, 0], [BA^\sigma, 0], [AB^\sigma, 0]\rangle$ cannot stabilise a point.

Suppose that $H_0$ is dihedral, an easy calculation shows that we always have $AB^\sigma \neq BA^\sigma$. It follows that one of $AB^\sigma$, $BA^\sigma$ is an involution outside the cyclic part of $H_0$ and hence inverts one of $AA^\sigma$ and $BB^\sigma$. If conjugation by $AB^\sigma$ is considered, the conditions $(B^\sigma)^{-1}A^{-1}XX^\sigma AB^\sigma = (X^\sigma)^{-1}X^{-1}$ for $X \in \{A, B\}$ both reduce to $(B^{-1}A)^2 = I$ in $\mathrm{PSL}(2, p^{2f})$. A similar calculation shows that the condition of $XX^\sigma$ being inverted by conjugation by $BA^\sigma$ for $X \in \{A, B\}$ both reduce to $(B^{-1}A)^2 = I$ again. The condition $(B^{-1}A)^2 = I$ in $\mathrm{PSL}(2, p^{2f})$ is equivalent to $(B^{-1}A)^2 = \varepsilon I$ in $\mathrm{SL}(2, p^{2f})$ for some $\varepsilon \in \mathrm{GF}(p^{2f})$. If $B = \mathrm{dia}(\lambda, 1)$ and $b = c = 0$ in $A$, then $\lambda\lambda^\sigma = -1$ which, by part 3 of Theorem 1, means that the orders of both $[A, 1]$ and $[B, 1]$ would be equal to 4. If one of $b, c \neq 0$, then $\varepsilon = -1$ and $\lambda\lambda^\sigma = 1$, a contradiction as $\lambda \in N(F)$. An entirely similar conclusion is obtained if $B = \mathrm{off}(\lambda, 1)$: if $a = d = 0$ in $A$, then $\lambda^\sigma/\lambda = -1$ and $[A, 1]$, $[B, 1]$ have order 4 by part 4 of Theorem 1, and if one of $a, d \neq 0$, then $\varepsilon = -1$ and $\lambda^\sigma = \lambda$, a contradiction again. It follows that for all non-singular pairs $([A, 1], [B, 1])$, we have $(B^{-1}A)^2 \neq I$ in $\mathrm{PSL}(2, p^{2f})$, and so $H_0$ cannot be dihedral.

By Corollary 1, the order of every element in $G \backslash G_0$ is divisible by 4. Since the order of one of $[A, 1]$ and $[B, 1]$ is assumed to be greater than 4, one of the elements $[AA^\sigma, 0]$, $[BB^\sigma, 0]$ of $H_0$ has to have even order greater than 2. But as $A_4$ and $A_5$ do not contain elements of even order greater than 2, $H_0$ cannot be isomorphic to $A_4$ or $A_5$. The case $H_0 \cong S_4$ is excluded as follows. If there was an orientably-regular map with automorphism group $H$ such that $H_0 \cong S_4$, then the orders of the elements $[A, 1]$, $[B, 1]$ would have to be in the set $\{4, 8\}$ since the largest even order of an element in $S_4$ is 4. According to the list of [2], however, there is no orientably-regular map of type $\{4, 8\}$ or $\{8, 8\}$ with (orientation-preserving) automorphism group of order $2|S_4| = 48$.

We may now apply Proposition 4 to the subgroup $H_0$ to conclude that $H$ is conjugate in $\overline{G}$ to a subgroup isomorphic to $M(p^{2e})$ for some positive divisor $e$ of $f$ such that $f/e$ is odd, completing the proof. □

It follows that a pair $([A, 1], [B, 1])$ of elements of $G$ as above generates a twisted subgroup of $G$ if and only if the pair is non-singular.

## 8 Orbits of Non-singular Pairs: The Diagonal Case

We will identify representatives of $\overline{G}$-orbits of non-singular pairs $([A, 1], [B, 1])$, dealing with $B = \text{dia}(\lambda, 1)$ and $A$ as in (1) here and deferring the case $B = \text{off}(\lambda, 1)$ to the next section.

Instead of working with matrices, the form of $A$ in (1) suggests to look at the corresponding quadruples $(\lambda, b, c, u)$, also called *non-singular*, under the induced action of the stabiliser of $[B, 1]$ in $\overline{G}$. We recall that the values of $\lambda$ and identification of the stabiliser are in items 1 and 3 of Theorem 1. To simplify the notation in what follows, for any $\omega \in F$, we will use the symbol $\sqrt[r]{\omega}$ to denote the set of all $r$th roots of $\omega$ in $F = \text{GF}(q^2), q = p^f$. The analysis in the third part of the proof of Theorem 1 tells us that the stabiliser of $[B, 1]$ in $\overline{G}$ consists exactly of the following elements of $\overline{G}$:

$$
\begin{aligned}
&[P_1(\eta), 0]\,, \text{ where } P_1 = \text{dia}(\eta, 1) \text{ and } \eta \in F_0^*;\\
&[P_2(\eta), 1]\,, \text{ where } P_2 = \text{dia}(\eta\lambda, 1) \text{ and } \eta \in F_0^*;\\
&[P_3(\zeta), 0]\,, \text{ where } P_3 = \text{off}(\zeta, 1) \text{ if } \lambda \in \sqrt[q+1]{-1} \text{ and } \zeta \in \sqrt[q-1]{\lambda^{-2}};\\
&[P_4(\zeta), 1]\,, \text{ where } P_4 = \text{off}(\zeta/\lambda, 1) \text{ if } \lambda \in \sqrt[q+1]{-1} \text{ and } \zeta \in \sqrt[q-1]{\lambda^{-2}}.
\end{aligned}
$$

To find the corresponding orbit of $[A, 1]$, we first evaluate the products $[P_j(\eta), 0]^{-1}$ $[A, 1][P_j(\eta), 0]$ for $A$ as in (1) and $j \in \{1, 2, 3, 4\}$:

$$
\begin{aligned}
&[P_1(\eta), 0]^{-1}\,[A, 1][P_1(\eta), 0] = [C_1, 1], \text{ where } C_1 = \begin{pmatrix} -1 & b\eta^{-1} \\ c\eta & \lambda^\sigma \end{pmatrix};\\
&[P_2(\eta), 1]^{-1}\,[A, 1][P_2(\eta), 1] = [C_2, 1], \text{ where } C_2 = \begin{pmatrix} -1, & b^\sigma(\eta\lambda)^{-1} \\ c^\sigma\eta\lambda^\sigma & \lambda^\sigma \end{pmatrix};\\
&[P_3(\zeta), 0]^{-1}\,[A, 1][P_3(\zeta), 0] = [C_3, 1], \text{ where } C_3 = \begin{pmatrix} -1 & c\zeta/\lambda \\ b\lambda/\zeta & \lambda^\sigma \end{pmatrix}; \text{ and}\\
&[P_4(\zeta), 1]^{-1}\,[A, 1][P_4(\zeta), 1] = [C_4, 1], \text{ where } C_4 = \begin{pmatrix} -1 & -c^\sigma\zeta/\lambda^2 \\ b^\sigma/\zeta & \lambda^\sigma \end{pmatrix}.
\end{aligned}
$$

Let $\lambda = \xi^i$ for a fixed primitive element $\xi \in F$ and some odd $i$ such that $1 \le i \le (q-1)/2$; note that here $\lambda \in \sqrt[q+1]{-1}$ if and only if $i = (q-1)/2$. It follows that we have either $(q-1)/4$ such odd values of $i$ if $q \equiv 1 \bmod 4$ and all are smaller than $(q-1)/2$, or else $(q-3)/4$ such odd $i < (q-1)/2$ together with $i = (q-1)/2$ if $q \equiv -1 \bmod 4$.

In each of the above cases, if $i < (q-1)/2$, the stabiliser $H$ of $[B, 1]$ for $B = \text{dia}(\lambda, 1)$ has order $2(q-1)$. By the above calculations leading to the matrices $C_1$ and $C_2$, the orbit $O$ of the induced action of $H$ on a quadruple $(\lambda, b, c, u)$ satisfying $bc + \lambda^\sigma = u \in N(F)$ is formed by the quadruples $(\lambda, b\eta^{-1}, c\eta, u)$ for $\eta \in F_0^*$ and $(\lambda, b^\sigma(\eta\lambda)^{-1}, c^\sigma\eta\lambda^\sigma, u^\sigma\lambda^\sigma/\lambda)$ for $\eta \in F_0^*$. It is easy to check that all these $2(q-1)$ quadruples are mutually distinct and hence each such orbit $O$ has size $2(q-1)$. If $u = \lambda^\sigma$, then one of $b, c$ would have to be equal to zero, contrary to non-singularity. For each of our $\lfloor (q-1)/4 \rfloor$ choices of $i$, we therefore have a total of $(q^2-3)/2$ choices for $u \in N(F)$, $u \neq \lambda^\sigma$; for each of these, we can freely choose a non-zero $b \in F$ in $q^2 - 1$ ways and then $c$ is uniquely determined. This gives a total of $\lfloor (q-1)/4 \rfloor (q^2 - 1)(q^2 - 3)/2$ quadruples $(\lambda, b, c, u)$ and the number of $H$-orbits on these is given by dividing by $|H| = 2(q-1)$. Denoting by $n_1$, the number of $\overline{G}$-equivalence classes of non-singular quadruples $(\lambda, b, c, u)$ for $i < (q-1)/2$ for diagonal $B$, we obtain

$$n_1 = (q+1) \left\lfloor \frac{q-1}{4} \right\rfloor \frac{q^2-3}{4} . \tag{3}$$

The enumeration is trickier for $i = (q-1)/2$, that is, for $\lambda = \xi^{(q-1)/2}$ and $q \equiv -1 \mod 4$. This is the case when the stabiliser $H^*$ of $[B, 1]$ for $B = \mathrm{dia}(\lambda, 1)$ has order $4(q-1)$ and the orbit $O^*$ of the induced action of $H^*$ on a quadruple $(\lambda, b, c, u)$ satisfying $bc + \lambda^\sigma = u \in N(F)$ includes: (i) the $2(q-1)$ quadruples in $O$ listed in the previous paragraph, and (ii) by the form of the matrices $C_3$ and $C_4$ also the $(q-1)$ quadruples $(\lambda, b', c', u)$ for $b' = c\zeta/\lambda$, $c' = b\lambda/\zeta$ as well as the $(q-1)$ quadruples $(\lambda, b'', c'', u/\lambda^2)$ for $b'' = -c^\sigma\zeta/\lambda^2$, $c'' = b^\sigma/\zeta$, where $\zeta$ can be equal to any of the $q-1$ values of $\sqrt[q-1]{\lambda^{-2}}$. Clearly, either $|O^*| = 4(q-1)$ or $O^* = O$, and we will investigate the second possibility.

We have $O^* = O$ if and only if $b$ is equal to one of $b'$ or $b''$. For our $\lambda = \xi^{(q-1)/2}$, we find that $\sqrt[q-1]{\lambda^{-2}} = \{\xi^{-1}\omega;\ \omega \in F_0^*\}$. It can be checked that $b$ cannot be equal to any of the values of $b''$ but it may happen that $b = b' = c\zeta/\lambda = c\xi^{-1}\xi^{(1-q)/2}\omega'$ for some $\omega' \in F_0^*$. Taking into account the fact that we are only interested in non-zero $b, c$ it follows that $O^* = O$ if and only if $c/b = \xi^{(q+1)/2}\omega$ for some $\omega \in F_0^*$. But this is equivalent to $(c/b)^{q-1} = -1$, which can be rewritten in the form $bc^\sigma + b^\sigma c = 0$. Since $bc^\sigma + b^\sigma c$ is the trace of $AA^\sigma$ for $A$ as in (1), the fact that the trace is zero is equivalent to $[A, 1]^2$ having order 2 in $G$. This all means that $O^* = O$ if and only if the element $[A, 1]$ has order 4 in $G$. But for $\lambda = \xi^{(q-1)/2}$, the order of $[B, 1]$ in $G$ is then equal to 4 as well, a contradiction with non-singularity. It follows that $|O^*| = 4(q-1)$.

Observe that if $c/b = \xi^{(q+1)/2}\omega$ for some $\omega \in F_0^*$, then the determinant equation gives $u = bc + \lambda^\sigma = b^2\xi^{(q+1)/2}\omega + \lambda^\sigma$, and since $q \equiv -1 \mod 4$ this means that $u - \lambda^\sigma \in S(F)$. Conversely, if this holds, then for every $\omega \in F_0^*$ the equation $b^2\xi^{(q+1)/2}\omega = u - \lambda^\sigma$ has two values of $b$ as a solution. For a fixed $v \in N(F)$, let $N(v)$ denote the set of all $u \in N(F)\backslash\{v\}$ for which $u - v \in S(F)$. It is easy to see that the value of $|N(v)|$ does not depend on the choice of $v \in N(F)$. Let $n_F$ denote this common value; it was shown in [14] that $n_F = (q^2 - 1)/4$.

It follows that for every $u \in N(\lambda^\sigma)$, we have $2(q-1)$ values of $b$ satisfying $b^2\xi^{(q+1)/2}\omega = u - \lambda^\sigma$ for some $\omega \in F_0^*$, and these yield the singular pairs identified in the previous paragraph. Thus, for any of the $n_F$ values of $u \in N(\lambda^\sigma)$, we have $(q^2-1) - 2(q-1) = (q-1)^2$ values of $b \in F^*$ that furnish a non-singular pair $[A, 1], [B, 1]$ for $G$, giving a total of $n_F(q-1)^2$ quadruples of the form $(\xi^{(q-1)/2}, b, c, u)$. Any of the remaining $(q^2-3)/2 - n_F$ non-square values of $u$ then simply give $((q^2-3)/2 - n_F)(q^2-1)$ non-singular quadruples $(\xi^{(q-1)/2}, b, c, u)$ that we have to consider. As established earlier, the induced action of the group $H^*$

on these quadruples is semi-regular. Hence, letting $n_2$ be the number of $\overline{G}$-equivalence classes of non-singular quadruples $(\lambda, b, c, u)$ for $\lambda = \xi^{(q-1)/2}$ and $q \equiv -1 \bmod 4$ and for diagonal $B$, we have

$$n_2 = \frac{n_F(q-1)^2 + ((q^2-3)/2 - n_F)(q^2-1)}{4(q-1)} = \frac{1}{8}\left((q+1)(q^2-3) - 4n_F\right) . \tag{4}$$

## 9 Orbits of Non-singular Pairs: The Off-Diagonal Case

We continue looking for representatives of $\overline{G}$-orbits of non-singular pairs ($[A, 1]$, $[B, 1]$) as in the previous section but this time for $B = \text{off}(\lambda, 1)$ and $A$ as in (2), we again reduce the task of finding representatives of the corresponding quadruples $(\lambda, a, d, u)$, called *non-singular* as well, under the induced action of the stabiliser of $[B, 1]$ in $\overline{G}$, with $\lambda$ and the stabiliser being determined by items 2 and 4 of Theorem 1. The orbit of $[A, 1]$ given by (2) under conjugation by elements stabilising $[B, 1]$ is found by calculating the conjugates by the following elements of $\overline{G}$ identified in the fourth part of the proof of Theorem 1:

$[P_1(\eta), 0]$, where $P_1 = \text{dia}(\eta, 1)$ and $\eta \in \sqrt[q+1]{1}$;
$[P_2(\eta), 1]$, where $P_2 = \text{off}(\eta\lambda, 1)$ and $\eta \in \sqrt[q+1]{1}$;
$[P_3(\zeta), 0]$, where $P_3 = \text{off}(\zeta, 1)$ if $\lambda \in \sqrt[q-1]{-1}$ and $\zeta \in \sqrt[q+1]{\lambda^2}$;
$[P_4(\zeta), 1]$, where $P_4 = \text{dia}(\zeta/\lambda, 1)$ if $\lambda \in \sqrt[q-1]{-1}$ and $\zeta \in \sqrt[q+1]{\lambda^2}$.

Evaluation of the products $[P_j(\eta), 0]^{-1}[A, 1][P_j(\eta), 0]$ for $A$ as in (2) and $j \in \{1, 2, 3, 4\}$ gives

$$[P_1(\eta), 0]^{-1}[A, 1][P_1(\eta), 0] = [D_1, 1], \text{ where } D_1 = \begin{pmatrix} a\eta^{-1} & \lambda^\sigma \\ -1 & d\eta \end{pmatrix};$$
$$[P_2(\eta), 1]^{-1}[A, 1][P_2(\eta), 1] = [D_2, 1], \text{ where } D_2 = \begin{pmatrix} -\eta^{-1}d^\sigma\lambda^\sigma/\lambda & \lambda^\sigma \\ -1 & -\eta a^\sigma \end{pmatrix};$$
$$[P_3(\zeta), 0]^{-1}[A, 1][P_3(\zeta), 0] = [D_3, 1], \text{ where } D_3 = \begin{pmatrix} d\zeta/\lambda & \lambda^\sigma \\ -1 & a\lambda/\zeta \end{pmatrix}; \text{ and}$$
$$[P_4(\zeta), 1]^{-1}[A, 1][P_4(\zeta), 1] = [D_4, 1], \text{ where } D_4 = \begin{pmatrix} -a^\sigma\zeta/\lambda & \lambda^\sigma \\ -1 & d^\sigma\lambda/\zeta \end{pmatrix}.$$

Let $\lambda = \xi^i$ for a fixed primitive element $\xi \in F$ and some odd $i$ such that $1 \le i \le (q+1)/2$; here, we have $\lambda \in \sqrt[q-1]{-1}$ if and only if $i = (q+1)/2$. This gives either $(q+1)/4$ such odd values of $i$ if $q \equiv -1 \bmod 4$, all smaller than $(q+1)/2$, or $(q-1)/4$ such odd $i < (q+1)/2$ together with $i = (q+1)/2$ if $q \equiv 1 \bmod 4$.

In each of these instances, if $i < (q+1)/2$, the stabiliser $H$ of $[B, 1]$ for $B = \mathrm{off}(\lambda, 1)$ has order $2(q+1)$. By the above calculations, the orbit $O$ of the induced action of $H$ on a quadruple $(\lambda, a, d, u)$ satisfying $ad + \lambda^\sigma = u \in N(F)$ consists of the quadruples $(\lambda, a\eta^{-1}, d\eta, u)$ and $(\lambda, -d^\sigma\lambda^\sigma(\eta\lambda)^{-1}, -a^\sigma\eta, u^\sigma\lambda^\sigma/\lambda)$ for $\eta \in F_0^*$. By non-singularity, at least one of $a, d$ is non-zero. This implies, by an easy computation, that all the above $2(q+1)$ quadruples are mutually distinct and hence each such orbit $O$ on our quadruples under the action of $H$ has size $2(q+1)$. If $u = \lambda^\sigma$, then exactly one of $a, d$ is equal to zero while the other one can assume any of the $q^2 - 1$ values in $F^*$. For each of the $\lfloor (q+1)/4 \rfloor$ choices of $i$ considered here, we have:

($\alpha$) $(q^2-3)/2$ choices of $u \in N(F)\backslash\{\lambda^\sigma\}$, each admitting a choice of $a \in F^*$ in $q^2-1$ ways, uniquely determining $d$ and making $\lfloor (q+1)/4 \rfloor (q^2-1)(q^2-3)/2$ quadruples; and

($\beta$) for $u = \lambda^\sigma$, we have $2(q^2-1)$ choices of non-zero $a$ (for $d = 0$) and non-zero $d$ (for $a = 0$), giving $\lfloor (q+1)/4 \rfloor \cdot 2(q^2-1)$ quadruples.

Since the number of $H$-orbits on quadruples in ($\alpha$) and ($\beta$) is given by dividing by $|H| = 2(q+1)$, we arrive at a total of $n_3$ $\overline{G}$-equivalence classes of non-singular quadruples $(\lambda, a, d, u)$ for $i < (q+1)/2$ and off-diagonal $B$, where

$$n_3 = (q-1)\left\lfloor \frac{q+1}{4} \right\rfloor \frac{q^2+1}{4}\,. \tag{5}$$

The enumeration is again more tricky for $i = (q+1)/2$, that is, for $\lambda = \xi^{(q+1)/2}$ and $q \equiv 1 \bmod 4$. The stabiliser $H^*$ of $[B, 1]$ for $B = \mathrm{off}(\lambda, 1)$ has in this case order $4(q+1)$ and the orbit $O^*$ of the induced action of $H^*$ on a quadruple $(\lambda, a, d, u)$ with values fulfilling the equation $ad + \lambda^\sigma = u \in N(F)$ contains the $2(q+1)$ quadruples in $O$ listed in the previous paragraph together with the $2(q+1)$ distinct quadruples $(\lambda, d\zeta/\lambda, a\lambda/\zeta, u)$ and $(\lambda, -a^\sigma\zeta/\lambda, d^\sigma\lambda/\zeta, -u^\sigma)$, where $\zeta$ can be any of the $q+1$ values of $\sqrt[q+1]{\lambda^2}$. As before, either $|O^*| = 4(q+1)$ or $O^* = O$, and we will study the second eventuality in detail.

By inspection of the orbits $O$ and $O^*$ in this case, the orbit $O^*$ coincides with $O$ if and only if either

($\gamma$) $a\eta^{-1} = d\zeta/\lambda$, or

($\delta$) $a\eta^{-1} = -a^\sigma\zeta/\lambda$ and $d\eta = d^\sigma\lambda/\zeta$,

all for some $\eta \in \sqrt[q+1]{1}$. For our calculations, we may without loss of generality assume that $\zeta = \xi$, as this is one of the $(q+1)$th roots of $\lambda^2 = \xi^{q+1}$. Note that we now have $\lambda^\sigma = -\lambda$, $\eta^\sigma = \eta^{-1}$, $\zeta^\sigma = \lambda^2/\zeta$, and therefore also $(\zeta\eta/\lambda)^\sigma = -(\zeta\eta/\lambda)^{-1}$.

The case ($\gamma$): From $a = d(\zeta\eta/\lambda)$ and its $\sigma$-image $a^\sigma = d^\sigma(\zeta\eta/\lambda)^\sigma = -d^\sigma(\zeta\eta/\lambda)^{-1}$ we obtain $aa^\sigma + dd^\sigma = 0$. But $aa^\sigma + dd^\sigma$ is the trace of $AA^\sigma$, which means that the order of both $[A, 1]$ and $[B, 1]$ is 4. It follows that ($\gamma$) cannot occur for a non-singular pair.

The case ($\delta$): From $a = -a^\sigma(\zeta\eta/\lambda)$ and its $\sigma$-image $a^\sigma = -a(\zeta\eta/\lambda)^\sigma = a(\zeta\eta/\lambda)^{-1}$, we have $a = -a$, that is, $a = 0$. Similarly, from $d = d^\sigma\lambda/(\zeta\eta)$, one concludes that $d = 0$, contrary to non-singularity.

We can now complete the enumeration of non-singular quadruples $(\lambda, a, d, u)$ in the case $B = \mathrm{off}(\lambda, 1)$ for $\lambda = \xi^{(q+1)/2}$. As we saw above, the case ($\delta$) is excluded. The case ($\gamma$) does not yield non-singular pairs but we still need more details about the corresponding quadruples to be excluded. We saw that ($\gamma$) occurs if and only if $d = a\lambda/(\zeta\eta)$, and this gives $u = ad + \lambda^\sigma = a^2\xi^{(q-1)/2}\xi^{i(q-1)} + \lambda^\sigma$, $i \in \{0, 1, \ldots, q\}$. Since now $q \equiv 1 \bmod 4$, it follows that $u - \lambda^\sigma \in S(F)$, that is, $u \in N(\lambda^\sigma)$. Conversely, if $u \in N(\lambda^\sigma)$, then for each of the $q + 1$ values of $i \in \{0, 1, \ldots, q\}$, we obtain two solutions for $a$ as above (and for each such $a$ a unique $d$), not giving a non-singular pair. It follows that for each of the $n_F$ elements $u \in N(\lambda^\sigma)$, we can choose $(q^2 - 1) - 2(q + 1) = (q + 1)(q - 3)$ values $a \in F^*$ (and for each such $a$ a unique $d$) such that $ad + \lambda^\sigma = u$, so that the order of $[A, 1]$ is not equal to 4. This gives a total of $n_F(q + 1)(q - 3)$ quadruples $(\xi^{(q+1)/2}, a, d, u)$ collected so far. Any of the remaining $(q^2 - 3)/2 - n_F$ non-square values of $u$ distinct from $\lambda^\sigma$ then give $((q^2 - 3)/2 - n_F)(q^2 - 1)$ quadruples $(\xi^{(q+1)/2}, a, d, u)$ that we have to consider for non-singular pairs. Finally, if $u = \lambda^\sigma$, then either $a = 0$ with $q^2 - 1$ choices for $d$, or $d = 0$ with $q^2 - 1$ choices for $a$, giving further $2(q^2 - 1)$ quadruples. Since we have shown earlier that the induced action of the group $H^*$ on these quadruples is semi-regular with orbits of length $4(q + 1)$, the number $n_4$ of $\overline{G}$-equivalence classes of non-singular quadruples $(\lambda, a, d, u)$ for off-diagonal $B$ and $\lambda = \xi^{(q+1)/2}$, $q \equiv 1 \bmod 4$, is

$$n_4 = \frac{n_F(q + 1)(q - 3) + ((q^2 - 3)/2 - n_F)(q^2 - 1) + 2(q^2 - 1)}{4(q + 1)},$$

which simplifies to

$$n_4 = \frac{1}{8}\left((q - 1)(q^2 + 1) - 4n_F\right). \tag{6}$$

## 10 Enumeration of Orientably-Regular Maps on $M(q^2)$

We have seen in Sect. 7 that a pair $([A, 1], [B, 1])$ of elements of $G$, with diagonal $B$ and $A$ given by (1) or with off-diagonal $B$ and $A$ given by (2), and with product of order two, generates a twisted subgroup of $G$ if and only if the pair is non-singular. In the previous two sections we have counted orbits of non-singular pairs in $G$ under conjugation in $\overline{G}$, with no regard to subgroups the pairs generate. The number of these orbits turns out to be $n_1 + n_3 + n_4$ if $q \equiv 1 \bmod 4$ and $n_1 + n_2 + n_3$ if $q \equiv -1 \bmod 4$; in both cases, the sum is equal to $(q^2 - 1)(q^2 - 2)/8$. We state this as a separate result.

**Proposition 10** *The number of $\overline{G}$-orbits of non-singular pairs in $G = M(q^2)$ is equal to $(q^2 - 1)(q^2 - 2)/8$.* □

We will now refine our considerations and take into account subgroups generated by non-singular pairs. For our group $G = G_{2f} = M(p^{2f})$ and for any positive divisor

$e$ of $f$ such that $f/e$ is odd we let $G_{2e}$ denote the canonical copy of $M(p^{2e})$ in $G$. In Lemma 2, we saw that the automorphism $\sigma$ of $F = F_{2f} = \mathrm{GF}(p^{2f})$ of order two restricts to an automorphism $\sigma_{2e}$ of order two of the subfield $F_{2e} = \mathrm{GF}(p^{2e})$ of $F$. We recall that $\overline{G} = \overline{G}_{2f} = G_{2f}\langle\sigma\rangle$ and we similarly introduce $\overline{G}_{2e}$ for every $e$ as above by letting $\overline{G}_{2e} = G_{2e}\langle\sigma_{2e}\rangle$.

Let $\mathrm{orb}_f(e)$ denote the number of $\overline{G}_{2f}$-orbits of non-singular pairs $([A, 1], [B, 1])$ of $G$ that generate a subgroup of $G$ isomorphic to $M(p^{2e})$. At the same time, let $\mathrm{orb}(e)$ be the number of orbits of non-singular pairs of $G_{2e}$ which generate $G_{2e}$. The two quantities are, in fact, equal which is fundamental for our enumeration.

**Proposition 11** *For each positive divisor $e$ of $f$ with $f/e$ odd, we have* $\mathrm{orb}_f(e) = \mathrm{orb}(e)$.

***Proof*** It is clear that every $\overline{G}_{2e}$-orbit of a non-singular pair in the canonical copy $G_{2e} \cong M(p^{2e})$ in $G_{2f}$ is contained in a $\overline{G}_{2f}$-orbit of the same pair. In the reverse direction, let a non-singular pair in $G$ generate a subgroup isomorphic to $M(p^{2e})$. Since, by the important Proposition 3, all such subgroups are $\overline{G}_{2f}$-conjugate in $G_{2f}$, we may assume that the non-singular pair is contained in $G_{2e}$. But then the $\overline{G}_{2f}$-orbit of this pair obviously contains a $\overline{G}_{2e}$-orbit of the same pair. The result will now be a consequence of the following claim:

*Let* $([A, 1], [B, 1])$ *and* $([A', 1], B', 1])$ *be two non-singular pairs of* $G_{2e}$ *both generating* $G_{2e}$ *and lying in the same* $\overline{G}_{2f}$*-orbit of* $G_{2f}$. *Then, the two pairs are contained in the same* $\overline{G}_{2e}$*-orbit of* $G_{2e}$.

We have $B = \mathrm{dia}(\lambda, 1)$ and $B' = \mathrm{dia}(\lambda', 1)$, or $B = \mathrm{off}(\lambda, 1)$ and $B' = \mathrm{off}(\lambda', 1)$, in both cases for some $\lambda, \lambda' \in F_{2e}$. The assumption that the two pairs are contained in the same $\overline{G}_{2f}$-orbit means, by Proposition 6, that there exists an element $P \in \mathrm{GL}(2, p^{2f})$ such that $AP^\sigma = \varepsilon P A'^{(\sigma)}$ and $BP^\sigma = \delta P B'^{(\sigma)}$ for some $\varepsilon, \delta \in F^*$. Propositions 7 and 8 applied to the equation $BP^\sigma = \delta P B'^{(\sigma)}$ tell us that $P = \mathrm{dia}(\omega, 1)$ or $P = \mathrm{off}(\omega, 1)$ for some $\omega \in F^*$. We will show that for all choices of $A$, $A'$ and $P$ the equation $AP^\sigma = \varepsilon P A'^{(\sigma)}$ implies that $\omega \in F_{2e}$, which will prove that our two non-singular pairs are in one $G_{2e}$-orbit.

There are two possibilities for $A$ and $A'$, given in (1) and (2), and two possibilities for $P$, the diagonal and the off-diagonal case, to substitute in the equation $AP^\sigma = \varepsilon P A'^{(\sigma)}$; in all the four cases, we need to conclude that $\omega \in F_{2e}$. We demonstrate it on the case when $P = \mathrm{off}(\omega, 1)$ and when $A$, $A'$ are given by (2), that is, when $A$ has rows $(a, \lambda^\sigma)$, $(-1, d)$ and $A'$ has rows $(a', \lambda'^\sigma)$, $(-1, d')$ for some $a, d, \lambda, a', d', \lambda' \in F_{2e}$. The above equation then reads

$$\begin{pmatrix} a & \lambda^\sigma \\ -1 & d \end{pmatrix} \begin{pmatrix} 0 & \omega^\sigma \\ 1 & 0 \end{pmatrix} = \varepsilon \begin{pmatrix} 0 & \omega \\ 1 & 0 \end{pmatrix} \begin{pmatrix} a' & \lambda'^\sigma \\ -1 & d' \end{pmatrix}^{(\sigma)},$$

giving the equations

$$\lambda^\sigma = -\varepsilon\omega\,, \quad \omega^\sigma a = \varepsilon\omega d'^{(\sigma)}\,, \quad d = \varepsilon a'^{(\sigma)}\,, \quad -\omega = \varepsilon^\sigma \lambda'^{(\sigma)}\,.$$

If $a', d \neq 0$, then the third equation in combination with the first gives $\omega = -\lambda^\sigma/\varepsilon = -\lambda^\sigma a'^{(\sigma)}/d$, showing that $\omega \in F_{2e}$. If $a' = d = 0$, we must have $a, d' \neq 0$ and then a combination of the first two equations gives $\omega^\sigma = \varepsilon\omega d'^{(\sigma)}/a = -\lambda^\sigma d'^{(\sigma)}/a$, showing again that $\omega \in F_{2e}$. The remaining three cases are similar (and easier) to check. This completes the proof. □

For positive integers $x$, let us define a function $h$ by $h(x) = (p^{2x} - 1)(p^{2x} - 2)/8$. In terms of $h$ and the numbers $\mathrm{orb}_f(e)$, Proposition 10 simply says that $\sum_e \mathrm{orb}_f(e) = h(f)$, where summation is taken over all positive divisors $e$ of $f$ such that $f/e$ is odd. By Proposition 11, we may replace $\mathrm{orb}_f(e)$ with $\mathrm{orb}(e)$ and obtain $\sum_e \mathrm{orb}(e) = h(f)$, with the same summation convention. This miniature but important detail enables us to make a substantial advance in the enumeration.

Let $f = 2^\alpha o$ where $o$ is an odd integer and let $e$ be a divisor of $f$ such that $f/e$ is odd; equivalently, $e = 2^\alpha d$ where $d$ is a positive (and necessarily odd) divisor of $o$. Taking the above notes into account, Proposition 10 may then be restated as follows:

$$\sum_{d|o} \mathrm{orb}(2^\alpha d) = h(2^\alpha o)\,. \tag{7}$$

Using the Möbius inversion, we obtain $\mathrm{orb}(f) = \mathrm{orb}(2^\alpha o) = \sum_{d|o} \mu(o/d)h(2^\alpha d)$, where $\mu$ is the classical number-theoretic Möbius function $\mu$ on positive integers. We thus arrive at our first main result.

**Theorem 2** *Let $q = p^f$ for an odd prime $p$, let $G = M(q^2)$, and let $f = 2^\alpha o$ with $o$ odd. The number of $\overline{G}$-orbits of non-singular generating pairs of $G$ is equal to*

$$\sum_{d|o} \mu(o/d)h(2^\alpha d)\,, \quad \text{where } h(x) = (p^{2x} - 1)(p^{2x} - 2)/8\,.$$

The last step is to study conjugacy of non-singular pairs of $M(q^2)$ under the action of the group $\mathrm{Aut}(M(q^2))$ which, as we know by Proposition 1, is isomorphic to $\mathrm{P\Gamma L}(2, q^2)$. Since for $q = p^f$, we have $\mathrm{P\Gamma L}(2, q^2) \cong \mathrm{PGL}(2, q^2) \rtimes Z_{2f} \cong \overline{G} \rtimes Z_f$, it is sufficient to investigate the induced action of the Galois automorphisms $\sigma_j : z \mapsto z^{p^j}$ for $z \in F = \mathrm{GF}(p^{2f})$ and $1 \le j \le f-1$ on the $\overline{G}$-orbits of our non-singular pairs $([A, 1], [B, 1])$. We will use the natural notation $O^{\sigma_j}$ for the $\sigma_j$-image of a $\overline{G}$-orbit $O$ of a pair $([A, 1], [B, 1])$ of elements of $G$. Note that $\sigma_f = \sigma$, and we also have $O^{\sigma_f} = O$, by the remark made at the beginning of Sect. 7. Clearly, if $O^{\sigma_j} \cap O \neq \emptyset$, then $O^{\sigma_j} = O$.

**Proposition 12** *Let $O$ be the orbit of a non-singular pair $([A, 1], [B, 1])$ of elements of $G$ under conjugation in $\overline{G}$ and let $j$ be the smallest positive integer for which $O^{\sigma_j} = O$. If $[A, 1]$ and $[B, 1]$ generate $G$, then $j = f$.*

***Proof*** We may assume that $f \ge 2$, otherwise the result is trivial. Suppose that the pair $([A, 1], [B, 1])^{\sigma_j} = ([A, 1]^{\sigma_j}, [B, 1]^{\sigma_j})$ is $\overline{G}$-conjugate to the pair $([A, 1], [B, 1])$, that is, there exists some $C \in \mathrm{GL}(2, q^2)$ and $i \in Z_2$ such that $[A, 1]^{\sigma_j} =$

$[C,i]^{-1}[A,1][C,i]$ and $[B,1]^{\sigma_j}=[C,i]^{-1}[B,1][C,i]$. It follows that for every $[X,1]\in\langle[A,1],[B,1]\rangle$, we have $[X,1]^{\sigma_j}=[C,i]^{-1}[X,1][C,i]$. Using our assumption that $\langle[A,1],[B,1]\rangle=G$, we conclude that the above is valid also for $X=\mathrm{dia}(\xi,1)$, where $\xi$ is a primitive element of $F=\mathrm{GF}(p^{2f})$. Letting $C$ have elements $\alpha,\beta,\gamma,\delta$ in the usual order, the equivalent form $[C,i][\mathrm{dia}(\xi^{p^j},1),1]=[\mathrm{dia}(\xi,1),1][C,i]$ of the above equation yields

$$\begin{pmatrix}\alpha & \beta\\ \gamma & \delta\end{pmatrix}\begin{pmatrix}(\xi^{p^j})^{(\sigma)} & 0\\ 0 & 1\end{pmatrix}=\varepsilon\begin{pmatrix}\xi & 0\\ 0 & 1\end{pmatrix}\begin{pmatrix}\alpha^\sigma & \beta^\sigma\\ \gamma^\sigma & \delta^\sigma\end{pmatrix}$$

for some $\varepsilon\in F^*$; here, we used the $(\sigma)$-convention introduced at the end of Sect. 4. This gives the system of equations

$$\alpha(\xi^{p^j})^{(\sigma)}=\varepsilon\xi\alpha^\sigma\ ,\quad \beta=\varepsilon\xi\beta^\sigma\ ,\quad \gamma(\xi^{p^j})^{(\sigma)}=\varepsilon\gamma^\sigma\ ,\ \delta=\varepsilon\delta^\sigma\ .$$

Consider first the case $\delta\neq 0$; without loss of generality, we then may assume $\delta=1$. Then $\varepsilon=1$, and the equation for $\alpha$ gives $(\xi^{p^j})^{(\sigma)}\xi^{-1}=\alpha^\sigma\alpha^{-1}$, or, equivalently, $\xi^{p^{j+if}-1}=\alpha^{p^f-1}$. It follows that $p^f-1$ is a divisor of $p^{j+if}-1$, which implies that $f$ divides $j+if$ and hence $f$ divides $j$, which, since $j\le f$, shows that $j=f$. If $\delta=0$ then, without loss of generality, $\beta=1$ and so $\varepsilon=\xi^{-1}$. The equation for $\gamma$ now implies $\xi^{p^{j+if}+1}=\gamma^{p^f-1}$. It follows that $p^f-1$ divides $p^{j+if}+1$ and hence also $p^{2(j+if)}-1$ and so $f$ must divide $2(j+if)$. Thus, $f$ is a divisor of $2j$ and as $j\le f$, we have either $j=f$ or $j=f/2$ (assuming $f$ is even). But the last case is easily seen to be impossible since $p^f-1$ is not a divisor of $p^{f/2}+1$ or $p^{3f/2}+1$. This completes the proof. □

Proposition 12 tells us that if a non-singular pair $([A,1],[B,1])$ of elements of $G$ actually generates $G$ and gives rise to an orbit $O$ under conjugation in $\overline{G}$, then the action of the group $\mathrm{Aut}(M(q^2))$ fuses the $f$ orbits $O^{\sigma_j}$ for $j\in\{0,1,\ldots,f-1\}$ into a single orbit. Recalling the one-to-one correspondence between isomorphism classes of orientably-regular maps supported by the group $G=M(q^2)$ and orbits of (necessarily non-singular) generating pairs of $G$ under conjugation by $\mathrm{Aut}(G)$, Theorem 2 then immediately implies our second main result.

**Theorem 3** *Let $q=p^f$ for an odd prime $p$ and let $f=2^\alpha o$ with $o$ odd. The number of orbits of non-singular generating pairs of $M(q^2)$ under the action of the group* $\mathrm{Aut}(M(q^2))$*, and hence the number of isomorphism classes of orientably-regular maps $\mathcal{M}$ with* $\mathrm{Aut}^+(\mathcal{M})\cong M(q^2)$*, is equal to*

$$\frac{1}{f}\sum_{d|o}\mu(o/d)h(2^\alpha d)\ ,\quad \text{where}\ \ h(x)=(p^{2x}-1)(p^{2x}-2)/8\ .$$

## 11 Enumeration of Reflexible Maps

A map is called *reflexible* if it admits an automorphism reversing the orientation of the surface. For orientably-regular maps represented by triples $(G, x, y)$ as indicated in the Introduction, reflexibility is equivalent to the existence of an automorphism $\theta$ of the group $G$ such that $\theta(x) = x^{-1}$ and $\theta(y) = y^{-1}$. Note that if such a $\theta$ exists, then $\theta^2 = id$.

In the specific situation considered in this paper, namely, when $G = M(q^2)$ for $q = p^f$, we established in Proposition 1 that $\mathrm{Aut}(G) \cong \mathrm{P\Gamma L}(2, q^2)$. Moreover, it is well known that every automorphism in $\mathrm{P\Gamma L}(2, q^2) \cong \mathrm{PGL}(2, q^2) \rtimes Z_{2f}$ is a composition of conjugation by some element of $\mathrm{PGL}(2, q^2)$ and power of the Frobenius automorphism $z \mapsto z^p$ of the Galois field $F = GF(p^{2f})$. It follows that an *involutory* automorphism $\theta$ of $G = M(q^2)$ is a composition of conjugation as above with $\sigma^i$ for $i \in \{0, 1\}$, where $\sigma$ is the automorphism of $F$ sending $z$ to $z^q$. By the remark at the beginning of Sect. 7, however, the action of $\sigma$ is equivalent to conjugation in $\overline{G} = G\langle\sigma\rangle$ by the element $[I, 1]$. Consequently, an orientably-regular map on the group $G = M(q^2)$ generated by a pair of elements $x = [A, 1]$ and $y = [B, 1]$ is reflexible if and only if the ordered pairs $(x, y)$ and $(x^{-1}, y^{-1})$ are conjugate by an *involutory* element of $\overline{G}$.

In this section, we will count the number of reflexible orientably-regular maps on $M(q^2)$. In particular, we will see that not all orientably-regular maps with automorphism group $M(q^2)$ are reflexible, in contrast with the position for $PGL(2, q^2)$, see e.g. [3]. We will use techniques similar to the main enumeration in previous sections and structure our explanations accordingly.

### *11.1 Conjugating Involutions*

In view of the above findings, an orientably-regular map on $M(q^2)$ given by the generating pair $([A, 1], [B, 1])$ is reflexible if and only if there is an involution $[C, i] \in \overline{G}$ for some $i \in \{0, 1\}$ such that $[C, i][A, 1][C, i] = [A, 1]^{-1}$ and $[C, i][B, 1][C, i] = [B, 1]^{-1}$. As before, we will be dealing separately with the diagonal and off-diagonal cases for matrix $B$.

**The diagonal case** $B = \mathrm{dia}(\lambda, 1)$ We can write $(B^\sigma)^{-1} = \begin{pmatrix} \varepsilon\lambda^{-q} & 0 \\ 0 & \varepsilon \end{pmatrix}$ for some $\varepsilon \neq 0$. We seek a conjugating involution $[C, i]$ as above and begin with $i = 0$. For $[C, 0]$ to be an involution, we must have $\mathrm{tr}(C) = 0$ and we can without loss of generality write $C = \begin{pmatrix} 1 & \beta \\ \gamma & -1 \end{pmatrix}$ or $C = \begin{pmatrix} 0 & \beta \\ 1 & 0 \end{pmatrix}$.

For the first form of $C$, the conjugation equation becomes:

$$\begin{pmatrix} 1 & \beta \\ \gamma & -1 \end{pmatrix} \begin{pmatrix} \lambda & 0 \\ 0 & 1 \end{pmatrix} \begin{pmatrix} 1 & \beta^q \\ \gamma^q & -1 \end{pmatrix} = \begin{pmatrix} \varepsilon\lambda^{-q} & 0 \\ 0 & \varepsilon \end{pmatrix}$$

This leads to the system of equations $\lambda + \beta\gamma^q = \varepsilon\lambda^{-q}$, $\lambda\beta^q - \beta = 0$, $\gamma\lambda - \gamma^q = 0$, and $\beta^q\gamma\lambda + 1 = \varepsilon$. If $\beta = 0$, then $\varepsilon = 1$ and so $\lambda = \lambda^{-q}$ or $\lambda \in \sqrt[q+1]{1}$. If $\beta \neq 0$, then $\lambda = \beta^{1-q}$. In either case, $\lambda \in S(F)$ contrary to the construction of $B$. Hence, there are no conjugating matrices $C$ of this form.

For the second possibility, we have:

$$\begin{pmatrix} 0 & \beta \\ 1 & 0 \end{pmatrix}\begin{pmatrix} \lambda & 0 \\ 0 & 1 \end{pmatrix}\begin{pmatrix} 0 & \beta^q \\ 1 & 0 \end{pmatrix} = \begin{pmatrix} \varepsilon\lambda^{-q} & 0 \\ 0 & \varepsilon \end{pmatrix}$$

This leads to $\beta = \varepsilon\lambda^{-q}$ and $\lambda\beta^q = \varepsilon$, giving potential solutions of this form with $\beta$ satisfying $\beta^{q-1} = \lambda^{q-1}$.

Now, consider the case $i = 1$. Since $[C, 1]$ is an involution, it cannot be in $G$; it follows that $CC^\sigma = \varepsilon I$ for some $\varepsilon \neq 0$ and $\det(C) \in S(F)$. Letting $C = \begin{pmatrix} \alpha & \beta \\ \gamma & \delta \end{pmatrix}$, we may either have $\gamma = 0$ in which case we can take $\alpha = 1$, or $\gamma = 1$, furnishing again two possible forms for $C$.

The first form of $C$ for $i = 1$ we will consider is $C = \begin{pmatrix} 1 & \beta \\ 0 & \delta \end{pmatrix}$ for some $\delta \in S(F)$. Since $CC^\sigma = \varepsilon I$, we must have $\varepsilon = 1$ and $\delta \in \sqrt[q+1]{1}$. In this case, the matrix equations are

$$\begin{pmatrix} 1 & \beta \\ 0 & \delta \end{pmatrix}\begin{pmatrix} \lambda^q & 0 \\ 0 & 1 \end{pmatrix}\begin{pmatrix} 1 & \beta \\ 0 & \delta \end{pmatrix} = \begin{pmatrix} \varepsilon\lambda^{-q} & 0 \\ 0 & \varepsilon \end{pmatrix}$$

It follows that $\lambda^q = \varepsilon\lambda^{-q}$ and $\delta^2 = \varepsilon$, implying $\delta^2 = \lambda^{2q}$, but as $\delta \in S(F)$ and $\lambda \in N(F)$, this is impossible. We conclude that the first form cannot occur.

The second form of $C$ for $i = 1$ to consider is $C = \begin{pmatrix} \alpha & \beta \\ 1 & \delta \end{pmatrix}$ with $\alpha\delta - \beta \in S(F)$. The requirement for $[C, 1]$ to have order two gives

$$\begin{pmatrix} \alpha & \beta \\ 1 & \delta \end{pmatrix}\begin{pmatrix} \alpha^q & \beta^q \\ 1 & \delta^q \end{pmatrix} = \begin{pmatrix} \varepsilon & 0 \\ 0 & \varepsilon \end{pmatrix}$$

Thus, $\delta = -\alpha^q$ and $C$ has the form $\begin{pmatrix} \alpha & \beta \\ 1 & -\alpha^q \end{pmatrix}$. The conjugation equation then becomes

$$\begin{pmatrix} \alpha & \beta \\ 1 & -\alpha^q \end{pmatrix}\begin{pmatrix} \lambda^q & 0 \\ 0 & 1 \end{pmatrix}\begin{pmatrix} \alpha & \beta \\ 1 & -\alpha^q \end{pmatrix} = \begin{pmatrix} \varepsilon\lambda^{-q} & 0 \\ 0 & \varepsilon \end{pmatrix}$$

Now, $\alpha\lambda^q - \alpha^q = 0$. If $\alpha \neq 0$, then $\lambda^q = \alpha^{q-1} \in S(F)$, which is impossible. So $\alpha = 0$ and $C$ must have the form $\begin{pmatrix} 0 & \beta \\ 1 & 0 \end{pmatrix}$ for some $\beta$. Since $CC^\sigma = \varepsilon I$, we must have $\beta = \beta^q$ and hence $\beta \in F'^*$.

**The off-diagonal case** $B = \mathrm{off}(\lambda, 1)$ We can write $(B^\sigma)^{-1} = \begin{pmatrix} 0 & \varepsilon \\ \varepsilon\lambda^{-q} & 0 \end{pmatrix}$ for some $\varepsilon \neq 0$. We begin with the case $i = 0$, and as before we can without loss of generality set $C = \begin{pmatrix} 1 & \beta \\ \gamma & -1 \end{pmatrix}$ or $C = \begin{pmatrix} 0 & \beta \\ 1 & 0 \end{pmatrix}$. For the first form of $C$, we have

$$\begin{pmatrix} 1 & \beta \\ \gamma & -1 \end{pmatrix} \begin{pmatrix} 0 & \lambda \\ 1 & 0 \end{pmatrix} \begin{pmatrix} 1 & \beta^q \\ \gamma^q & -1 \end{pmatrix} = \begin{pmatrix} 0 & \varepsilon \\ \varepsilon\lambda^{-q} & 0 \end{pmatrix}$$

It follows that $\beta + \gamma^q\lambda = 0$, $\beta^{q+1} - \lambda = \varepsilon$, $\gamma^{q+1}\lambda - 1 = \varepsilon\lambda^{-q}$ and $-\beta^q - \gamma\lambda = 0$. If $\beta = 0$, then $\gamma = 0$ and so $\varepsilon = -\lambda$ and $\lambda^{1-q} = 1$. If $\beta, \gamma \neq 0$, then $\beta + \gamma^q\lambda^q = 0$, giving $\lambda = \lambda^q$. In either case, we conclude that $\lambda \in S(F)$, contrary to the construction of $B$. Consequently, there are no matrices $C$ of this form.

For the possibility $C = \begin{pmatrix} 0 & \beta \\ 1 & 0 \end{pmatrix}$, we have

$$\begin{pmatrix} 0 & \beta \\ 1 & 0 \end{pmatrix} \begin{pmatrix} 0 & \lambda \\ 1 & 0 \end{pmatrix} \begin{pmatrix} 0 & \beta^q \\ 1 & 0 \end{pmatrix} = \begin{pmatrix} 0 & \varepsilon \\ \varepsilon\lambda^{-q} & 0 \end{pmatrix}$$

This leads to $\beta^{q+1} = \varepsilon$, $\lambda = \varepsilon\lambda^{-q}$ and we have potential solutions of this form with $\beta$ satisfying $\beta^{q+1} = \lambda^{q+1}$.

Now, consider the second case $i = 1$. As before the first possible form for $C$ is $\begin{pmatrix} 1 & \beta \\ 0 & \delta \end{pmatrix}$ for some $\delta \in \sqrt[q+1]{1}$. In this case, the matrix equations are

$$\begin{pmatrix} 1 & \beta \\ 0 & \delta \end{pmatrix} \begin{pmatrix} 0 & \lambda^q \\ 1 & 0 \end{pmatrix} \begin{pmatrix} 1 & \beta \\ 0 & \delta \end{pmatrix} = \begin{pmatrix} 0 & \varepsilon \\ \varepsilon\lambda^{-q} & 0 \end{pmatrix}$$

We must have $\beta = 0$ and $\delta = \varepsilon\lambda^{-q}$, giving potential solutions of this form for any $\delta \in \sqrt[q+1]{1}$.

The second possible form of $C$ for $i = 1$ is $C = \begin{pmatrix} \alpha & \beta \\ 1 & -\alpha^q \end{pmatrix}$ for which the equations are

$$\begin{pmatrix} \alpha & \beta \\ 1 & -\alpha^q \end{pmatrix} \begin{pmatrix} 0 & \lambda^q \\ 1 & 0 \end{pmatrix} \begin{pmatrix} \alpha & \beta \\ 1 & -\alpha^q \end{pmatrix} = \begin{pmatrix} 0 & \varepsilon \\ \varepsilon\lambda^{-q} & 0 \end{pmatrix}$$

Here, we have $\alpha\beta + \alpha\lambda^q = 0$, $\beta^2 - \alpha^{q+1}\lambda^q = \varepsilon$, $-\alpha^{q+1} + \lambda^q = \varepsilon\lambda^{-q}$, and $-\alpha^q\beta - \alpha^q\lambda^q = 0$. If $\alpha \neq 0$, then $\lambda^q = -\beta$ but $\beta \in F'^*$, so $-\beta \in S(F)$ and this is impossible. It follows that $\alpha = 0$ but then $\varepsilon = \beta^2 = \lambda^{2q}$, while at the same time, $\beta \in S(F)$ and $\lambda^q \in N(F)$, a contradiction. Hence, there are no matrices $C$ of this form.

**Summary of possible conjugating elements** In the case $B = \mathrm{dia}(\lambda, 1)$, the possible conjugating elements can have the following forms:

$$[C, 0]; \quad C = \begin{pmatrix} 0 & \beta \\ 1 & 0 \end{pmatrix}; \quad \beta^{q-1} = \lambda^{q-1} \tag{8}$$

$$[C,1];\quad C=\begin{pmatrix}0&\beta\\1&0\end{pmatrix};\quad \beta\in F'^{*} \tag{9}$$

In the case $B=\text{off}(\lambda,1)$, the possible conjugating elements can have the following forms.

$$[C,0];\quad C=\begin{pmatrix}0&\beta\\1&0\end{pmatrix};\quad \beta^{q+1}=\lambda^{q+1} \tag{10}$$

$$[C,1];\quad C=\begin{pmatrix}1&0\\0&\delta\end{pmatrix};\quad \delta\in \sqrt[q+1]{1} \tag{11}$$

## 11.2 Enumeration

We now proceed to the actual enumeration and follow the same strategy as we used for general maps, namely, counting orbits in the diagonal and off-diagonal cases for $B$ as in Sects. 8 and 9 and then deriving the final enumeration result using the Möbius inversion formula as in Sect. 10.

**Counting orbits: the case $B=\text{dia}(\lambda,1)$** In this case, we have $A=\begin{pmatrix}-1&b\\c&\lambda^q\end{pmatrix}$. We can write $[A,1]^{-1}$ as $[(A^\sigma)^{-1},1]$ where $(A^\sigma)^{-1}$ has the form $\varepsilon\begin{pmatrix}-\lambda&b^q\\c^q&1\end{pmatrix}$ for some $\varepsilon\neq 0$.

The first possibility is that the conjugating element has the form $[C,0]$ with $C=\begin{pmatrix}0&\beta\\1&0\end{pmatrix}$ for some $\beta$ such that $\beta^{q-1}=\lambda^{q-1}$. In this case, the equations are

$$\begin{pmatrix}0&\beta\\1&0\end{pmatrix}\begin{pmatrix}-1&b\\c&\lambda^q\end{pmatrix}\begin{pmatrix}0&\beta^q\\1&0\end{pmatrix}=\varepsilon\begin{pmatrix}-\lambda&b^q\\c^q&1\end{pmatrix}$$

After some manipulation, the resulting equations lead to $\beta=-b^q/c$; $\beta^{q-1}=\lambda^{q-1}$. Another way to express the second equation is $(bc)^{1-q}=\lambda^{q-1}$ or $(bc\lambda)^{q-1}=1$ so that $bc\lambda\in F'^{*}$.

Since $u=bc+\lambda^q\in N(F)\setminus\{\lambda^q\}$, we can count the possible values of $u$ by noting that $u\lambda=bc\lambda+\lambda^{q+1}$. So $u\lambda\in F'^{*}$ but we must have $u\neq\lambda^q$ and therefore there are exactly $q-2$ choices for $u$. For each choice of $u$, we then set $bc=u-\lambda^q$ and then there are $q^2-1$ choices for $b$ and $c$ is determined.

For given $\lambda=\xi^i$, the number of matrices $A$ satisfying the equations is then $(q-2)(q^2-1)$. The $\overline{G}$-orbits have length $2(q-1)$ if $i<(q-1)/2$ and $4(q-1)$ if $i=(q-1)/2$. So the number of orbits of generating pairs of this form for each $i$ is

$$r_{1,i}=\begin{cases}(q-2)\dfrac{q+1}{2} & \text{for } \lambda=\xi^i,\ i<(q-1)/2\\[2mm] (q-2)\dfrac{q+1}{4} & \text{for } \lambda=\xi^i,\ i=(q-1)/2\end{cases} \tag{12}$$

The total number of orbits is therefore

$$r_1 = \sum_{i \le (q-1)/2} r_{1,i} = \frac{(q^2-1)(q-2)}{8} \tag{13}$$

The second possibility is that the conjugating element has the form $[C, 1]$ with $C = \begin{pmatrix} 0 & \beta \\ 1 & 0 \end{pmatrix}$ for some $\beta \in F'^*$. In this case, the equations are

$$\begin{pmatrix} 0 & \beta \\ 1 & 0 \end{pmatrix} \begin{pmatrix} -1 & b^q \\ c^q & \lambda \end{pmatrix} \begin{pmatrix} 0 & \beta \\ 1 & 0 \end{pmatrix} = \varepsilon \begin{pmatrix} -\lambda & b^q \\ c^q & 1 \end{pmatrix}$$

The resulting equations become $\beta = -b^q/c^q$; $\beta \in F'^*$; in particular, $b/c \in F'^*$. Now note that $bc = c^2 b/c \in S(F)$ and so the number of possible values for $u = bc + \lambda^q$ is just $n_F = (q^2-1)/4$ from before. It follows that there are $(q^2-1)/4$ possible values for the product $bc$ and we can count the number of ways we can choose $b$.

Since $b/c \in F'^*$ we can write $x = b/c,\ y = bc$ and there are $q-1$ choices for $x$. For each choice, we need $b$ to satisfy the equations $x = b/c,\ y = bc$ so $xy = b^2$. Knowing that $xy \in S(F)$, there are exactly two choices for $b$ and so the total number of matrices $A$ satisfying the equations for a given $\lambda$ is $(q-1)(q^2-1)/2$.

Then, given the orbit lengths as before, the number of orbits of pairs of this form for each $i$ is

$$r_{2,i} = \begin{cases} \dfrac{q^2-1}{4} & \text{for } \lambda = \xi^i, i < (q-1)/2 \\ \dfrac{q^2-1}{8} & \text{for } \lambda = \xi^i, i = (q-1)/2 \end{cases} \tag{14}$$

$$r_2 = \sum_{i \le (q-1)/2} r_{2,i} = \frac{(q^2-1)(q-1)}{16} \tag{15}$$

Note also that in the set of matrices counted by $r_1$ above, we have $bc \in N(F)$ and for those counted by $r_2$, we have $bc \in S(F)$, so that the sets are disjoint.

**Counting orbits: the case $B = \text{off}(\lambda, 1)$** In this case, we have $A = \begin{pmatrix} a & \lambda^q \\ -1 & d \end{pmatrix}$. We can write $[A, 1]^{-1}$ as $[(A^\sigma)^{-1}, 1]$ where $(A^\sigma)^{-1}$ has the form $\varepsilon \begin{pmatrix} d^q & -\lambda \\ 1 & a^q \end{pmatrix}$ for some $\varepsilon \neq 0$.

The first possibility is that the conjugating element has the form $[C, 0]$ with $C = \begin{pmatrix} 0 & \beta \\ 1 & 0 \end{pmatrix}$ for some $\beta$ such that $\beta^{q+1} = \lambda^{q+1}$. In this case, the equations are

$$\begin{pmatrix} 0 & \beta \\ 1 & 0 \end{pmatrix} \begin{pmatrix} a & \lambda^q \\ -1 & d \end{pmatrix} \begin{pmatrix} 0 & \beta^q \\ 1 & 0 \end{pmatrix} = \varepsilon \begin{pmatrix} d^q & -\lambda \\ 1 & a^q \end{pmatrix}$$

It follows that $\beta d = \varepsilon d^q$, $-\beta^{q+1} = -\varepsilon\lambda$, $\lambda^q = \varepsilon$, and $a\beta^q = \varepsilon a^q$. So either $ad = 0$ or $(\lambda ad)^{q-1} = 1$. If $ad \neq 0$ then as before there are exactly $q-2$ choices for $u = ad + \lambda^q$, each with $q^2 - 1$ choices for $a$ and then $d$ is determined. If $ad = 0$ there are $q^2 - 1$ choices for $a$ if $d = 0$ and $q^2 - 1$ choices for $d$ if $a = 0$.

Thus for a given $\lambda = \xi^i$, there are $q(q^2-1)$ matrices $A$ satisfying the equations. The orbit lengths are $2(q+1)$ if $i < (q+1)/2$ and $4(q+1)$ if $i = (q+1)/2$. So the number of orbits of this form for each $i$ is

$$r_{3,i} = \begin{cases} \dfrac{q(q-1)}{2} & \text{for } \lambda = \xi^i, i < (q+1)/2 \\ \dfrac{q(q-1)}{4} & \text{for } \lambda = \xi^i, i = (q+1)/2 \end{cases} \tag{16}$$

$$r_3 = \sum_{i \le (q+1)/2} r_{3,i} = \frac{q(q^2-1)}{8} \tag{17}$$

The remaining possibility is that the conjugating element has the form $[C, 1]$ with $C = \begin{pmatrix} 1 & 0 \\ 0 & \delta \end{pmatrix}$ for some $\delta$ such that $\delta^{q+1} = 1$. In this case, the equations are

$$\begin{pmatrix} 1 & 0 \\ 0 & \delta \end{pmatrix} \begin{pmatrix} a^q & \lambda \\ -1 & d^q \end{pmatrix} \begin{pmatrix} 1 & 0 \\ 0 & \delta \end{pmatrix} = \varepsilon \begin{pmatrix} d^q & -\lambda \\ 1 & a^q \end{pmatrix}$$

This leads to $a^q = \varepsilon d^q$, $\lambda\delta = -\lambda\varepsilon$, $-\delta = \varepsilon$, and $\delta^2 d^q = \varepsilon a^q$. So $\delta = -a^q/d^q$ and $\delta \in \sqrt[q+1]{1}$. Now $ad \in S(F)$ since $a^q/d^q = -\delta \in S(F)$, which implies that the number of possible $u = ad + \lambda^q$ is, as before, $n_F = (q^2-1)/4$. To proceed, write $x = a/d$, $y = ad$ and then $x^{q(q+1)} = 1$. This gives $x \in \sqrt[q+1]{1}$ and so there are $q+1$ choices for $x$ and for each $x$, there are $(q^2-1)/4$ choices for $y$, with still having to choose $d$ to satisfy $d^2 = y/x$. Since we know $y/x \in S(F)$, there are two choices for $d$.

Thus, for a given $\lambda = \xi^i$, there are $(q+1)(q^2-1)/2$ matrices $A$ satisfying the above equations. The orbit lengths are as before, so the number of orbits of this form for each $i$ is

$$r_{4,i} = \begin{cases} \dfrac{q^2-1}{4} & \text{for } \lambda = \xi^i, i < (q+1)/2 \\ \dfrac{q^2-1}{8} & \text{for } \lambda = \xi^i, i = (q+1)/2 \end{cases} \tag{18}$$

$$r_4 = \sum_{i \le (q+1)/2} r_{4,i} = \frac{(q+1)(q^2-1)}{16} \tag{19}$$

Again, the sets of matrices counted by $r_3$ and $r_4$ are disjoint since in the first case $ad \in N(F)$ and in the second $ad \in S(F)$.

**Summary of counting orbits** For $B = \text{dia}(\lambda, 1)$, the total number of $\overline{G}$-orbits of generating pairs for reflexible maps is

$$R_1 = r_1 + r_2 = \frac{(q^2 - 1)(3q - 5)}{16} \tag{20}$$

For $B = \text{off}(\lambda, 1)$, the total number of such orbits is

$$R_2 = r_3 + r_4 = \frac{(q^2 - 1)(3q + 1)}{16} \tag{21}$$

The total number of orbits is therefore

$$R = R_1 + R_2 = \frac{(q^2 - 1)(3q - 2)}{8} \tag{22}$$

### 11.3 Counting Reflexible Maps

We may enumerate the orientably-regular reflexible maps on $M(q^2)$ by using the above calculations in conjunction with the logic of Sect. 10. Since details of this process are exactly as in Sect. 10 except for using the input on counting orbits from Sect. 11.2, we present just the final result.

**Theorem 4** *Let $q = p^f$ be an odd prime power, with $f = 2^\alpha o$ where $o$ is odd. The number of orientably-regular reflexible maps $\mathcal{M}$ with $\text{Aut}^+(\mathcal{M}) \cong M(q^2)$ is, up to isomorphism, equal to*

$$\frac{1}{f} \sum_{d|o} \mu(o/d)\tilde{h}(2^\alpha d) ,$$

*where $\tilde{h}(x) = (p^{2x} - 1)(3p^x - 2)/8$ and $\mu$ is the Möbius function.*

## 12 Remarks

As stated in the Introduction, orientably-regular maps have been enumerated for very sparse classes of non-trivial groups, and in terms of *all* such maps (not just maps of restricted types) only for almost-Sylow-cyclic groups [4] and the linear fractional groups PSL$(2, q)$ and PGL$(2, q)$ [3, 19]. It should be noted, however, that the available results for PSL$(2, q)$ and PGL$(2, q)$ are more detailed by giving ‘closed formulae’ for the number of orientably-regular maps of every given type, whereas our main results in Theorems 3 and 4 contain formulae for the total number of such maps.

In order to obtain a refined version of our enumeration of orientably-regular maps with automorphism group isomorphic to a twisted linear fractional group $G = M(q^2)$, one could follow [13], which requires setting up both a character table for $G$ and the Möbius function for the lattice of subgroups of $G$. The number of orientably-regular maps on the group $G$ is then obtained as a combination of a character-theoretic formula for counting solutions of the equation $xyz = 1$ for $x, y, z$ in given conjugacy classes of $G$ (a special case of a general formula of Frobenius [7]) combined with Möbius inversion, which is a forthcoming project of the authors. Whether the project will return a 'nice' formula, however, is not clear due to another significant difference between the family of orientably-regular maps on $M(q^2)$ compared to those on $\mathrm{PGL}(2, q)$. Namely, in the case of $\mathrm{PGL}(2, q)$, for any even $k, \ell \geq 4$ not both equal to 4 there is an orientably-regular map for infinitely many values of $q$, cf. [3]. Our next result shows that this fails to hold in the case of $M(q^2)$.

**Proposition 13** *If* $k, \ell \equiv 0 \pmod 8$ *and* $k \not\equiv \ell \pmod{16}$ *then there is no orientably-regular map of type* $(k, \ell)$ *on* $M(q^2)$ *for any* $q$.

***Proof*** By Theorem 1 orders of elements in $G \backslash G_0$ are $o_i = 2(q-1)/\gcd\{q-1, i\}$ and $o'_i = 2(q+1)/\gcd\{q+1, i\}$ for odd $i$ such that $1 \leq i \leq (q-1)/2$ and $1 \leq i \leq (q+1)/2$, respectively. Note that if $o_i \equiv 0 \mod 8$, then $o'_i \equiv 4 \mod 8$ and vice versa. Further, if $o_i \equiv 8 \mod 16$, then $q-1 \equiv 4 \mod 8$ since $i$ is odd, and if $o_i \equiv 0 \mod 16$ then $q-1 \equiv 0 \mod 8$. It follows that for a given $q$ we cannot have a non-singular generating pair of orders $o_i \equiv 0 \mod 16$ and $o_j \equiv 8 \mod 16$. The argument for orders of the form $o'_i$ is similar. □

Besides reflexibility, another frequently studied property of orientably-regular maps is self-duality. In general, an oriented map is *positively self-dual* if it is isomorphic to its dual with the same orientation, and *negatively self-dual* if it is isomorphic to its oppositely oriented dual map. In terms of orientably-regular maps represented by triples $(G, x, y)$, positive and negative self-duality is equivalent to the existence of an (involutory) automorphism of $G$ sending the ordered pair $(x, y)$ onto $(y, x)$ and $(y^{-1}, x^{-1})$, respectively. By the same arguments as in the second paragraph of Sect. 11, one concludes that for our group $G = M(q^2)$, an orientably-regular map defined by a generating pair $[A, 1], [B, 1]$ will be positively self-dual if and only if there exists an involution $[C, i] \in \overline{G}$ conjugating the two generators, and the map will be negatively self-dual if there is such an involution conjugating $[A, 1]$ to $[B, 1]^{-1}$.

Setting up the corresponding matrix equations for such conjugations, however, leads to enormously complicated formulae from which we were not able to extract 'nice' closed formulae. Clearly, in a self-dual map, we have $k = \ell$ so that the orders of $[A, 1]$ and $[B, 1]$ must be equal. We used GAP [21] to construct all the regular maps on $M(q^2)$, for small values of $q$, with the generators of equal order, and then tested for self-duality by determining if a conjugating element $[C, i]$ as above exists. The results of this computation are given in Table 1, showing the numbers #($k{=}\ell$) of maps that have generators of equal orders, those which are positively or negatively self-dual, and those which are both.

**Table 1** Numbers of self-dual maps on $M(q^2)$

| $q$ | $B = \text{dia}(\lambda, 1)$ | | | | $B = \text{off}(\lambda, 1)$ | | | |
|---|---|---|---|---|---|---|---|---|
| | #($k$=$\ell$) | + self-dual | − self-dual | Both | #($k$=$\ell$) | + self-dual | − self-dual | Both |
| 3 | 0 | 0 | 0 | 0 | 3 | 3 | 3 | 3 |
| 5 | 15 | 15 | 5 | 5 | 10 | 10 | 6 | 6 |
| 7 | 28 | 28 | 8 | 8 | 78 | 42 | 14 | 14 |
| 9 | 95 | 45 | 9 | 9 | 68 | 36 | 10 | 10 |
| 11 | 276 | 132 | 24 | 24 | 265 | 165 | 33 | 33 |
| 13 | 469 | 273 | 39 | 39 | 666 | 234 | 42 | 42 |
| 17 | 2556 | 612 | 68 | 68 | 1312 | 544 | 72 | 72 |
| 19 | 1960 | 760 | 80 | 80 | 2799 | 855 | 95 | 95 |

Note that the computational evidence suggests that a negatively self-dual map on $M(q^2)$ is also positively self-dual.

**Acknowledgements** The authors thank Marston Conder for independently checking our computational results displayed in Table 1 using the MAGMA package. The last two authors gratefully acknowledge the support of this work by the Slovak Research Grants APVV-15-0220, APVV-17-0428, VEGA 1/0142/17 and VEGA 1/0238/19.

## References

1. P.J. Cameron, G.R. Omidi, B. Tayfeh-Rezaie, 3-designs from PGL$(2, q)$. Electron. J. Combin. **13**, Art. R50 (2006)
2. M.D.E. Conder, List of orientably-regular chiral and reflexible maps up to genus 301 (2012), https://www.math.auckland.ac.nz/~conder/
3. M.D.E. Conder, P. Potočnik, J. Širáň, Regular hypermaps over projective linear groups. J. Australian Math. Soc. **85**, 155–175 (2008)
4. M.D.E. Conder, P. Potočnik, J. Širáň, Regular maps with almost Sylow-cyclic automorphism groups, and classification of regular maps with Euler characteristic $-p^2$. J. Algebra **324**, 2620–2635 (2010)
5. M.D.E. Conder, T. Tucker, Regular Cayley maps for cyclic groups. Trans. Am. Math. Soc. **366**(7), 3585–3609 (2014)
6. H.S.M. Coxeter, W.O.J. Moser, *Generators and Relations for Discrete Groups* (Springer, Berlin, 1980)
7. F.G. Frobenius, *Über Gruppencharaktere* (Sitzber. Königlich Preuss. Akad. Wiss. Berlin, 1896), pp. 985–1021
8. M. Giudici, Maximal subgroups of almost simple groups with socle PSL$(2, q)$ (2007), arXiv:math/0703685
9. G.A. Jones, Ree groups and Riemann surfaces. J. Algebra **165**, 41–62 (1994)
10. G.A. Jones, S.A. Silver, Suzuki groups and surfaces. J. Lond. Math. Soc. **48**(2), 117–125 (1993)
11. G.A. Jones, D. Singerman, Belyĭ functions, hypermaps and Galois groups. Bull. London Math. Soc. **28**(6), 561–590 (1996)

12. G.A. Jones, D. Singerman, Theory of maps on orientable surfaces. Proc. Lond. Math. Soc. **3–37**(2), 273–307 (1978)
13. G.A. Jones, Combinatorial categories and permutation groups. Ars Math. Contemp. **10**(2), 237–254 (2016)
14. J.B. Kelly, A characteristic property of quadratic residues. Proc. Amer. Math. Soc. **5**, 38–46 (1954)
15. D. Leemans, J. de Saadeleer, On the rank two geometries of the groups PSL$(2, q)$: part I. Ars Math. Contemp. **3**, 177–192 (2010)
16. A.M. Macbeath, Generators of the linear fractional groups, in *1969 Number Theory (Proc. Sympos. Pure Math., Vol. XII, Houston, Tex.), Am. Math. Soc.* (Providence, R.I. 1967), pp. 14–32
17. A. Malnič, R. Nedela, M. Škoviera, Regular maps with nilpotent automorphism groups. Eur. J. Combin. **33**(8), 1974–1986 (2012)
18. R. Nedela, Regular maps—combinatorial objects relating different fields of mathematics. J. Korean Math. Soc. **38**, 1069–1105 (2001)
19. C.H. Sah, Groups related to compact Riemann surfaces. Acta Math. **123**, 13–42 (1969)
20. J. Širáň, How symmetric can maps on surfaces be? in *Surveys in Combinatorics*, vol. 409. London Mathematical Society Lecture Note Series (Cambridge University Press, Cambridge, 2013), pp. 161–238
21. The GAP Group, GAP—Groups, Algorithms, and Programming, Version 4.7.8 (2015), http://www.gap-system.org
22. H. Zassenhaus, Über endliche Fastkörper. Abh. Math. Sem. Univ. Hamburg **11**, 187–220 (1936)

# From Schur Rings to Constructive and Analytical Enumeration of Circulant Graphs with Prime-Cubed Number of Vertices

**Victoria Gatt, Mikhail Klin, Josef Lauri and Valery Liskovets**

**Abstract** Two methods, structural (constructive) and multiplier (analytical), of exact enumeration of undirected and directed circulant graphs of orders 27 and 125 are elaborated and represented in detail here together with intermediate and final numerical data and generating functions. The first method is based on the classification of circulant graphs in terms of $S$-rings and results in exhaustive listing (with the aid of COCO and GAP) of all corresponding $S$-rings of the indicated orders. The latter method is based on a general theoretical approach developed earlier for counting circulant graphs of prime-power orders. It is a Redfield–Pólya type of enumeration based on an isomorphism criterion for circulant graphs of such orders. In particular, five intermediate enumeration subproblems arise, which are refined further into eleven subproblems of this type. We give a brief survey of some background theory of the results which form the basis of our computational approach. Some curious and rather unexpected identities are established between intermediate valency-specified enumerators and their validity is conjectured for arbitrary cubed odd primes $p^3$. We conclude with the enumeration of self-complementary circulant graphs of orders

Mikhail Klin and Valery Liskovets—Supported by the Scientific Grant Agency of the Slovak Republic (see Acknowledents).

V. Gatt
University of Malta, Msida, Malta
e-mail: victoria.gatt@um.edu.mt

M. Klin
Ben-Gurion University of the Negev, 84105 Beer Sheva, Israel
e-mail: klin@cs.bgu.ac.il

Matej Bel University, Banská Bystrica, Slovakia

J. Lauri (✉)
University of Malta, Msida, Malta
e-mail: josef.lauri@um.edu.mt

V. Liskovets
National Academy of Sciences of Belarus, Minsk, Belarus
e-mail: liskov@im.bas-net.by



G. A. Jones et al. (eds.), *Isomorphisms, Symmetry and Computations in Algebraic Graph Theory*, Springer Proceedings in Mathematics & Statistics 305,
https://doi.org/10.1007/978-3-030-32808-5_2

27 and 125. We believe that this research can serve as the crucial step towards the explicit uniform enumeration formulae for circulant graphs of orders $p^3$ for arbitrary primes $p > 2$.

**Keywords** Circulant graph · Cyclic group · $S$-ring · Constructive enumeration · Multiplier · Enumeration under group action · Graph isomorphism · Pólya's method · Self-complementary graphs · Combinatorial identity

**Mathematic Subject Classifications** 05C30 · 05C25 · 05C20

## 1 Introduction

The present research is carried out in the framework of the general program outlined in the paper [10] for counting circulant graphs of prime-power orders. We refer to this paper for details concerning two approaches to the exact enumeration of circulant graphs, namely, constructive and analytical. The former enables us, in principle at least, to list the non-isomorphic circulant graphs we are counting. The analytic approach is based on the isomorphism theorem [12] for circulant graphs of prime-power orders, and we are guided also here by the subsequent adaptation of this theorem to the enumeration of circulant graphs as developed in [19].

We restrict ourselves to orders 27 and 125 only. There are several arguments for our choice. First of all, there are almost no numerical results for the number of isomorphism classes of circulant graphs of prime-cubed orders, including $p = 3$ and 5. There exist huge numbers of circulant graphs even of these orders so that it hardly makes much sense to enumerate constructively circulant graphs of larger orders. In principle, no such obstacle arises for analytical enumeration, but even here these two orders require much effort. Presumably, the main difficulties of analytical enumeration should become apparent already on these least prime-cubed orders. Moreover, our aim is also to compare both approaches on the same classes of objects and to obtain confirmation of numerical results obtained in both ways using COCO and GAP and also partially by brute force.

The importance and difficulty of counting circulant graphs stems from the falsity of a very natural conjecture of Ádám giving a condition on the connecting sets for isomorphism to hold. However, it turned out to be false: Ádám's condition is sufficient for isomorphism but not necessary. The falsity of this conjecture led to some beautiful results which characterised completely the conditions on the order of the circulant graph for the conjecture to hold. Ádám's condition for isomorphism extended to general Cayley graphs led to the Cayley Isomorphism Problem, a very important thread of research in algebraic graph theory.

In this paper, since we are enumerating circulant graphs, we use results from the first line of research as well as the structural approach based more directly on Schur rings. We describe these two methods below, giving more detail on the multiplier approach. We point out several relations arising between the intermediate terms which form the generating functions which enumerate these circulants by valency, and we conjecture that the relations which emerge from our generating functions for $k = 3$ and $p = 3, 5$ hold for all odd prime $p$. We then conclude with the enumeration of self-complementary circulant graphs.

Fuller details including all case-by-case analysis, all generating functions produced and all the GAP programmes used can be found in [6, 7].

For standard graph theoretic terms, we refer the reader to the two texts [14, 15].

## 1.1 First Definitions

A circulant graph is a Cayley graph over a cyclic group. That is, let $H$ be a cyclic group (which we shall represent as the group $\mathbb{Z}_n$ of integers with addition modulo $n$, the order of the group) and let $S \subseteq H$ (called the *connecting set* of the Cayley graph) such that $0 \notin S$. Then a circulant is a Cayley graph $\mathrm{Cay}(H, S)$ which has $H$ as vertex-set and two vertices $g, h$ are adjacent if $g = h + s$ for some $s \in S$. If the set $S$ generates $H$ then the circulant graph $\mathrm{Cay}(H, S)$ is connected. In the special case when $-S = S$ (that is, $s \in S$ if and only if $-s \in S$), the circulant graph is also referred to as an *undirected graph*. For brevity, we shall sometimes refer to "circulants" instead of "circulant graphs".

The *valency* of a vertex $v$ in a directed graph is equal to the number of arcs of the form $(v, x)$; for an undirected graph this is equal to the number of edges containing that vertex. In our generating functions, we usually denote valency by the letter $r$.

Finally, $I_{(G,X)}$ will denote the cycle index of the permutation group $G$ acting on the set $X$.

## 1.2 The Structural Approach: An Introduction

The group ring $\langle \mathbb{Z}[\mathbb{Z}_n]; +, \cdot \rangle$ of $\mathbb{Z}_n$ over $\mathbb{Z}$, consists of the set of all formal linear combinations of elements of $\mathbb{Z}_n$ with integral coefficients, that is, all formal sums $\sum_{h \in \mathbb{Z}_n} \alpha_h \underline{h}$ with $\alpha_h \in \mathbb{Z}$, $h \in \mathbb{Z}_n$, together with addition

$$\sum_{h \in \mathbb{Z}_n} \alpha_h \underline{h} + \sum_{h \in \mathbb{Z}_n} \beta_h \underline{h} := \sum_{h \in \mathbb{Z}_n} (\alpha_h + \beta_h) \underline{h}$$

and formal multiplication

$$\Big(\sum_{h \in \mathbb{Z}_n} \alpha_h \underline{h}\Big) \cdot \Big(\sum_{k \in \mathbb{Z}_n} \beta_k \underline{k}\Big) := \sum_{h,k \in \mathbb{Z}_n} \alpha_h \beta_k \Big(\underline{h + k}\Big) = \sum_{h \in \mathbb{Z}_n} \Big(\sum_{k \in \mathbb{Z}_n} \alpha_{h-k} \beta_k\Big) \underline{h}.$$

Note that we are writing $\underline{h}$ for $h \in \mathbb{Z}_n$ in order to distinguish clearly between elements of $\mathbb{Z}_n$ and $\mathbb{Z}$.

For the elements of $\mathbb{Z}[\mathbb{Z}_n]$ the *Schur-Hadamard product* is also defined as follows:

$$\Big(\sum_{h\in\mathbb{Z}_n}\alpha_h\underline{h}\Big)\circ\Big(\sum_{h\in\mathbb{Z}_n}\beta_h\underline{h}\Big):=\sum_{h\in\mathbb{Z}_n}(\alpha_h\beta_h)\underline{h}.$$

Therefore, for $T, T' \subseteq \mathbb{Z}_n$ we have $\underline{T}\circ\underline{T'}=\underline{T\cap T'}$.

The $\mathbb{Z}$-submodule of $\mathbb{Z}[\mathbb{Z}_n]$ generated by elements $\lambda_1,\ldots,\lambda_r \in \mathbb{Z}[\mathbb{Z}_n]$ will be denoted by $\langle\lambda_1,\ldots,\lambda_r\rangle$. Therefore the $\mathbb{Z}$-submodule $\langle\lambda_1,\ldots,\lambda_r\rangle$, consists of all linear combinations of $\lambda_1,\ldots,\lambda_r$.

Assume $T\subseteq\mathbb{Z}_n$, $T=\{t_1,t_2,\ldots,t_r\}$. Elements of the form $\underline{T}:=\sum_{h\in T}\underline{h}$ are called *simple quantities* of $\mathbb{Z}[\mathbb{Z}_n]$. One can consider $\underline{T}$ as the formal sum $\sum_{h\in\mathbb{Z}_n}\alpha_h\underline{h}$ with $\alpha_h=1$ if and only if $h\in T$ and $\alpha_h=0$ otherwise. In other words, we identify a simple quantity in which every entry has multiplicity 1 with the corresponding set. For $T=\{t_1,t_2,\ldots,t_r\}$ we use the notation $\underline{t_1,\ldots,t_r}$ instead of $\underline{\{t_1,\ldots,t_r\}}$.

A subring $\mathfrak{S}$ of a group ring $\mathbb{Z}[\mathbb{Z}_n]$ is called a *Schur ring* or *S*-ring over $\mathbb{Z}_n$, of rank $r$ if the following conditions hold:

1. $\mathfrak{S}$ is closed under addition and multiplication with elements from $\mathbb{Z}$ (i.e. $\mathcal{S}$ is a $\mathbb{Z}$-module);
2. Simple quantities $\underline{T}_0,\underline{T}_1,\ldots,\underline{T}_{r-1}$ exist in $\mathfrak{S}$ such that every element $\sigma\in\mathfrak{S}$ has a unique representation $\sigma=\sum_{i=0}^{r-1}\sigma_i\underline{T}_i$;
3. $\underline{T}_0=\underline{0}$, $\sum_{i=0}^{r-1}\underline{T}_i=\underline{\mathbb{Z}_n}$, that is, $\{T_0,T_1,\ldots,T_{r-1}\}$ is a partition of $\mathbb{Z}_n$;
4. For every $i\in\{0,1,2,\ldots,r-1\}$ there exists a $j\in\{0,1,2,\ldots,r-1\}$ such that $\underline{T}_j=\underline{-T}_i(=\underline{\{n-x : x\in T_i\}})$;
5. For $i,j\in\{1,\ldots,r\}$, there exist non-negative integers $p_{ij}^k$ called structure constants, such that $\underline{T}_i\cdot\underline{T}_j=\sum_{k=1}^{r}p_{ij}^k\underline{T}_k$.

The simple quantities $\underline{T}_0,\underline{T}_1,\ldots,\underline{T}_{r-1}$ form a standard basis for $\mathfrak{S}$ and their corresponding sets $T_i$ are *basic sets* of the *S*-ring. The circulant graphs $\Gamma_i=\mathrm{Cay}(\mathbb{Z}_n,T_i)$, where $0\le i\le r-1$, are called *basic circulant graphs* [14]. The following notation will denote an *S*-ring generated by its basic sets $\underline{T}_0,\underline{T}_1,\ldots\underline{T}_{r-1}$:

$$\mathfrak{S}=\langle\underline{T}_0,\underline{T}_1,\ldots\underline{T}_{r-1}\rangle.$$

Note that both $\mathbb{Z}[\mathbb{Z}_n]$ and $\langle\underline{0},\underline{\mathbb{Z}_n-\{0\}}\rangle$ are Schur rings over $\mathbb{Z}_n$ which we call the trivial Schur rings over $\mathbb{Z}_n$.

A permutation $g:\mathbb{Z}_n\to\mathbb{Z}_n$ is called an automorphism of an *S*-ring $\mathfrak{S}$, if it is an automorphism of every graph $\Gamma_i$. Equivalently, the intersection of the automorphism groups of the basic circulant graphs of an *S*-ring $\mathfrak{S}=\langle\underline{T}_0,\underline{T}_1,\ldots,\underline{T}_{r-1}\rangle$, gives the automorphism group of the *S*-ring, that is,

$$Aut(\mathfrak{S}):=\bigcap_{i=0}^{r-1}Aut(\Gamma_i) \tag{1}$$

The structural approach to the enumeration of circulants on $n$ vertices is based on the lattice $\mathfrak{L}(n)$ of all Schur rings over $\mathbb{Z}_n$ which, together with information on the automorphism groups of the Schur rings, suffices to carry out the enumeration. This enumeration scheme has already been described in [10], so we give here only a brief summary.

We first use the lattice of Schur rings to count the number of labelled circulant graphs, as follows:

1. Construct the lattice $\mathfrak{L}(n)$ of all Schur rings as a sequence $\mathfrak{L}(n) = (\mathfrak{S}_1, \mathfrak{S}_2, \ldots \mathfrak{S}_s)$ such that $\mathfrak{S}_j \subseteq \mathfrak{S}_i$ implies $j \leq i$;
2. For directed circulants, let $\tilde{d}_{ir}$ be the number of $r$-element basic sets of the $\mathcal{S}$-ring $\mathfrak{S}_i$, different from the basic set $T_0 = \{0\}$, that is,

$$\tilde{d}_{ir} := |\{T_{(x)} \in \mathfrak{S}_i \,|\, x \neq 0 \text{ and } |T_{(x)}| = r\}|;$$

3. For undirected circulants, let $d_{ir}$ be the number of $r$-element symmetrized (that is closed under taking of inverses) basic sets of $\mathfrak{S}_i$, different from $T_0$. That is,

$$d_{ir} := |\{T_{(x)}^{sym} \,|\, x \neq 0 \text{ and } |T_{(x)}^{sym}| = r\}|;$$

4. Enumeration of all labelled directed and undirected circulant graphs which belong to the Schur ring $\mathfrak{S}_i$ may then be carried out by making use of generating functions $\tilde{f}_i(t)$ and $f_i(t)$ respectively, given by

$$\begin{aligned} \tilde{f}_i(t) &:= \sum_{r=0}^{n-1} \tilde{f}_{ir} t^r := \prod_{r=1}^{n-1} (1 + t^r)^{\tilde{d}_{ir}} \\ f_i(t) &:= \sum_{r=0}^{n-1} f_{ir} t^r := \prod_{r=1}^{n-1} (1 + t^r)^{d_{ir}} \end{aligned} \qquad (2)$$

Substituting $t = 1$ in the generating functions, would give us the number of all labelled directed and undirected circulant graphs in $\mathfrak{S}_i$. In addition, the graph corresponding to $T \in \mathfrak{S}_i$ is of valency $r$ if $T$ has $r$ elements.

The link between the number of labelled and unlabelled circulant graphs is given by Lemma 1.

**Lemma 1** ([10]) *Let $G_i = Aut\ (\mathfrak{S}_i)$, let $N(G_i) = N_{S_n}(G_i)$ be the normalizer of the group $G_i$ in $S_n$, and let $\Gamma$ be a circulant graph belonging to $\mathfrak{S}_i$. Then*

*(a) $Aut\ (\Gamma) = G_i \iff \Gamma$ generates $\mathfrak{S}_i$.*
*(b) If $Aut\ (\Gamma) = G_i$ then there are exactly $[N(G_i) : G_i]$ (that is, equal to the number of cosets of $G_i$ in $N(G_i)$) distinct circulant graphs which are isomorphic to $\Gamma$.*

**Remark** In (a), a Cayley graph $\Gamma$ is identified with its connecting set. This is why we are allowed to say that $\Gamma$ generates $\mathfrak{S}_i$.

So, let the generating function for the number of non-isomorphic undirected circulant graphs with automorphism group $G_i$ be given by $g_i(t) = \sum_{r=0}^{n-1} g_{ir} t^r$ and let the generating function for the number of non-isomorphic directed circulant graphs with automorphism group $G_i$ be given by $\tilde{g}_i(t) = \sum_{r=0}^{n-1} \tilde{g}_{ir} t^r$ (In all our generating functions, the coefficient of $t^r$ equals the number of circulants under consideration in which all vertices have valency $r$.)

Moreover, let $g(t) = g(n, t)$ and $\tilde{g}(t) = \tilde{g}(n, t)$ denote the generating functions for the number of non-isomorphic undirected and directed circulant graphs, respectively, with $n$ vertices. The values $g(1)$ and $\tilde{g}(1)$ therefore give the numbers of all non-isomorphic undirected and directed circulant graphs, respectively, with $n$ vertices. These generating functions are then given by the following theorem whose proof is based on the inclusion-exclusion principle.

**Theorem 1** ([10])

$$
\begin{aligned}
g_i(t) &= \frac{|G_i|}{|N(G_i)|}\left(f_i(t) - \sum_{\mathfrak{G}_j \subseteq \mathfrak{G}_i} \frac{|N(G_j)|}{|G_j|} g_j(t)\right),\\
\tilde{g}_i(t) &= \frac{|G_i|}{|N(G_i)|}\left(\tilde{f}_i(t) - \sum_{\mathfrak{G}_j \subseteq \mathfrak{G}_i} \frac{|N(G_j)|}{|G_j|} \tilde{g}_j(t)\right), \qquad (3)\\
g(t) &= \sum_{i=1}^{s} g_i(t), \quad \tilde{g}(t) = \sum_{i=1}^{s} \tilde{g}_i(t).
\end{aligned}
$$

In Sect. 3 we shall use this approach to enumerate the number of non-isomorphic undirected circulants of order $3^3$.

## 1.3 The Multiplier Approach: An Introduction

Let $\mathbb{Z}_n^*$ be the multiplicative group consisting of all the units in $\mathbb{Z}_n$ (when $n$ is prime, $\mathbb{Z}_n^* = \mathbb{Z}_n - \{0\}$). It is clear that if $\Gamma_1 = \mathrm{Cay}(\mathbb{Z}_n, S)$ and $\Gamma_2 = \mathrm{Cay}(\mathbb{Z}_n, T)$ are circulants such that there exists an $m \in \mathbb{Z}_n^*$ with $mS = \{ms : s \in S\} = T$, then $\Gamma_1$ and $\Gamma_2$ are isomorphic. In this case we say that the connecting sets are *equivalent*. In [1] Ádám conjectured that the converse is also true, that is, two isomorphic circulant graphs have equivalent connecting sets. This conjecture turned out to be false.

The principal theorem which gives the most correct version of Ádám's Conjecture is the following due to Muzychuk [20], with a more general situation which is considered in [21].

**Theorem 2** *Let $\Gamma_1$ and $\Gamma_2$ be two circulant graphs on n vertices, and suppose that n is square-free. Then $\Gamma_1$, $\Gamma_2$ are isomorphic if and only if their connecting sets are equivalent.*

The easiest square-free case occurs when $n$ is prime, and this was solved by Elspas and Turner [3] by means of a clever use of Pólya's enumeration theorem applied to the multiplicative action of $\mathbb{Z}_p^*$ on $\mathbb{Z} - \{0\}$. The most natural non-square-free cases to consider would be when the order $n$ is a power $k$ of a prime, that is, $n = p^k$, for $k \geq 2$. But to enumerate circulant graphs of such an order requires some multiplicative relations between the connecting sets of two circulant graphs which are necessary and sufficient for them to be isomorphic, that is, we require the correct version of Ádám's Conjecture for $n = p^k$. We call this method of enumerating non-isomorphic circulants the multiplier approach. We shall consider in some detail the multiplier approach for $p^3$ with $p = 3, 5$ in Sect. 4.

## 2 Automorphism Groups of Prime-Cubed Circulants

We pause here to provide a brief survey of some background concepts and results from algebraic graph theory which are needed for a full justification of the theoretical results at the basis of our computational approach. As a rule, in this section, we avoid giving rigorous and precise proofs of the results involved. Our modest aim is to help the reader achieve a satisfactory intuitive feel for these results and after that the more interested reader may be able to obtain a full understanding of how these proofs are obtained. Such an understanding is, however, not required in order to follow the arguments presented in the subsequent sections of the paper.

We denote by $(\mathbb{Z}_n, \mathbb{Z}_n)$ the regular cyclic permutation group acting on the set $\{0, 1, \ldots, n-1\}$ and generated by the cyclic shift $(0\ 1\ 2\ \ldots\ n-1)$. Recall that the automorphism group $\mathrm{Aut}(\mathbb{Z}_n)$ of the group $\mathbb{Z}_n$ is $\mathbb{Z}_n^*$, the multiplicative group, modulo $n$, of units of $\mathbb{Z}_n$. It has order $\varphi(n)$, where $\varphi$ is the famous Euler function. The group $\mathbb{Z}_n^*$ acts on $\mathbb{Z}_n$ by multiplication modulo $n$. For our goals, it is enough to consider the case $n = p^k$ where $p$ is prime and $k$ is mainly 1, 2 or 3. It is well-known that $\varphi(p^k) = (p-1)p^{k-1}$. For these values of $n$ the group $\mathbb{Z}_n^*$ is also cyclic.

More generally, let $(G, \Omega)$ be a finite permutation group acting on the set $\Omega$. Denote by 2-orb$(G, \Omega)$ the set of all 2-orbits of $(G, \Omega)$ (in the sense of Wielandt), that is, the set of the orbits of the action $(G, \Omega^2)$ induced naturally as follows: for $g \in G$ and $(\alpha, \beta) \in \Omega^2$ g acts on $\Omega \times \Omega$, as $(\alpha, \beta)^g = (\alpha^g, \beta^g)$, where $x^g$ is the image of the element $x \in \Omega$ under the action of $g \in G$. Let 2-orb$(G, \Omega) = \{R_i : 0 \leq i \leq r-1\}$. Each pair $\Gamma_i = (\Omega, R_i)$ is regarded as a directed graph with vertex-set $\Omega$. Then the group $(G^{(2)}, \Omega)$, where

$$G^{(2)} = \cap_{i=0}^{r-1} \mathrm{Aut}(\Gamma_i),$$

is called the 2-*closure* of $(G, \Omega)$. The group $(G, \Omega)$ is called 2-*closed* if $(G^{(2)}, \Omega) = (G, \Omega)$.

A classical (almost trivial) result by Wielandt claims that each regular permutation group is 2-closed. Thus, in particular, $(\mathbb{Z}_n, \mathbb{Z}_n)$ is 2-closed. (Note that later on the notation $(\mathbb{Z}_n, \mathbb{Z}_n)$ may be reduced to $\mathbb{Z}_n$ if is it clear from the context that we mean the regular action of $\mathbb{Z}_n$.)

The lattice of all 2-closed overgroups of the group $(\mathbb{Z}_n, \mathbb{Z}_n)$ plays an important part in this theory. Here, by an overgroup of $\mathbb{Z}_n$ we understand a subgroup $G$ of $S_n$, the symmetric group on $\{0, 1, 2, 3, \ldots, n-1\}$, which contains $(\mathbb{Z}_n, \mathbb{Z}_n)$. Note that in such a context $Z_n$ appears in two different roles: the set of elements $\{0, 1, 2, \ldots, n-1\}$ and the set of permutations.

It turns out that for the case $n = p^k$ there exists an anti-isomorphism between the lattice of all 2-closed overgroups of $\mathbb{Z}_n$ in $S_n$ and the lattice $\mathfrak{S}_n$ of all $S$-rings over $\mathbb{Z}_n$.

For arbitrary values of $n$, establishing such a bijection between overgroups and $S$-rings turns out to be more sophisticated: one has to consider only so-called Schurian $S$-rings. The good news for the case $n = p^k$ is that here all $S$-rings are Schurian, a result due to Pöschel [24].

For all values of $n$ there are two trivial overgroups, the minimal $(\mathbb{Z}_n, \mathbb{Z}_n)$ and the maximal $(S_n, \mathbb{Z}_n)$. All other overgroups appear between these two extremal objects.

## 2.1 Wreath Products

Let $(G_1, M_1)$ and $(G_2, M_2)$ be two permutation groups and let $M = M_1 \times M_2$. Let $G$ be the set of all mappings $g : M \to M$ such that, for $x = (x_1, x_2) \in M$, $x^g = (x_1, x_2)^g = (f_1(x_1, x_2), f_2(x_1, x_2))$, where the following conditions hold:

(a) $f_1$ depends only on coordinate $x_1$;
(b) the mapping $x_1 \mapsto f_1(x_1, x_2)$ is a permutation in $G_1$;
(c) for every $x_1 \in M_1$, the mappings $x_2 \mapsto f_2(x_1, x_2)$ are permutations which belong to $G_2$.

In this case, for brevity, the notation $g = [g_1, g_2(x_1)]$ is used and is called the *table* of $g$. By definition, we we have

$$x^g = (x_1, x_2)^g = (x_1^{g_1}, x_2^{g_2(x_1)}).$$

It is easy to check that here $G$ is a permutation group $G \leq Sym(M)$. The group $(G, M)$ is called the *wreath product* of $(G_1, M_1)$ and $(G_2, M_2)$ and will be denoted by $(G_1 \wr G_2, M_1 \times M_2)$ or $(G_1, M_1) \wr (G_2, M_2)$.

The wreath product is a group of order $|G_1| \cdot |G_2|^{M_1}$ (= the number of all tables); sometimes, $G_1$ and $G_2$ are called active and passive factors of $G$, respectively. Note also that we are using here the so-called orthodox notation for the wreath product, due to L. A. Kalužnin.

Let $\Gamma_1 = (V_1, E_1)$, $\Gamma_2 = (V_2, E_2)$ be two graphs (directed or undirected). Then the graph $\Gamma = (V, E)$ defined by $V = V_1 \times V_2$, $E = \{((x_1, x_2), (x_1', x_2')) : (x_1, x_2') \in E_1 \vee (x_1 = x_1' \wedge (x_2, x_2') \in E_2)\}$ is called the *composition* of $\Gamma_1$ and $\Gamma_2$ and is usually denoted by $\Gamma_1[\Gamma_2]$ (substitute $\Gamma_2$ for each vertex of $\Gamma_1$ and connect all vertices of the corresponding components according to the connections in $\Gamma_1$).

One of the traditional questions in graph theory was to study conditions under which the automorphism group of the composed graph $\Gamma_1[\Gamma_2]$ equals the wreath product of $\mathrm{Aut}(\Gamma_1)$ with $\mathrm{Aut}(\Gamma_2)$.

In cases where we agree to consider transitive permutation groups only, this question, in the context of the current presentation finds a very suitable solution. Namely it turns out that

$$(G_1 \wr G_2, M)^{(2)} = (G_1, M_1)^{(2)} \wr (G_2, M_2)^{(2)}$$

where $M = M_1 \times M_2$.

In order to explain the significance of this equality in the context of Schur rings we need one more definition. An $S$-ring $\mathfrak{S} = \langle \underline{T_1}, \underline{T_2}, \ldots, \underline{T_d} \rangle$ over $\mathbb{Z}_{p^k}$ is called *wreath decomposable* (or briefly *decomposable*) if there exists a non-trivial proper subgroup $K \leq \mathbb{Z}_{p^k}$ such that for each basic set either $T_i \subseteq K$ or $T_i$ is a union of suitable cosets of $\mathbb{Z}_{p^k}/K$ (that is, $T_i = \cup_{x \in T_i} K + x$). In particular, one has $\underline{K} \in \mathfrak{S}$. The $S$-ring $\mathfrak{S}$ is called *wreath indecomposable* (briefly *indecomopsable*) if it is not wreath decomposable.

Using these concepts and facts, one can prove that, for an arbitrary $S$-ring $\mathfrak{S}$ over $\mathbb{Z}_{p^k}$, $\mathfrak{S}$ is wreath decomposable if and only if $\mathrm{Aut}(\mathfrak{S})$ can be represented as the wreath product of the automorphism groups of an $S$-ring over $\mathbb{Z}_{p^{k-i}}$ and an $S$-ring over $\mathbb{Z}_{p^i}$, where $1 \leq i < k$.

**Example 1** Let $n = 9$, $\mathfrak{S}_1 = \langle \underline{0}, \underline{3}, \underline{6}, \underline{1, 4, 7}, \underline{2, 5, 8} \rangle$. Then $\mathrm{Aut}(\mathfrak{S}_1) = \mathbb{Z}_3 \wr S_3 = \mathrm{Aut}(\Gamma_1)$ is a group of order $3 \cdot (3!)^3 = 2^3 \cdot 3^4 = 648$. Here $\Gamma_1 = \vec{C}_3[E_3]$ is the composition of a directed cycle $\vec{C}_3$ with the empty 3-vertex graph $E_3$. The graph $\Gamma_1$ is a Cayley graph $\mathrm{Cay}(\mathbb{Z}_9, \{1, 4, 7\})$, that is, a 9-vertex circulant. The three copies of $E_3$ are joined cyclically by means of three sets of nine arcs: each such set of arcs joining the three vertices of one copy of $E_3$ to the next copy of $E_3$.

In what follows, indecomposable $S$-rings over groups $\mathbb{Z}_{p^i}$ will be called *atoms*. It turns out that if we understand the structure of all atoms then we understand the automorphism groups of all $S$-rings over $\mathbb{Z}_{p^k}$.

## 2.2 Affine Overgroups of $(\mathbb{Z}_n, \mathbb{Z}_n)$

Let $\mathrm{Aff}(1, n)$ be the group of all one-dimensional linear transformations over $\mathbb{Z}_n$, that is,

$$\mathrm{Aff}(1, n) := \{\mathcal{M}_{a,b} : a \in \mathbb{Z}_n^*, b \in \mathbb{Z}_n\},$$

where the *affine transformation* $\mathcal{M}_{a,b}$ of $\mathbb{Z}_n$ is given by

$$\mathcal{M}_{a,b} : x \mapsto ax + b \quad (x \in \mathbb{Z}_n).$$

The following well-known facts prove to be very helpful in our context.

1. Every $\mathcal{M}_{a,b} \in \mathrm{Aff}(1, n)$ is a permutation on $\mathbb{Z}_n$;
2. $(\mathrm{Aff}(1, n), \mathbb{Z}_n)$ is a permutation group;

3. $|\mathrm{Aff}(1, n)| = n \cdot \varphi(n)$;
4. $\mathbb{Z}_n \cong \{\mathcal{M}_{1,b} : b \in \mathbb{Z}_n\}$ is a normal subgroup of $\mathrm{Aff}(1, n)$.

Here we restrict our considerations to the case $n = p^k$; moreover, only small values of $k$ will actually be required. Therefore $|\mathbb{Z}^*_{p^k}| = \varphi(p^k) = p^{k-1}(p-1)$ and $|\mathrm{Aff}(1, p^k)| = p^{2k-1}(p-1)$. Note that the permutation $\mathcal{M}_{1,1}$ is nothing else than the standard cycle $(0\ 1\ \ldots\ n-1)$. Every subgroup $(G, \mathbb{Z}_n)$ of $\mathrm{Aff}(1, n)$ which contains the cycle $\mathcal{M}_{1,1}$ will be called an *affine overgroup* of $(\mathbb{Z}_n, \mathbb{Z}_n)$ or simply an affine group. It is convenient to call $\mathrm{Aff}(1, n)$ the *complete affine group*.

Clearly, each affine group $G$ can be represented as a semidirect product

$$G = \mathbb{Z}_n \rtimes L,$$

where $L$ is a subgroup of $\mathbb{Z}^*_n$, while in the group $G$, the subgroup $L$ is the stabiliser $G_0$ of the element $0 \in \mathbb{Z}$.

This is why orbits of $L$ on the set $\mathbb{Z}_n$ form an $S$-ring over $\mathbb{Z}_n$. Such an $S$-ring which stems from a suitable affine group $G$ is called an *affine S-ring* over $\mathbb{Z}_n$.

By the given definitions for an affine $S$-ring $\mathfrak{S}$, which is obtained from $(G, \mathbb{Z}_n)$, the group $G$ is a subgroup of $\mathrm{Aut}(\mathfrak{S})$. A significant issue is to understand the full group $\mathrm{Aut}(\mathfrak{S})$, or, in other words, the 2-closure of $(G, \mathbb{Z}_n)$.

For the case $n = p$, the full affine group of order $p(p-1)$ is 2-transitive and its 2-closure is the symmetric group $S_p$. All other affine groups are *uniprimitive*, that is, primitive but not 2-transitive.

One of the classical results in the theory of permutation groups (due to Burnside and Schur) is that each uniprimitive permutation group of prime degree $p$ is affine and 2-closed.

For $n = p^k, k > 1$ all affine groups are imprimitive! It turns out that here the description of the 2-closure is becoming a more involved task. Recall that a transitive permutation group $(G, M)$ is called a *Frobenius group* if each non-identical permutation $g \in G$ has at most one fixed point in $M$.

**Proposition 1** *Every imprimitive Frobenius group is 2-closed.*

Note that, in general, the proposition fails for primitive permutation groups, though it remains valid for some restricted classes, like the above-mentioned uniprimitive permutation groups of degree $p$.

For $k \geq 2$, Proposition 1 allows us to detect a one-parameter family of affine 2-closed groups of order $t \cdot p^k$, where $t$ is any divisor of $p - 1$. These imprimitive Frobenius groups form one class of $p^k$-atoms in the process of classification of the automorphism groups of circulants.

There exists also an efficient criterion for the indecomposability of an affine $S$-ring over $\mathbb{Z}_{p^k}$. To avoid technical complications, this criterion will not be formulated explicitly here, however, it will be exploited implicitly in a further presentation.

At this stage, we are sufficiently prepared to go ahead towards formulating the main results in this section.

## 2.3 Main Results

Let us denote by $u_n$ the number of subgroups of the multiplicative group $\mathbb{Z}_n^*$. It is easy to understand that for $n = p^k$, $p$ odd prime, there is the equality $u_n = k \cdot d$, where $d$ is the number of all natural divisors of $p - 1$.

Thus, in the context of the current paper, the following values are mostly significant:

$$u_3 = 2, \quad u_9 = 4, \quad u_{27} = 6, \quad u_5 = 3 \quad u_{25} = 6, \quad u_{125} = 9.$$

**Proposition 2** *There are exactly $u_p$ 2-closed overgroups of $\mathbb{Z}_p$:*

*(a) The symmetric group $S_p$; and*
*(b) Frobenius uniprimitive groups $\mathfrak{F}_p^s$ of order $sp$, where $s$ is a proper divisor of $p-1$, (that is, $s < p-1$).*

All the groups which appear in the formulation of Proposition 2 play the role of $p$-atoms in the recursive description of the 2-closed overgroups of $\mathbb{Z}_{p^k}$, $k > 1$.

**Proposition 3** *Every 2-closed overgroup of $\mathbb{Z}_{p^2}$ is of one of the following types:*

*(a) wreath product of $p$-atoms;*
*(b) $S_{p^2}$;*
*(c) Frobenius group $\mathfrak{F}_{p^2}^s$, where $s$ is any divisor of $p-1$.*

Groups of types (b) or (c) in the above proposition will be called $p^2$-atoms.

**Corollary 1** *There are exactly $1 + u_p + (u_p)^2$ 2-closed overgroups of $\mathbb{Z}_{p^2}$.*

The first difficulty in the description of the 2-closure of affine overgroups of $\mathbb{Z}_{p^k}$ appears in the case when $k = 3$. Each affine group $G$ over $\mathbb{Z}_{p^3}$ can be presented in the form $G = \mathbb{Z}_{p^3} \rtimes L$, where $L \leq \mathbb{Z}_{p^3}^*$.

It turns out that we have to classify all affine groups into three classes; according to the appearance of elements $(p+1)$ and $(p^2+1)$ in $L$.

If $(p+1) \notin L$ and $(p^2+1) \notin L$, then $L$ is a Frobenius (imprimitive) group, which is 2-closed.

If $(p+1) \in L$ and $(p^2+1) \in L$, then the orbits of $L$ define a decomposable $S$-ring. Thus $G$ is not 2-closed, however, $G^{(2)}$ can be described with the aid of the iterated wreath product.

**Example 2** Here $n = 27$, $L = \{1, 4, 7, 10, 13, 16, 19, 22, 25\}$. Clearly both $1+3$ and $1+3^2$ are in $L$. The group $G = \mathbb{Z}_{27} \rtimes L$ defines an $S$-ring $\mathfrak{S}_2$ with the basic elements $T_1 = \underline{1, 4, 7, 10, 13, 16, 19, 22, 25}$, $2T_1$, $T_3 = \underline{3, 12, 21}$, $2T_3$, $T_5 = \underline{9}$, $2T_5$, $\underline{0}$.

Analysing the lattice of $S$-rings over $\mathbb{Z}_{27}$, it is possible to observe that $\mathrm{Aut}(\mathfrak{S}_2) = \mathrm{Aut}(\Gamma_1) \cap \mathrm{Aut}(\Gamma_3) \cap \mathrm{Aut}(\Gamma_5)$, where the $\Gamma_i$ are the circulants defined by $T_i$.

In [7] we give a schematic figure depicting graphs $\Gamma_1$, $\Gamma_3$ and $\Gamma_5$ together which should help the reader conceptualise the structure of $\mathrm{Aut}(\mathfrak{S}_2)$; it turns out to be $\mathbb{Z}_3 \wr \mathbb{Z}_3 \wr \mathbb{Z}_3$. (Note that the operation of wreath product is associative.) Thus $|\mathrm{Aut}(\mathfrak{S}_2| = 3 \cdot (3 \cdot 3^3)^3 = 3^{13}$.

Finally, the third most sophisticated case of affine overgroups of $\mathbb{Z}_{p^3}$ is when $(p+1) \notin L$, but $(p^2+1) \in L$. In this case, the affine group $G \rtimes L$ defines an indecomposable $S$-ring, however the group $G$ is not 2-closed.

Again we will try to understand this more sophisticated situation with the aid of an example.

**Example 3** Here again $n = 27$. Consider $L = \{1, 10, 19\}$. Note that $(3+1) \notin L$, however $(3^2+1) \in L$, thus we indeed face the third case which we declared to be the most difficult one. The group $G = \mathbb{Z}_{27} \rtimes L$ defines the $\mathfrak{S}_3$ with the basic quantities of lengths 3 and 1 as follows:

$$\mathfrak{S}_3 = \langle \underline{0}, \underline{1, 10, 19}, \underline{2, 11, 20}, \underline{8, 17, 26}, \underline{7, 16, 25}, \underline{4, 13, 22}, \underline{5, 14, 23}, \underline{3}, \underline{6}, \underline{9}, \underline{12}, \underline{15}, \underline{18}, \underline{21}, \underline{24} \rangle.$$

It is possible to prove that in this case $\mathrm{Aut}(\mathfrak{S}_3) = \cap_{i=1}^{3} \mathrm{Aut}(\Gamma_i)$, where $\Gamma_1 = \mathrm{Cay}(\mathbb{Z}_{27}, \{1, 10, 19\})$, $\Gamma_2 = \mathrm{Cay}(\mathbb{Z}_{27}, \{9\})$ and $\Gamma_3 = \mathrm{Cay}(\mathbb{Z}_{27}, \{3\})$. These circulant graphs $\Gamma_1$ and $\Gamma_2$ are depicted together in Fig. 1 where solid black arrows denote nine arcs from one directed triangle to the next one: each vertex in the first triangle joined by three arcs to each vertex in the second triangle.

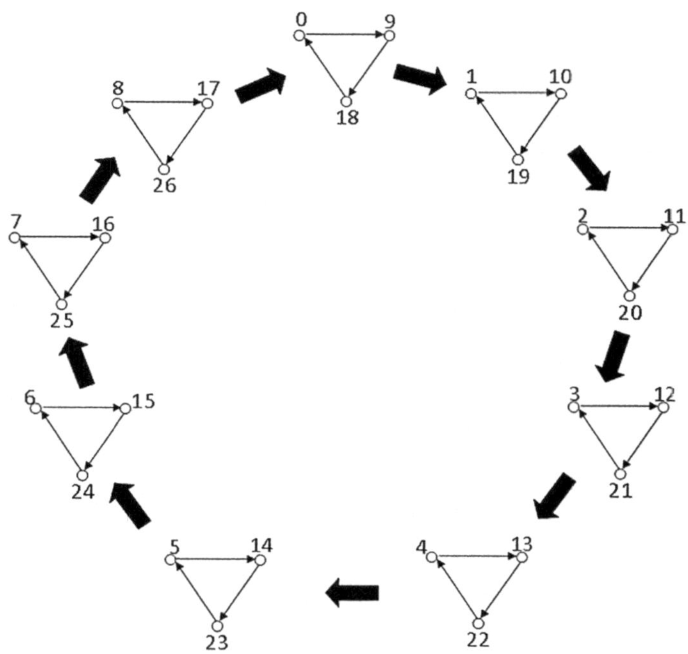


**Fig. 1** The circulant graphs $\Gamma_1$ (thin arcs) and $\Gamma_2$ (solid arcs)

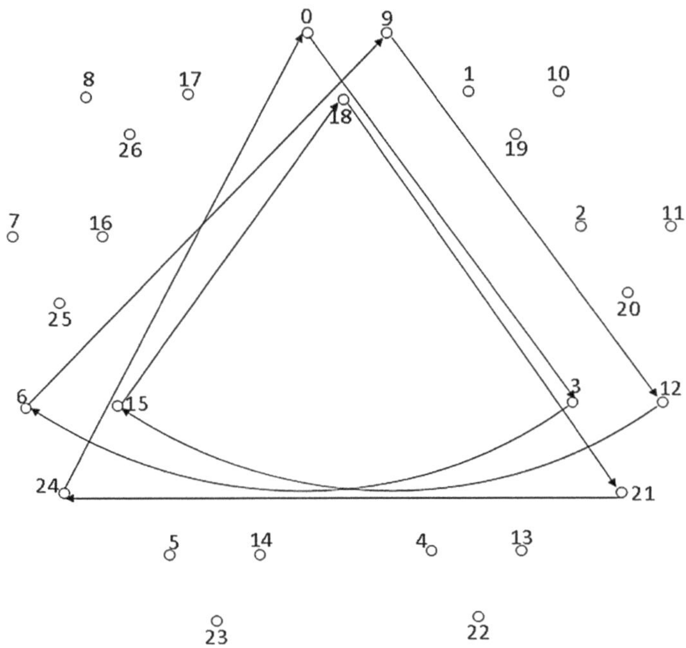


**Fig. 2** One of the three connected components of the graph $\Gamma_3$

The situation with graph $\Gamma_3$ is more complicated provided we want to arrange its vertices to afford the viewer a "correct" visualisation: that is to consider $\Gamma_3$ together with the two previous graphs in order to visualise the action of the group $\mathrm{Aut}(\mathfrak{S}_3)$.

In principle, the graph $\Gamma_3$ has a very simple structure of the form $3\vec{C}_9$, that is, the disjoint union of three directed cycles of length 9. Nevertheless, we intentionally prefer to depict it in a more sophisticated "skew" manner, as it appears in Fig. 2.

In this figure, we have deliberately shown only one of the three connected components of $\Gamma_3$ (it is the one spanning those vertices which are multiples of 3 in $\mathbb{Z}_{27}$). The arcs of the two other isomorphic connected components are omitted.

From Fig. 2 it immediately follows that $\mathrm{Aut}(\Gamma_1) \cap \mathrm{Aut}(\Gamma_2) = \mathbb{Z}_9 \wr \mathbb{Z}_3$. Clearly, $\mathrm{Aut}(\Gamma_3) = \mathrm{Aut}(3\vec{C}_9) = S_3 \wr \mathbb{Z}_3$. Thus, in principle, $G^{(2)} = \mathrm{Aut}(\mathfrak{S}_3) = (\mathbb{Z}_9 \wr \mathbb{Z}_3) \cap (S_3 \wr \mathbb{Z}_3)$. However, this abstract formula is useful only if we take into account the actual form of the graph $\Gamma_3$ relative to the other two graphs. The reader is welcome to check that the group $G^{(2)}$ has the form $(\mathbb{Z}_3)^3 \cdot \mathbb{Z}_9$ (note that this is not a semidirect product!) and

$$G^{(2)} = \langle h, h_0, h_1, h_2 \rangle.$$

Here $h = \mathcal{M}_{1,1}$ is a generating element of $\mathbb{Z}_{27}$, while

$$h_0 = (0\ 9\ 18)(3\ 12\ 21)(6\ 15\ 24) \tag{4}$$

$$h_1 = (1\ 10\ 19)(4\ 13\ 22)(7\ 16\ 25) \tag{5}$$
$$h_2 = (2\ 11\ 20)(5\ 14\ 23)(8\ 17\ 26). \tag{6}$$

Indeed, $h_0$ preserves the copy of $\vec{C}_9$ depicted in Fig. 2, fixing the two other copies of $\vec{C}_9$, preserving simultaneously graphs $\Gamma_1$ and $\Gamma_2$. The permutations $h_1$ and $h_2$ have similar interpretations.

Finally, we wish to stress that while $|G| = |\mathbb{Z}_{27}| \cdot |L| = 27 \cdot 3 = 81$, for the 2-closure we get $|G^{(2)}| = 3^3 \cdot 9 = 243$. In other words, the order of the 2-closure $G^{(2)}$ is $p$ times larger in comparison with the order of the affine group $G = \mathrm{Aut}(\mathfrak{S}_3)$. Note also that the group $G^{(2)}$ is not wreath decomposable, that is, it cannot be represented as a wreath product of suitable smaller atoms. According to Klin and Pöschel, the group $G^{(2)}$ is called a subwreath product (see the discussion below).

Now we are prepared for the consideration of the next theoretical claim, which again will be presented here without its formal justification.

**Theorem 3** *Every 2-closed overgroup of $\mathbb{Z}_{p^3}$ belongs to one of the following types:*

*(a) Wreath product of $p$ atoms and $p^2$-atoms or vice-versa;*
*(b) wreath product of three $p$-atoms;*
*(c) $S_{p^3}$;*
*(d) A Frobenius group $\mathbb{Z}_{p^3} \rtimes L$ such that $(p+1) \notin L$ and $(p^2+1) \notin L$;*
*(e) The permutation group $(G^{(2)}, \mathbb{Z}_{p^3})$, where $G^{(2)}$ is the 2-closure of $G = \mathbb{Z}_{p^3} \rtimes L$, such that $(p^2+1) \in L$, but $(p+1) \notin L$.*

**Corollary 2** *There exist exactly $1 + 4u_p + 2(u_p)^2 + (u_p)^3$ different 2-closed overgroups of $\mathbb{Z}_{p^3}$.*

As has been explained in the previous section, in order to apply the structural approach for the enumeration of circulants, besides knowledge of all 2-closed overgroups of $\mathbb{Z}_{p^k}$, the orders of their normalisers (in $S_{p^k}$) are also required. In general, for the enumeration purposes of this paper, we did not try to obtain this information on a theoretical level since for small values of $p$ this information may be obtained with the aid of computer algebra packages such as GAP.

The alternative, so-called multiplier approach, relies on some implicit information regarding the behaviour of the normalisers of 2-closed overgroups of $\mathbb{Z}_{p^k}$, namely, the knowledge of necessary and sufficient conditions of isomorphism of circulants is enough for this approach. For the case $n = p^k$, $p$ odd, these conditions were formulated in terms very suitable for the purposes of analytical information. For general $n$, the problem of analytical enumeration of the $n$-vertex circulants might become a subject of future special attention.

## 2.4 Brief Historical Summary

The crucial concept of a Schur ring, which is used in this paper, goes back to the seminal paper by Schur [25, 26]. For more than two decades this work was known only to a few colleagues and followers of Schur and it was revitalised by Wielandt, in particular, due to a special chapter on $S$-rings in his classic book [30]. Nowadays, $S$-rungs are discussed in many modern textbooks on group theory, such as [2, 27].

The paper [29] is a direct predecessor of the current text. It relies on [13] and presents first attempt to create (with the aid of a computer) a full catalogue of $S$-rings over the cyclic group $\mathbb{Z}_{125}$.

Already at that stage, it became clear that in the problem of the description of automorphism groups of circulants, a crucial step is to move from $n = p^2$ to $n = p^3$. Main ideas in this direction were announced in the abstract [9].

Another pioneering text [10] dealt with analytical enumeration of prime-squared circulants.

A more complete discussion of the historical background to the concepts in this paper can be found in our preprint [7].

## 3 The Structural Approach: The Case $p^3$ for $p = 3$

We shall now use the structural approach to enumerate the *undirected* circulant graphs of order 27. The number of such circulant graphs has already been determined by Brendan McKay and listed in [11], but we shall here also obtain the generating function by valency for the number of these circulant graphs.

A list of symmetric Schur rings over $\mathbb{Z}_{27}$ was obtained using the package COCO (see [4]). (By a "symmetric Schur-Ring", we mean one in which every basic set $T$ satisfies $-T = T$. This is sufficient for our purpose of enumerating undirected circulant graphs.) This list is given below. In this list, we can observe that $\mathfrak{S}_1$ is the finest Schur ring with the smallest automorphism group, while $\mathfrak{S}_8$ has the largest automorphism group. Therefore $\mathfrak{S}_1$ contains all the other Schur rings. We may now construct the lattice of Schur rings. This is given in Fig. 3.

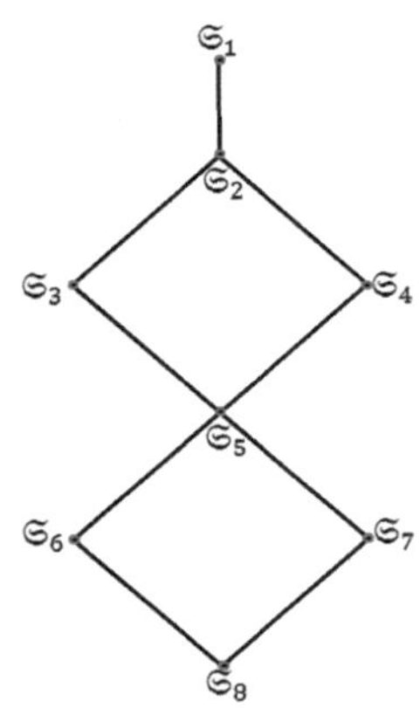


**Fig. 3** Lattice of all $\mathcal{S}$-rings for $n = 27$ (symmetric)

$$\mathfrak{S}_1 = \langle \underline{0}, 1, 26, 2, 25, 3, 24, 4, 23, 5, 22, 6, 21, 7, 20, 8, 19, 9, 18, 10, 17, 11, 16, 12, 15, 13, 14\rangle,$$
$$\mathfrak{S}_2 = \langle \underline{0}, 1, 26, 8, 19, 10, 17, 2, 25, 7, 20, 11, 16, 3, 24, 4, 23, 5, 22, 13, 14, 6, 21, 9, 18, 12, 15\rangle,$$
$$\mathfrak{S}_3 = \langle \underline{0}, 1, 26, 2, 25, 4, 23, 5, 22, 7, 20, 8, 19, 10, 17, 11, 16, 13, 14, 3, 24, 6, 21, 9, 18, 12, 15\rangle,$$
$$\mathfrak{S}_4 = \langle \underline{0}, 1, 26, 8, 19, 10, 17, 2, 25, 7, 20, 11, 16, 3, 24, 6, 21, 12, 15, 4, 23, 5, 22, 13, 14, 9, 18\rangle,$$
$$\mathfrak{S}_5 = \langle \underline{0}, 1, 26, 2, 25, 4, 23, 5, 22, 7, 20, 8, 19, 10, 17, 11, 16, 13, 14, 3, 24, 6, 21, 12, 15, 9, 18\rangle,$$
$$\mathfrak{S}_6 = \langle \underline{0}, 1, 26, 2, 25, 4, 23, 5, 22, 7, 20, 8, 19, 10, 17, 11, 16, 13, 14, 3, 24, 6, 21, 9, 18, 12, 15\rangle,$$
$$\mathfrak{S}_7 = \langle \underline{0}, 1, 26, 2, 25, 3, 24, 4, 23, 5, 22, 6, 21, 7, 20, 8, 19, 10, 17, 11, 16, 12, 15, 13, 14, 9, 18\rangle,$$
$$\mathfrak{S}_8 = \langle \underline{0}, 1, 2, 3, 4, 5, 6, 7, 8, 9, 10, 11, 12, 13, 14, 15, 16, 17, 18, 19, 20, 21, 22, 23, 24, 25, 26\rangle$$

Using (2) we can obtain the generating functions $f_i(t)$. These are as follows

$$\begin{aligned}
f_1(t) &= (1+t^2)^{13},\\
f_2(t) &= (1+t^6)^3(1+t^2)^4,\\
f_3(t) &= (1+t^{18})(1+t^2)^4,\\
f_4(t) &= (1+t^6)^4(1+t^2),\\
f_5(t) &= (1+t^{18})(1+t^6)(1+t^2),\\
f_6(t) &= (1+t^{18})(1+t^8),\\
f_7(t) &= (1+t^{24})(1+t^2),\\
f_8(t) &= (1+t^{26}).
\end{aligned}$$

Table 1 gives a list of the orders of the automorphism groups and their normalizers. These were again obtained using GAP.

We may now determine $g_i = g_i(t)$ for $i = 1, 2, \ldots, 8$, using Eq. (3) and Fig. 3.

$$\begin{aligned}
g_8 &= f_8 = 1 + t^{26},\\
g_7 &= f_7 - g_8 = t^{24} + t^2,\\
g_6 &= f_6 - g_8 = t^{18} + t^8,\\
g_5 &= f_5 - (g_8 + g_7 + g_6) = t^{20} + t^6,
\end{aligned}$$

**Table 1** Sizes of automorphism groups and their normalizers for $n = 27$

| $G_i$ | $\lvert G_i\rvert$ | $\lvert N(G_i)\rvert$ | $\frac{\lvert G_i\rvert}{\lvert N(G_i)\rvert}$ |
|---|---|---|---|
| $G_1$ | 54 | 486 | 1/9 |
| $G_2$ | 486 | 4374 | 1/9 |
| $G_3$ | 34992 | 104976 | 1/3 |
| $G_4$ | 181398528 | 544195584 | 1/3 |
| $G_5$ | 13060694016 | 13060694016 | 1 |
| $G_5$ | 286708355039232000 | 286708355039232000 | 1 |
| $G_7$ | 3656994324480 | 3656994324480 | 1 |
| $G_8$ | 10888869450418352160768000000 | 10888869450418352160768000000 | 1 |

$$g_4 = (f_4 - g_8 - g_7 - g_6 - g_5)/3 = t^{20} + t^{18} + 2t^{14} + 2t^{12} + t^8 + t^6,$$
$$g_3 = (f_3 - g_8 - g_7 - g_6 - g_5)/3 = t^{24} + 2t^{22} + t^{20} + t^6 + 2t^4 + t^2,$$
$$g_2 = (f_2 - g_8 - g_7 - g_6 - g_5 - 3g_4 - 3g_3)/9 = t^{18} + 2t^{16} + t^{14} + t^{12} + 2t^{10} + t^8,$$
$$g_1 = (f_1 - g_8 - g_7 - g_6 - g_5 - 3g_4 - 3g_3 - 9g_2)/9,$$

which finally gives us

$$g_1 = t^{24} + 8t^{22} + 31t^{20} + 78t^{18} + 141t^{16} + 189t^{14} + 189t^{12} + 141t^{10} + 78t^8 + 31t^6 + 8t^4 + t^2.$$

Therefore $g(t) = g_1 + g_2 + \cdots + g_8$ which gives

$$t^{26} + 3t^{24} + 10t^{22} + 34t^{20} + 81t^{18} + 143t^{16} + 192t^{14} + 192t^{12} + 143t^{10} + 81t^8 + 34t^6 + 10t^4 + 3t^2 + 1$$

This gives the same generating function as that obtained below using the multiplier method. It confirms McKay's old result [unpublished, 1995] that there are exactly 928 non-isomorphic, undirected circulant graphs on 27 vertices.

# 4 The Multiplier Approach for $n = p^3$

## *4.1 The Main Isomorphism Theorem*

Since Ádám's Conjecture does not hold for $n = p^3$ we need the next result which tells us, in terms of their connecting sets, when two circulant graphs of this order are isomorphic. This isomorphism criterion will require us to partition the elements of the connecting sets into *layers*, which we now define.

First consider the set $\mathbb{Z}'_{p^3} = \mathbb{Z}'_{p^3} - \{0\}$ and divide its elements into three layers, namely, $Y_0$, $Y_1$ and $Y_2$, where $Y_0$ will contain those elements which do not have $p$ as a factor, $Y_1$ will contain those elements which do have $p$ as a factor but not $p^2$, and $Y_2$ will contain those elements which have $p^2$ as a factor. A connecting set $X$ is then given by $X = X_0 \dot{\cup} X_1 \dot{\cup} X_2$, where $X_i = X \cap Y_i$. The layer $X_0$ is a subset of $\mathbb{Z}^*_{p^3}$ while the layer $X_1$ is a subset of $p\mathbb{Z}^*_{p^2}$ and $X_2$ is a subset of $p^2\mathbb{Z}^*_p$. In addition, when these layers are acted upon (multiplicatively) by elements of $\mathbb{Z}^*_n$, wherein this case $n = p^3$, these layers are invariant.

**Theorem 4** (Main Isomorphism Theorem) *Let $n = p^3$ ($p$ an odd prime) and let $\Gamma$ and $\Gamma'$ be two $p^3$-circulants with the connecting sets $X$ and $X'$, respectively. Then $\Gamma$ and $\Gamma'$ are isomorphic if and only if their respective layers are multiplicatively equivalent, that is,*

$$X'_0 = m_0 X_0, \quad X'_1 = m_1 X_1 \quad X'_2 = m_2 X_2, \tag{$M_3$}$$

*for an arbitrary set of multipliers* $m_0, m_1, m_2 \in \mathbb{Z}_{p^3}^*$. *Moreover, in the above, one must have*

*(i)*

$$m_1 \equiv m_0 (\text{mod } p^2) \text{ and } m_2 \equiv m_1 (\text{mod } p) \quad (E_{00})$$

*whenever*

$$(1 + p^2)X_0 \neq X_0, \quad (R_{00})$$

*(ii)*

$$m_1 \equiv m_0 (\text{mod } p) \quad (E_{01})$$

*whenever*

$$(1 + p)X_0 \neq X_0, \quad (R_{01})$$

*(iii)*

$$m_2 \equiv m_1 (\text{mod } p) \quad (E_{10})$$

*whenever*

$$(1 + p)X_1 \neq X_1. \quad (R_{10})$$

This isomorphism theorem for $p^3$ involves three multipliers and $\binom{3}{2}$ cases coming from the non-invariance conditions $R_{ij}$, (the easier case for enumeration of circulants of order $p^2$ is explained in detail in [7, 10]). What makes the enumeration problem particularly difficult is not only that there are multipliers for the separate layers of the connecting sets, but that, depending on non-invariance conditions, some multipliers must be equal in certain cases. Moreover, the intersection between the conditions makes this case even more difficult. The three non-invariance relations $R_{00}$, $R_{01}$, $R_{10}$ below, will break up into five cases which will eventually give eleven enumeration subproblems, as we shall see below.

## 4.2 Representation and Computational Implementation of the Main Isomorphism Theorem, $p = 3, 5$

Liskovets and Pöschel in [19] manage, in the case when $n = p^3$, to partition the conditions of the Main Theorem into five parts which makes their use in enumeration much easier. These authors take into consideration all combinations of non-invariance conditions ($R_{ij}$), together with the remaining invariance conditions

$$(1 + p^{k-i-j-1})X_i = X_i \quad (\neg R_{ij})$$

and make use of a number of results, in order to obtain the subproblem list for counting circulants of order $p^k$. For details of how this list has been generated from the Main

**Table 2** The conditions for isomorphism of circulants of order $p^3$

| Subproblem | Non-invariance conditions | Invariance conditions | Conditions on multipliers |
|---|---|---|---|
| $A_1$ | $\emptyset$ | $\neg R_{01}, \neg R_{10}$ | no restriction |
| $A_2$ | $R_{00}$ | $\emptyset$ | $m_2 = m_1 = m_0$ |
| $A_3$ | $R_{01}$ | $\neg R_{00}, \neg R_{10}$ | $m_1 = m_0$ |
| $A_4$ | $R_{10}$ | $\neg R_{01}$ | $m_2 = m_1$ |
| $A_5$ | $R_{01}, R_{10}$ | $\neg R_{00}$ | $m_2 = m_1$ and $m_1 \equiv m_0 (\text{mod } p)$ |

Theorem using results from number theory and walks through a rectangular lattice, the reader is referred to [19]. The necessary information required for the enumeration of $p^3$ circulants is listed in Table 2, which we, therefore, take to be a rewording of Theorem 4. This has been obtained from Table 1 in [19].

The five subcases $A_1$ to $A_5$ shown in Table 2, give conditions on the three multipliers $m_0$, $m_1$ and $m_2$ for two circulants of order $p^3$ to be isomorphic. The re-interpretation of the Main Isomorphism Theorem by Liskovets and Pöschel says that the three multipliers must satisfy at least one of the five sets of conditions for two ciruclants of order $p^3$ to be isomorphic. So, for example, condition $A_3$ means that if the non-invariance condition $R_{01}$ holds, together with the invariance conditions $\neg R_{00}$ and $\neg R_{10}$, then $m_1 = m_0$ but $m_2$ can be independent.

If we let $A$ denote set of all non-isomorphic circulants of order $p^3$ and let $A_i$, for $i = 1, \ldots, 5$, also denote the set of circulants which are non-isomorphic under the respective condition of Table 2, then, the result of Liskovets and Pöschel says that

$$A = A_1 \cup A_2 \cup A_3 \cup A_4 \cup A_5.$$

In addition to this information given in Table 2, we shall also use, without explicit mention, the following observations [19].

$$(R_{ij}) \Rightarrow (R_{ij'}) \text{ whenever } j' \geq j$$

Therefore

$$\neg(R_{ij'}) \Rightarrow \neg(R_{ij})$$

As a result we have that

$$\neg(R_{01}) \Rightarrow \neg(R_{00}) \tag{7}$$

In addition,

$$(E_{ij}) \Rightarrow (E_{i'j}) \text{ whenever } i' \geq i$$

Therefore

$$\neg(E_{i'j}) \Rightarrow \neg(E_{ij})$$

that is,

$$\neg(E_{10}) \Rightarrow \neg(E_{00}) \tag{8}$$

As we explained before, when the subproblem in question includes one or more non-invariance conditions, these are changed to invariance conditions and then the result is subtracted from the total number. Therefore, in order to count under a given non-invariance relation $R_{ij}$, we first

(i) Determine the count under the action assuming the invariance relation $\neg(R_{ij})$,
(ii) Determine the count under the action without any (non)-invariance relations,
(iii) Subtract the result of (i) from (ii).

This procedure is often complicated by having both invariance and non-invariance conditions. For example, to count the number of non-isomorphic circulants in case $A_3$ of Table 2, we

(i) First count under the conditions $m_1 = m_0, \neg(R_{00}), \neg(R_{10})$.
(ii) Then count under the conditions $m_1 = m_0, \neg(R_{01}), \neg(R_{00}), \neg(R_{10})$.
(iii) Then subtract the result of (ii) from (i).

Having counted the number of non-isomorphic circulants under each of the five isomorphism conditions we then need to calculate $|A|$ and therefore we would need to consider the intersections between the $A_i$. It, however, transpires that these intersections are empty. This can be seen by considering the invariance and non-invariance relations. It, therefore, follows that

$$|A| = |A_1| + |A_2| + |A_3| + |A_4| + |A_5| \qquad (*)$$

This is the main reason why it is easier to do enumeration using the formulation of Theorem 4 as in Table 2.

We shall now briefly explain how these five problems lead to eleven subproblems using the case of undirected circulant graphs of order $p^3$ for $p = 3$. Roughly speaking, our goal is to solve each concrete subproblem by considering a suitable group of multipliers acting on a suitable combination of layers and applying the standard enumeration technique of Pólya-Redfield. A more detailed explanation can be found in [7].

Let $Y_0, Y_1, Y_2$, be the three layers of $\mathbb{Z}'_{27}$. If $X$ is the connecting set of the circulant graph then, by Theorem 4, the non-invariance conditions in this case are

$$\begin{aligned} R_{00} :&\quad 10X_0 \neq X_0 \\ R_{01} :&\quad 4X_0 \neq X_0 \\ R_{10} :&\quad 4X_1 \neq X_1, \end{aligned}$$

where $X_0 = X \cap Y_0$, $X_1 = X \cap Y_1$ and $X_2 = X \cap Y_2$.

When we enumerate under an invariance condition, such as $10X_0 = X_0$, we must take $X_0$ from whole subsets of $Y_0$ which are invariant under $10Y_0 = Y_0$. These subsets partition $Y_0$, therefore, under the condition $10X_0 = X_0$, the set $X_0$ must be a union of these parts. Therefore, the multiplicative action is taken on these parts or blocks.

We shall denote the partitioned set corresponding to the invariance condition $10Y_0 = Y_0$ by $Y_0^*$, that corresponding to $4Y_0 = Y_0$ by $Y_0^{**}$, and that corresponding to the invariance condition $4Y_1 = Y_1$ by $Y_1^*$.

For the sake of illustration, we shall discuss $A_4$, one of the problems which splits into two subproblems. Here we have the conditions $R_{10}$ and $\neg R_{01}$ when $m_2 = m_1$. Therefore we shall now consider $A_{41}$ and $A_{42}$ as follows:

$A_{41}$ is the set of non-isomorphic circulants resulting from the action with $\neg R_{01}$, that is with layers arising from $4X_0 = X_0$. Therefore in this action $X_0$ must be a union of parts in $Y_0^{**}$, therefore we shall use $Y_0^{**}$ instead of $Y_0$.

$A_{42}$ is the set of non-isomorphic circulants resulting from the action with $\neg R_{01}$ and $\neg R_{10}$. This means the set $X_0$ must be a union of the layers in $Y_0^{**}$ and $X_1$ a union of layers in $Y_1^*$.

The required result for $|A_4|$ will then be $|A_{41}| - |A_{42}|$. This example also gives us an opportunity to illustrate our direct use of the cycle indices of the relevant group of multipliers. Since $m_2 = m_1$ while $m_0$ is independent, we require the cycle index $I_{(\mathbb{Z}_{27}^*, Y_1 \cup Y_2)} \times I_{(\mathbb{Z}_{27}^*, Y_0)}$, blocked as required. Therefore we have these cycle indices: $I_{(\mathbb{Z}_{27}^*, Y_1 \cup Y_2)} \times I_{(\mathbb{Z}_{27}^*, Y_0^{**})}$ for $A_{41}$ and $I_{(\mathbb{Z}_{27}^*, Y_1^* \cup Y_2)} \times I_{(\mathbb{Z}_{27}^*, Y_0^{**})}$ for $A_{42}$.

Finally, we should also briefly explain how $A_5$ splits into four subproblems. Once again, $A_5$ will be divided into the problems $A_{51}$ and $A_{52}$. Although $A_{51}$ is determined in a manner similar to $A_{41}$, one should be cautious when determining $A_{52}$, since this time we have two non-invariance conditions. This means that we have that $A_{52}$ is the set of non-isomorphic circulants resulting from the action with $\neg(R_{01}$ and $R_{10})$ and $\neg R_{00}$. Now, by de Morgan's laws, we have that

$$\begin{aligned}\neg(R_{01} \text{ and } R_{10}) \text{ and } \neg R_{00} =& (\neg R_{01} \text{ or } \neg R_{10}) \text{ and } \neg R_{00}\\ =& (\neg R_{01} \text{ and } \neg R_{00}) \text{ or } (\neg R_{10} \text{ and } \neg R_{00})\end{aligned}$$

Now $\neg(R_{01}) \Rightarrow \neg(R_{00})$, therefore for $A_{52}$ we have

$$\begin{aligned}|\neg(R_{01} \text{ and } R_{10}) \text{ and } \neg R_{00}| =& |\neg(R_{01})| + |(\neg(R_{10}) \text{ and } \neg(R_{00}))| -\\ & |(\neg(R_{01}) \text{ and } \neg(R_{10}))|.\end{aligned}$$

Therefore we split $A_{52}$ into 3 enumeration subproblems, with the first problem enumerating under the condition $\neg(R_{01})$, the second under $\neg(R_{10})$ and $\neg(R_{00})$ and the last under the invariance conditions $\neg(R_{01})$ and $\neg(R_{10})$. These are the subproblems $A_{521}$, $A_{522}$, $A_{523}$, respectively, giving that $|A_5| = |A_{51}| - |A_{521}| - |A_{522}| + |A_{523}|$.

These methods finally give us an 11-term formula for $|A|$ in terms of its subproblems:

$$\begin{aligned}|A| = & |A_1| + |A_{21}| - |A_{22}| + |A_{31}| - |A_{32}| + |A_{41}| - |A_{42}|\\ & + |A_{51}| - |A_{521}| - |A_{522}| + |A_{523}|. \qquad (**)\end{aligned}$$

We observe that (**) consists of Ádám's term $A_{21}$, the enumerator up to Cayley isomorphism, together with ten less obvious small correcting terms.

The reason why we have eleven subproblems arising from the five terms in (*) following Table 2 is because $11 = 1 \cdot 1 + 3 \cdot 2 + 1 \cdot 4$, which gives the 3rd (little) Schröder number (see, for example, sequence A001003 in [28]). Note that, in this context, 5 is the third Catalan number.

The above analysis can be carried out in an analogous way for $p = 5$, and for directed or undirected graphs. To generalise a bit our notation, let us define $A[s; p^3]$ for $s = u$, undirected, $s = d$, directed, and $p = 3, 5$ to be the number of undirected/directed circulant graphs on $p^3$ vertices. We similarly define $A_i[s; p^3]$, $A_{ij}[s; p^3]$ and $A_{ijk}[s; p^3]$ to be the number of undirected/directed circulant graphs on $p^3$ vertices making up the corresponding intermediate terms $|A_i|$, $|A_{ij}|$ or $|A_{ijk}|$, respectively. With a slight abuse of notation, we also let $A_{\mathbf{w}}(t) := A_{\mathbf{w}}[s; p^3](t)$ (where $\mathbf{w}$ represents the subscript $i$, $ij$ or $ijk$) denote generically the generating function by valency.

## 4.3 Results for the Cases $p = 3, 5$, Undirected and Directed

We first give our main results in Table 3 which shows the number of circulant (di)graphs on 27 and 125 vertices regardless of valency.

Next, Table 4 contains the values of the intermediate terms, which, as defined and described in the previous section, jointly, in accordance with (*) and (**), yield the values in Table 3. We note that it is Ádám's term $A_{21}$, which is the greatest contributor to these values.

Then, in Table 5 we give the final and all the intermediate generating functions appearing in the directed case of order 27. For three other cases, the final and some of the intermediate terms are represented in our preprint [7]. The full list of the generating functions can be found in Appendix C in [6].

We also observe that the number of non-isomorphic undirected circulants on 27 vertices is now verified by Matan Ziv-Av's brute-force methods as well as that of McKay's from 1995, and by the structural and multiplier approaches described here. The generating functions for these circulants are now also verified by the structural (for the undirected case of order 27) and multiplier approaches and also by Ziv-Av's methods. The value $A[d; 27] = 3728891$ has recently appeared in [28] (Sequence A04929); we also note that $A_{21}[d; 27] = 3730584$ is represented there as well, in A056391.

**Table 3** The number of undirected and directed $p^3$-circulant graphs, $p = 3, 5$

| | Undirected | Directed |
|---|---|---|
| n = 27 | 928 | 3, 728, 891 |
| n = 125 | 92, 233, 720, 411, 499, 283 | 212, 676, 479, 325, 586, 539, 710, 725, 989, 876, 778, 596 |

**Table 4** The number of undirected and directed $p^3$-circulant graphs, $p = 3, 5$: intermediate contributors and totals

| Term | Undir n = 27 | Undir n = 125 | Dir n = 27 | Dir n = 125 |
|---|---|---|---|---|
| $A_1$ | 8 | 27 | 27 | 216 |
| $A_{21}$ | 944 | 92233720411833168 | 3730584 | ($a$) |
| $A_{22}$ | 48 | 419664 | 2776 | 879609512976 |
| $A_2 = A_{21} - A_{22}$ | 896 | 92233720411413504 | 3727808 | ($b$) |
| $A_{31}$ | 16 | 1272 | 156 | 5034768 |
| $A_{32}$ | 8 | 30 | 30 | 420 |
| $A_3 = A_{31} - A_{32}$ | 8 | 1242 | 126 | 5034348 |
| $A_{41}$ | 16 | 1272 | 156 | 5034768 |
| $A_{42}$ | 8 | 30 | 30 | 420 |
| $A_4 = A_{41} - A_{42}$ | 8 | 1242 | 126 | 5034348 |
| $A_{51}$ | 32 | 86592 | 1168 | 175943379264 |
| $A_{521}$ | 16 | 1680 | 200 | 13423440 |
| $A_{522}$ | 16 | 1680 | 200 | 13423440 |
| $A_{523}$ | 8 | 36 | 36 | 1044 |
| $(A_{52} = A_{521} + A_{522} - A_{523})$ | 24 | 3324 | 364 | 26845836 |
| $A_5 = A_{51} - A_{521} - A_{522} + A_{523}$ | 8 | 83268 | 804 | 175916533428 |
| $A = A_1 + A_2 + A_3 + A_4 + A_5$ | 928 | 92233720411499283 | 3728891 | ($c$) |

($^a$) 212676479325586539710726693559689232
($^b$) 212676479325586539710725813950176256
($^c$) 212676479325586539710725989876778596

## 4.4 Discussion of Results: Unexpected Patterns and Corollaries

**Numerical observations and some identities** A direct phenomenological analysis of the main and intermediate analytical formulae shown in these tables reveals some hidden patterns that need to be explained in general, either combinatorially, algebraically or analytically. First of all, in the four cases we are studying (undirected / directed, $p = 3$ / $p = 5$) the following three 'coincidences' are observed:

$$A_{31}[s;\, p^3] = A_{41}[s;\, p^3] \tag{9}$$

$$A_{32}[s;\, p^3] = A_{42}[s;\, p^3] \tag{10}$$

$$A_{521}[s;\, p^3] = A_{522}[s;\, p^3] \tag{11}$$

and (as a corollary of the first two)

$$A_3[s;\, p^3] = A_4[s;\, p^3]. \tag{12}$$

for $p = 3, 5$ and $s = u, d$ (where $A_4[s;\, p^3] := A_{41}[s;\, p^3] - A_{42}[s;\, p^3]$).

**Table 5** The generating functions for directed circulant graphs of order 27

| | |
|---|---|
| $A_1$ | $t^{26}+t^{25}+t^{24}+t^{23}+t^{22}+t^{21}+t^{20}+t^{19}+t^{18}+t^{17}+t^{16}+t^{15}+t^{14}+t^{13}+t^{12}+t^{11}+t^{10}+t^9+t^8+t^7+t^6+t^5+t^4+t^3+t^2+t+1$ |
| $A_{21}$ | $t^{26}+3t^{25}+23t^{24}+152t^{23}+850t^{22}+3680t^{21}+12850t^{20}+36606t^{19}+86919t^{18}+173701t^{17}+295311t^{16}+429388t^{15}+536810t^{14}+577996t^{13}+536810t^{12}+429388t^{11}+295311t^{10}+173701t^9+86919t^8+36606t^7+12850t^6+3680t^5+850t^4+152t^3+23t^2+3t+1$ |
| $A_{22}$ | $t^{26}+2t^{25}+6t^{24}+11t^{23}+22t^{22}+38t^{21}+65t^{20}+92t^{19}+129t^{18}+172t^{17}+214t^{16}+235t^{15}+263t^{14}+276t^{13}+263t^{12}+235t^{11}+214t^{10}+172t^9+129t^8+92t^7+65t^6+38t^5+22t^4+11t^3+6t^2+2t+1$ |
| $A_2=A_{21}-A_{22}$ | $t^{25}+17t^{24}+141t^{23}+828t^{22}+3642t^{21}+12785t^{20}+36514t^{19}+86790t^{18}+173529t^{17}+295097t^{16}+429153t^{15}+536547t^{14}+577720t^{13}+536547t^{12}+429153t^{11}+295097t^{10}+173529t^9+86790t^8+36514t^7+12785t^6+3642t^5+828t^4+141t^3+17t^2+t$ |
| $A_{31}$ | $t^{26}+t^{25}+t^{24}+2t^{23}+2t^{22}+2t^{21}+6t^{20}+6t^{19}+6t^{18}+10t^{17}+10t^{16}+10t^{15}+14t^{14}+14t^{13}+14t^{12}+10t^{11}+10t^{10}+10t^9+6t^8+6t^7+6t^6+2t^5+2t^4+2t^3+t^2++t+1$ |
| $A_{32}$ | $t^{26}+t^{25}+t^{24}+t^{23}+t^{22}+t^{21}+t^{20}+t^{19}+t^{18}+t^{17}+t^{16}+t^{15}+2t^{14}+2t^{13}+2t^{12}+t^{11}+t^{10}+t^9+t^8+t^7+t^6+t^5+t^4+t^3+t^2+t+1$ |
| $A_3=A_{31}-A_{32}$ | $t^{23}+t^{22}+t^{21}+5t^{20}+5t^{19}+5t^{18}+9t^{17}+9t^{16}+9t^{15}+12t^{14}+12t^{13}+12t^{12}+9t^{11}+9t^{10}+9t^9+5t^8+5t^7+5t^6+t^5+t^4+t^3$ |
| $A_{41}$ | $t^{26}+2t^{25}+6t^{24}+10t^{23}+14t^{22}+10t^{21}+6t^{20}+2t^{19}+t^{18}+t^{17}+2t^{16}+6t^{15}+10t^{14}+14t^{13}+10t^{12}+6t^{11}+2t^{10}+t^9+t^8+2t^7+6t^6+10t^5+14t^4+10t^3+6t^2+2t+1$ |
| $A_{42}$ | $t^{26}+t^{25}+t^{24}+t^{23}+2t^{22}+t^{21}+t^{20}+t^{19}+t^{18}+t^{17}+t^{16}+t^{15}+t^{14}+2t^{13}+t^{12}+t^{11}+t^{10}+t^9+t^8+t^7+t^6+t^5+2t^4+t^3+t^2+t+1$ |
| $A_4=A_{41}-A_{42}$ | $t^{25}+5t^{24}+9t^{23}+12t^{22}+9t^{21}+5t^{20}+t^{19}+t^{16}+5t^{15}+9t^{14}+12t^{13}+9t^{12}+5t^{11}+t^{10}+t^7+5t^6+9t^5+12t^4+9t^3+5t^2+t$ |
| $A_{51}$ | $t^{26}+2t^{25}+6t^{24}+11t^{23}+18t^{22}+20t^{21}+29t^{20}+38t^{19}+47t^{18}+64t^{17}+86t^{16}+91t^{15}+109t^{14}+124t^{13}+109t^{12}+91t^{11}+86t^{10}+64t^9+47t^8+38t^7+29t^6+20t^5+18t^4+11t^3+6t^2+2t+1$ |
| $A_{521}$ | $t^{26}+2t^{25}+6t^{24}+10t^{23}+14t^{22}+10t^{21}+6t^{20}+2t^{19}+t^{18}+t^{17}+4t^{16}+10t^{15}+20t^{14}+26t^{13}+20t^{12}+10t^{11}+4t^{10}+t^9+t^8+2t^7+6t^6+10t^5+14t^4+10t^3+6t^2+2t+1$ |
| $A_{522}$ | $t^{26}+t^{25}+t^{24}+2t^{23}+4t^{22}+2t^{21}+6t^{20}+10t^{19}+6t^{18}+10t^{17}+20t^{16}+10t^{15}+14t^{14}+26t^{13}+14t^{12}+10t^{11}+20t^{10}+10t^9+6t^8+10t^7+6t^6+2t^5+4t^4+2t^3+t^2++t+1$ |
| $A_{523}$ | $t^{26}+t^{25}+t^{24}+t^{23}+2t^{22}+t^{21}+t^{20}+t^{19}+t^{18}+t^{17}+2t^{16}-t^{15}+2t^{14}+4t^{13}+2t^{12}+t^{11}+2t^{10}+t^9+t^8+t^7+t^6+t^5+2t^4+t^3+t^2+t+1$ |
| $A_5=A_{51}-A_{521}$ $-A_{522}+A_{523}$ | $2t^{22}+9t^{21}+18t^{20}+27t^{19}+41t^{18}+54t^{17}+64t^{16}+72t^{15}+77t^{14}+76t^{13}+77t^{12}+72t^{11}+64t^{10}+54t^9+41t^8+27t^7+18t^6+9t^5+2t^4$ |
| $A=A_1+A_2$ $+A_3+A_4+A_5$ | $t^{26}+3t^{25}+23t^{24}+152t^{23}+844t^{22}+3662t^{21}+12814t^{20}+36548t^{19}+86837t^{18}+173593t^{17}+295172t^{16}+429240t^{15}+536646t^{14}+577821t^{13}+536646t^{12}+429240t^{11}+295172t^{10}+173593t^9+86837t^8+36548t^7+12814t^6+3662t^5+844t^4+152t^3+23t^2+3t+1$ |

For example $A_{31}[u;\,125] = A_{41}[u;\,125] = 1272$. Notice that their enumeration formulae are distinct. Moreover, refined by valencies, the corresponding pair of generating functions are also distinct. However, unexpectedly at first sight, the multisets of coefficients in these pairs of polynomials coincide. A more thorough analysis enabled us to reveal a simple pattern. Namely, in all four cases, we observe the following identities:

$$A_{31}[s;\,p^3](t) \equiv A_{41}[s;\,p^3](t^p) \pmod{t^{p^3-1}}, \tag{9$^t$}$$

$$A_{32}[s;\,p^3](t) \equiv A_{42}[s;\,p^3](t^p) \pmod{t^{p^3-1}}, \tag{10$^t$}$$

$$A_{522}[s;\,p^3](t) \equiv A_{521}[s;\,p^3](t^p) \pmod{t^{p^3-1}}, \tag{11$^t$}$$

and most spectacularly, as a corollary of the first two,

$$A_{3}[s;\,p^3](t) \equiv A_{4}[s;\,p^3](t^p) \pmod{t^{p^3-1}}. \tag{12$^t$}$$

Notice that the transformation $\eta_{p,3} : t \mapsto t^p$ modulo $t^{p^3-1}$ in the ring of polynomials of $t$ over the rationals is periodic of order 3. Thus,

$$A_{4}[s;\,p^3](t) \equiv A_{3}[s;\,p^3](t^{p^2}) \pmod{t^{p^3-1}},$$

etc. Besides, this operation fixes the terms $d \cdot t^{e(p^2+p+1)}$, $e = 0, 1, \ldots, p-1$.

Finally, more hidden identities of the same nature are valid in all four cases: $A_1[s;\,p^3](t)$ and $A_{523}[s;\,p^3](t)$ are invariant with respect to $\eta_{p,3}$, that is,

$$A_{1}[s;\,p^3](t) \equiv A_{1}[s;\,p^3](t^p) \pmod{t^{p^3-1}}, \tag{13$^t$}$$

and

$$A_{523}[s;\,p^3](t) \equiv A_{523}[s;\,p^3](t^p) \pmod{t^{p^3-1}}, \tag{14$^t$}$$

as can be seen from our tables. Of course, Equations (13$^t$) and (14$^t$) make no sense for valency-unspecified circulants ($t = 1$).

For the case of directed circulants of order 27 the reader can verify all these identities on the generating functions represented in Table 5.

We conjecture that the above identities hold in general.

**Conjecture** *Identities* (9$^t$) to (11$^t$), (13$^t$) *and* (14$^t$), *(and, consequently, identities* (12$^t$), *and* (9) *to* (12)) *are valid in general for all odd prime* $p$ and $s = u, d$.

If valid in general, this conjecture should have a simple formal analytical proof. For comparison, this is the case for two identities similar to (13$^t$) and (14$^t$) that are valid for intermediate enumerative polynomials for circulant graphs of prime-squared orders (see [7, 10, 18]). They promise a simple analytical proof and even suggest the existence of a direct bijective proof of combinatorial and/or algebraic nature (cf. [17]).

Hopefully the structural approach can help here (maybe even within the framework of the Isomorphism Theorem and related results?); in such a case the identities would gain some value for the structural theory of circulant graphs. Perhaps a link can be established between these formulae and the figures discussed in Sect. 2 whose valencies are multiples of $p^i, i = 1, 2$, while the graphs have clear homomorphic images of smaller size, though still belonging to the same variety of prime-power circulants.

Various formal identities are rather characteristic for the enumerators of circulant graphs of prime or prime-squared orders [10, 18]. However, the identities considered above are of a different nature: valency-violating. It is interesting to note that, rather unexpectedly, they have served as a hint for the discovery of similar valency-violating identities for intermediate classes of circulants of order $p^2,\ p \geq 3$. These results lie beyond the scope of the current paper. Details concerning the new identities and their analytical proof can be found in Appendix A of our preprint [7].

**The enumeration of self-complementary circulant graphs** The generating functions $A[s;\ p^3](t)$ make it possible to calculate easily the numbers of the corresponding self-complementary circulants (a graph is called *self-complementary* if it is isomorphic to its complementary graph).

**Proposition 4** ([16]) *For any odd prime $p$ and integer $k \geq 1$,*

$$A[u_{sc};\ p^k] = A[u;\ p^k](t)|_{t^2:=-1}, \tag{15}$$

$$A[d_{sc};\ p^k] = A[d;\ p^k](t)|_{t:=-1}, \tag{16}$$

*where $u_{sc}$ and $d_{sc}$ stand for undirected and directed self-complementary circulant graphs, resp.*

This can be shown using the well-known techniques (going back to de Bruijn and even to Redfield himself; see, for example, Sect. 6.2 in [8] and also [5, 23] used to count self-dual configurations by manipulating with the cycle index in a way which preserves variables corresponding to cycles of even length and excluding those corresponding to cycles of odd length.

**Remark** It was conjectured in [18], Conjecture 6.1, that the same assertation is valid for arbitrary odd orders $n$. The right-hand-side expressions in Identities (15) and (16) are the alternating sums of the coefficients, and we may reformulate Proposition 4 in terms of the general pattern (cf. A000171 in [28] and also [22])

$$sc(n) = e(n) - o(n), \tag{17}$$

where $e(n)$ and $o(n)$ stand for the number of non-isomorphic graphs of a certain class with even and odd number of edges, respectively, and $sc(n)$ stands for the number of self-complementary graphs (sc-graphs for short) of the same class.

**Table 6** The numbers of self-complementary undirected and directed circulant graphs of orders 27 and 125; subclasses

| Term (s-c) | Undir $n = 27$ | Undir $n = 125$ | Dir $n = 27$ | Dir $n = 125$ |
|---|---|---|---|---|
| $A_1$ | 0 | 1 | 1 | 8 |
| $A_{21}$ | 0 | 42949840 | 472 | 46116860227391504 |
| $A_{22}$ | 0 | 208 | 24 | 209936 |
| $A_2 = A_{21} - A_{22}$ | 0 | 42949632 | 448 | 46116860227181568 |
| $A_{31} = A_{41}$ | 0 | 8 | 4 | 432 |
| $A_{32} = A_{42}$ | 0 | 2 | 2 | 12 |
| $A_3 = A_4 = A_{31} - A_{32}$ | 0 | 6 | 2 | 420 |
| $A_{51}$ | 0 | 64 | 16 | 43328 |
| $A_{521} = A_{522}$ | 0 | 16 | 8 | 848 |
| $A_{523}$ | 0 | 4 | 4 | 20 |
| $A_5 = A_{51} - A_{521} - A_{522} + A_{523}$ | 0 | 36 | 4 | 41652 |
| $A = A_1 + A_2 + A_3 + A_4 + A_5$ | 0 | 42949681 | 457 | 46116860227224068 |

**Corollary 3** *The values for $A[s_{sc}; p^3]$ for $s = u, d$ and $p = 3, 5$ are given by*

$$\begin{aligned} A[u_{sc}; 27] &= 0, \\ A[u_{sc}; 125] &= 42949681, \\ A[d_{sc}; 27] &= 457, \\ A[d_{sc}; 125] &= 46116860227224068. \end{aligned}$$

The vanishing of $A[u_{sc}; 27]$ is obvious since $27 \times 26/4$ is fractional.

It makes sense to calculate the corresponding values for the intermediate generating functions (using the same substitutions as in Proposition 4): these are the numbers of self-complementary circulant graphs of the corresponding subclasses. The intermediate contributors for all four classes of circulants are represented in Table 6.

**Acknowledgements** The authors would like to thank Matan Ziv-Av for several helpful suggestions in the writing of GAP programs and for verifying some of our numerical results using diverse methods. The authors gratefully acknowledge support from the Scientific Grant Agency of the Slovak Republic under the number VEGA-1/0988/16.

## References

1. Á. Ádám, Research problem 2–10. J. Combin. Theory **2**, 393 (1967)
2. J.D. Dixon, B. Mortimer, *Permutation Groups*, vol. 163, Graduate Texts in Mathematics (Springer, New York, 1996)
3. B. Elspas, J. Turner, Graphs with circulant adjacency matrices. J. Combin. Theory **9**, 297–307 (1970)
4. I.A. Faradžev, M. Klin, Computer package for computation with coherent configurations. *Proceedings of ISSAC-91 (ACM Press, 1991)*, pp. 219–223
5. A. Farrugia, *Self-Complementary Graphs and Generalizations: a Comprehensive Reference Manual*. Master's thesis (University of Malta, 1999)
6. V. Gatt, *On the Enumeration of Circulant Graphs of Prime-Power Order: the Case of* $p^3$. *With Appendix C: Complete List of Generating Functions*. Master's thesis (University of Malta, 2013), https://arxiv.org/pdf/1703.06038v2
7. V. Gatt, M. Klin, J. Lauri, V. Liskovets, Constructive and analytic enumeration of circulant graphs with $p^3$ vertices; $p = 3, 5$ (2018), https://arxiv.org/pdf/1512.07744v4
8. F. Harary, E.M. Palmer, *Graphical Enumeration* (Academic Press, 1973), pp. 139, 140, 243
9. M. Klin, Automorphism groups of $p^m$ circulant graphs, in *The First China-Japan International Symposium on Algebraic Combinatorics* (Beijing, 1994), pp. 56–58
10. M. Klin, V. Liskovets, R. Pöschel, Analytic enumeration of circulant graphs with prime-squared number of vertices. Sém. Lotharing. Combin. B36d:36 (1996)
11. M. Klin, V. Liskovets, R. Pöschel, On the analytical enumeration of circulant graphs. *Technical Report MATH-AL-07-2003* (TU-Dresden, 2003)
12. M. Klin, R. Pöschel, The isomorphism problem for circulant digraphs with $p^n$ vertices. *Preprint P34/80*. Berlin , AdWDDR, ZIMM, 1980
13. M. Klin, R. Pöschel, The König problem, the isomorphism problem for cyclic graphs and the method of Schur rings, in *Algebraic methods in graph theory, Vol. I, II (Szeged, 1978)*, vol. 25. Colloq. Math. Soc. János Bolyai (North-Holland, Amsterdam-New York, 1981), pp. 405–434
14. M. Klin, C. Rücker, G. Rücker, G. Tinhofer, Algebraic combinatorics in mathematical chemistry. methods and algorithms. I. permutation groups and coherent (cellular) algebras. Match **40**, 7–138 (1999)
15. J. Lauri, R. Scapellato. *Topics in Graph Automorphisms and Reconstruction*. Number 432 in London Mathematical Society Lecture Notes Series (Cambridge University Press, Cambridge, 2016)
16. V. Liskovets, Enumeration of self-complementary circulant graphs of prime-power order: old and new results, in *The Second Malta Conference in Graph Theory and Combinatorics (2MCGTC., Qawra)* (University of Malta, Abstracts, 2017), pp. 85–86
17. V. Liskovets, New identities in the enumeration of circulant graphs of prime-squared orders. *Proc. XII Belarusian Math. Conf. (September 2016, Minsk)* **4**, 73–74; Inst. Math. Nat. Acad. Sci. Belarus
18. V. Liskovets, Some identities for enumerators of circulant graphs. J. Algebraic Combin. **18**(3), 189–209 (2003)
19. V. Liskovets, R. Pöschel, Counting circulant graphs of prime-power order by decomposing into orbit enumeration problems. Discrete Math. **214**, 173–191 (2000)
20. M. Muzychuk, Ádám's conjecture is true in the square-free case. J. Combin. Theory (Ser. A) **72**, 118–134 (1995)
21. M. Muzychuk, A solution of the isomorphism problem for circulant graphs. Proc. Lond. Math. Soc. **88**, 1–41 (2004)
22. A. Nakamoto, T. Shirakura, Sh Tazawa, An alternative enumeration of self-complementary graphs. Utilitas Mathematica **80**, 25–32 (2009)
23. E.M. Palmer, R.W. Robinson, Enumeration of self-dual configurations. Pac. J. Math. **110**(1), 203–221 (1984)
24. R. Pöschel, Untersuchungen von S-Ringen, insbensondere im Gruppenring von $p$-Gruppen. Math. Nach. **60**, 1–27 (1974)

25. I. Schur, On the theory of transitive permutation groups. Zap. Nauchn. Sem. S.-Peterburg. Otdel. Mat. Inst. Steklov. (POMI), 301(Teor. Predst. Din. Sist. Komb. i Algoritm. Metody. 9), 5–34, 243 (2003). Russian translation of [26]
26. I. Schur, Zur Theorie der einfach transitiven Permutationgruppen. S.-B. Preuss. Akad. Wiss. phys.-math. Kl. **18**(20), 598–623 (1933)
27. W.R. Scott, *Group Theory*, 2nd edn. (Dover Publications Inc, New York, 1987)
28. N.J.A. Sloane (ed.), *The On-Line Encyclopedia of Integer Sequences*, Published electronically at https://oeis.org
29. V.A. Vyšenskiĭ, M.H. Klin, N.I. Čeredničenko, Realization of an algorithm for construction of $S$-rings of cyclic groups of order $p^m$ and its application for solution of the problem of cataloguing $p^m$-vertex cyclic graphs, in *Computations in Algebra and Combinatorial Analysis (Russian)* (Akad. Nauk Ukrain. SSR, Inst. Kibernet., Kiev, 1978), pp. 73–86
30. H. Wielandt, *Finite Permutation Groups*. Translated from the German by R. Bercov (Academic Press, New York-London, 1964)

# A Note on a Problem of L. Martínez on Almost-Uniform Partial Sum Families

**Štefan Gyürki**

**Abstract** It is known that there is a close relation between directed strongly regular graphs and partial sum families. In this short note, we construct some partial sum families from known directed strongly regular graphs. As a byproduct, we give a partial answer to a question posed by Martínez in [4].

**Keywords** Partial sum families · Directed strongly regular graphs

## 1 Introduction

Martínez in [4] constructed several uniform and almost-uniform partial sum families and as a consequence, also proved the existence of some directed strongly regular graphs with related parameter sets. He also asked whether there exists almost-uniform partial sum families with prescribed parameters. In this paper, we partially solve his question and generalize the idea to partial sum families for other parameters.

## 2 Preliminaries

Strongly regular graphs play an important role in algebraic graph theory. Their possible generalization for directed graphs was given by Duval [2].

**Definition 1** A directed strongly regular graph (DSRG, for short) with parameters $(n, k, \mu, \lambda, t)$ is a regular directed graph on $n$ vertices with valency $k$, such that every vertex is incident with $t$ undirected edges, and the number of paths of length 2

Š. Gyürki (✉)
Department of Mathematics and Descriptive Geometry, Faculty of Civil Engineering, Slovak University of Technology, 810 05 Bratislava, Slovak Republic
e-mail: gyurki@math.sk

Mathematical Institute Slovak Academy of Sciences, 974 01 Banská Bystrica, Slovak Republic



G. A. Jones et al. (eds.), *Isomorphisms, Symmetry and Computations in Algebraic Graph Theory*, Springer Proceedings in Mathematics & Statistics 305,
https://doi.org/10.1007/978-3-030-32808-5_3

directed from a vertex $x$ to another vertex $y$ is $\lambda$, if there is an arc from $x$ to $y$, and $\mu$ otherwise. In particular, a DSRG with $t = k$ is an undirected strongly regular graph, and a DSRG with $t = 0$ is a doubly regular tournament.

**Definition 2** Let $H$ be a group of order $n$ and let $m$ be a positive integer. A family $\mathfrak{S} = \{S_{i,j}\}$, with $0 \leq i, j < m$, of subsets of $H$ is a $(m, n, k, \mu, \lambda, t)$-*partial sum family* (PSF, for short) if it simultaneously satisfies

1. For every $i$, it holds that $0 \notin S_{i,i}$, where 0 is the identity of $H$.
2. For every $i$, it holds that

$$\sum_{j=0}^{m-1} |S_{i,j}| = \sum_{j=0}^{m-1} |S_{j,i}| = k.$$

3. For every $i$ and $j$, it holds that

$$\sum_{l=0}^{m-1} \underline{S_{i,l}} \cdot \underline{S_{l,j}} = \delta_{i,j}(t - \mu) \cdot \underline{\{0\}} + (\lambda - \mu) \cdot \underline{S_{i,j}} + \mu \cdot \underline{H},$$

where $\delta_{i,j}$ is the Kronecker delta, $\underline{A}$ means the formal sum of all elements of the set $A$, and the equation is interpreted in the group ring $\mathbb{Z}H$.

Martínez and Araluze in [5] proved that the existence of a $(m, n, k, \mu, \lambda, t)$-PSF over a group $H$ is equivalent to the existence of a $(mn, k, \mu, \lambda, t)$-DSRG which admits a group of automorphisms isomorphic to $H$ acting semiregularly and having $m$ orbits. The usual direction is to construct DSRGs from PSFs, see [1, 4, 5]. Here, we go in the opposite way. We analyze the group of automorpshims of DSRGs with suitable parameter sets and try to construct PSF with related parameters.

Uniform PSFs were introduced in [1] as a slight generalization of PSFs, and finally, the term of almost-uniform PSFs appeared in [4].

**Definition 3** Let $H$ be a group of order $n \geq 2$ and $H'$ its subgroup of order $n'$. Let $m$ be a positive integer. Then an $(m, n, k, \mu, \lambda, t)$-PSF $\mathfrak{S} = \{S_{i,j}\}$, with $0 \leq i, j < m$, with the $S_{i,j}$ subsets of $H$, is *almost-uniform* with respect to $H'$ if it satisfies the following conditions:

1. The cardinalities of the diagonal blocks $S_{i,i}$ are all equal.
2. The cardinalities of the non-diagonal blocks $S_{i,j}$, $i \neq j$ are all equal.
3. The diagonal blocks form a partition of $H - H'$.

In particular, if the subgroup $H'$ is trivial, then $\mathfrak{S}$ is called *uniform*.

## 3 The Main Result

In [4], the author studies uniform and almost-uniform PSFs. With their aid, he constructs DSRGs. He proved (Proposition 7.3. in [4]) that if there exists an $(m, n, k, \mu, \lambda, t)$-almost-uniform PSF of a cyclic group $H$ with respect to a proper subgroup $H'$ and $m \geq 3$, then the possible parameter sets can have just three different forms. For one family of the feasible parameters (Family 1 in [4]), there are known several almost-uniform PSFs realizing them. This family has parameters

$$\begin{aligned} n &= (im - m + i)imr^2 + (1 - i)mr + ms, \\ k &= (im - m + i)imr^2 + (m - i - im)r + ms, \\ \mu &= (im - m + i)ir^2 - ri + s, \\ \lambda &= (im - m + i)ir^2 + (m - 3i)r + s, \\ t &= -ri + i^2r^2m + 2, \end{aligned}$$

where $i, r, s$ are suitable integers. For this family, he constructed DSRGs for $s = 1, 2, 3, 4$ with parameter sets $(9s + 9, 3s + 2, s, s + 1, s + 2)$ and $(9s + 9, 3s + 4, s + 2, s + 1, s + 4)$, respectively, from almost-uniform PSFs with corresponding integers $(m, i, r, s) = (3, 1, \pm 1, s)$ In the case when $m \in \{4, 5\}$ also a few DSRGs were found. Moreover, he asked whether there exists almost-uniform PSFs with $(m, i, r, s) = (3, 1, \pm 1, s)$ for arbitrary positive integer $s$.

On the other hand, Olmez and Song in [6], constructed among others, also DSRGs with parameter set which is complementary to $(9s + 9, 3s + 4, s + 2, s + 1, s + 4)$ using tactical configurations. We took these graphs and looked at their groups of automorphisms and orbits. Using GAP [3], we identified semi-regular subgroups in it and computed the corresponding almost-uniform PSFs. After a few experiments, we could guess the general situation, and proved the following theorem:

**Theorem 1** *Let $m, f > 1$ be positive integers such that* $\mathrm{GCD}(m, f) = 1$. *Let $H \cong (\mathbb{Z}_m \times \mathbb{Z}_f, \oplus)$ be the cyclic group of order $m \cdot f$. Denote its two normal subgroups by*

$$\begin{aligned} H' &= \mathbb{Z}_m \times \{0\} = \{(0, 0), (1, 0), \ldots, (m - 1, 0)\} \\ K &= \{0\} \times \mathbb{Z}_f = \{(0, 0), (0, 1), \ldots, (0, f - 1)\}. \end{aligned}$$

*Further, define sets $Q_i$ for $i = 0, 1, \ldots, m - 1$ as follows:*

$$Q_i = \{K \oplus (i, i)\} \setminus H' = \{(i, 1), (i, 2), \ldots, (i, f - 1)\}.$$

*Finally, for $i \neq j$ set $S_{i,j} = Q_j \cup \{(0, 0)\}$ and for $i = j$ let $S_{i,i} = Q_i$. Then $\mathfrak{S} = \{S_{i,j}\}_{0 \leq i, j < m}$ forms an almost-uniform $(m, mf, mf - 1, f - 1, m + f - 3, m + f - 2)$-PSF in $H$ with respect to $H'$.*

***Proof*** It is easy to see that all the axioms (but one concerning the equation in the group ring) are satisfied. So we need to prove that the equation

$$\sum_{l=0}^{m-1} \underline{S_{i,l}} \cdot \underline{S_{l,j}} = \delta_{i,j}(t-\mu)\underline{\{0\}} + (\lambda-\mu)\underline{S_{i,j}} + \mu\underline{H},$$

holds for all $i$ and $j$ in the group ring $\mathbb{Z}H$.

Let us start the proof with a few observations. Clearly, $\sum_{i=0}^{m-1} \underline{Q_i} = \underline{H - H'}$. Further,

$$\underline{Q_i} \cdot \underline{Q_j} = (f-2) \cdot \underline{Q_{i+j}} + (f-1) \cdot \underline{(i+j,0)}.$$

From the latter equation, one can derive the following ones:

– if $j \neq k \neq i$, then

$$\underline{S_{i,k}} \cdot \underline{S_{k,j}} = (f-2) \cdot \underline{Q_{j+k}} + (f-1) \cdot \underline{(j+k,0)} + 1 \cdot \underline{Q_j} + 1 \cdot \underline{Q_k} + 1 \cdot \underline{\{0\}};$$

– if $i \neq j$, then

$$\underline{S_{i,i}} \cdot \underline{S_{i,j}} = (f-2) \cdot \underline{Q_{i+j}} + (f-1) \cdot \underline{(i+j,0)} + 1 \cdot \underline{Q_i};$$

– and

$$\underline{S_{i,i}} \cdot \underline{S_{i,i}} = (f-2) \cdot \underline{Q_{i+i}} + (f-1) \cdot \underline{(i+i,0)}.$$

With the aid of the recently derived equations, we obtain

$$\sum_{k=0}^{m-1} \underline{S_{i,k}} \cdot \underline{S_{k,j}} = (m-2) \cdot \underline{S_{i,j}} + (f-1) \cdot \underline{H}$$

in the case when $i \neq j$;

$$\sum_{k=0}^{m-1} \underline{S_{i,k}} \cdot \underline{S_{k,i}} = (m-1) \cdot \underline{\{0\}} + (m-2) \cdot \underline{S_{i,i}} + (f-1) \cdot \underline{H}$$

when $i = j$, and the stated values for the parameters of the almost-uniform PSF follow. □

In Theorem 1, we showed that there always exists an almost-uniform PSF belonging to Family 1 when $m$ and $s+1$ are relatively prime and $i = r = 1$. Thus, we partially answered Martínez's question, but also proved a slightly stronger result, since in our case,4 the value of $m$ is an arbitrary integer.

**Example 1** Let us consider the case for $m = 4$ and $f = 3$. It is needless to recall the form of $Q_0, \ldots, Q_3$ and $\mathfrak{G}$ when working with $\mathbb{Z}_4 \times \mathbb{Z}_3$, but it can be interesting how to transfer this information to the group $\mathbb{Z}_{12} = \{0, 1, \ldots, 11\}$. A simple trick, which works here is to assign the element $a + 4b \in \mathbb{Z}_{12}$ to $(a, b) \in \mathbb{Z}_4 \times \mathbb{Z}_3$. We obtain $Q_0 = \{4, 8\}$, $Q_1 = \{1, 5\}$, $Q_2 = \{2, 10\}$, $Q_3 = \{7, 11\}$ and the corresponding almost-uniform (4, 12, 11, 2, 4, 5)-PSF of $\mathbb{Z}_{12}$ with respect to the subgroup $\mathbb{Z}_4 = \{0, 3, 6, 9\}$ is

| $S_{i,j}$ | $j = 1$ | $j = 2$ | $j = 3$ | $j = 4$ |
|---|---|---|---|---|
| $i = 1$ | $\{4, 8\}$ | $\{0, 1, 5\}$ | $\{0, 2, 10\}$ | $\{0, 7, 11\}$ |
| $i = 2$ | $\{0, 4, 8\}$ | $\{1, 5\}$ | $\{0, 2, 10\}$ | $\{0, 7, 11\}$ |
| $i = 3$ | $\{0, 4, 8\}$ | $\{0, 1, 5\}$ | $\{2, 10\}$ | $\{0, 7, 11\}$ |
| $i = 4$ | $\{0, 4, 8\}$ | $\{0, 1, 5\}$ | $\{0, 2, 10\}$ | $\{7, 11\}$ |

**Acknowledgements** The author gratefully acknowledges the contribution of the Scientific Grant Agency of Slovak Republic under the grant VEGA 1/0988/16 and the contribution of the Slovak Research and Development Agency under the projects APVV-0136-12 and APVV-0220-15.

## References

1. A. Araluze, I. Kovács, K. Kutnar, L. Martínez, D. Marušič, Partial sum quadruples and bi-Abelian digraphs. J. Combin. Theory Ser. A **119**, 1811–1831 (2012)
2. A.M. Duval, A directed graph version of strongly regular graphs. J. Combin. Theory Ser. A **47**, 71–100 (1988)
3. *GAP - Groups, Algorithms, Programming - a System for Computational Discrete Algebra*, https://www.gap-system.org
4. L. Martínez, Strongly regular $m$-Cayley circulant graphs and digraphs. Ars Mathem. Contemporanea **8**, 195–213 (2015)
5. L. Martínez, A. Araluze, New tools for construction of directed strongly regular digraphs: difference digraphs and partial sum families. J. Combin. Theory Ser. B **100**, 720–728 (2010)
6. O. Olmez, S.Y. Song, Some families of directed strongly regular graphs obtained from certain finite incidence structures. Graphs Combin. **30**, 1529–1549 (2014)

# The Paulus–Rozenfeld–Thompson Graph on 26 Vertices Revisited and Related Combinatorial Structures

**Štefan Gyürki, Mikhail Klin and Matan Ziv-Av**

**Abstract** This paper deals with the strongly regular graph $T$, having parameters $(26, 10, 3, 4)$ and the largest possible automorphism group $G = \mathrm{Aut}(T)$ of order 120. The group $G$ in its action on the vertex set of $T$ has two orbits of length 20 and 6. Many special features of the graph $T$ and its group $G$ make it a remarkable object in algebraic graph theory. The presentation, arranged in the style of a tutorial, describes the graph $T$ from many viewpoints of mathematics and computer algebra. Special attention is paid to the links of $T$ with such classical structures as the Petersen graph, semi-icosahedron, icosahedron, dodecahedron, two-graphs on 26 points, Paley graphs, inversive plane and generalized quadrangles of order 5 and others.

**Keywords** Algebraic graph theory · Strongly regular graph · Distance regular graph · Antipodal graph · Two-graph · Coherent configuration · Association scheme · Finite permutation group · Block design · Icosahedron · Semi-icosahedron · Dodecahedron · Paley graph · Weisfeiler–Leman stabilization · Computer algebra · GAP · COCO · Inversive plane · Generalized quadrangle.

Š. Gyürki
Department of Mathematics and Descriptive Geometry, Faculty of Civil Engineering,
Slovak University of Technology, 81005 Bratislava, Slovak Republic
e-mail: gypista@gmail.com

M. Klin (✉) · M. Ziv-Av
Department of Mathematics, Ben-Gurion University of the Negev, 8410501 Beer Sheva, Israel
e-mail: klin@cs.bgu.ac.il

M. Klin
Department of Mathematics, Faculty of Natural Sciences, Matej Bel University,
97401 Banská Bystrica, Slovakia

 
G. A. Jones et al. (eds.), *Isomorphisms, Symmetry and Computations in Algebraic Graph Theory*, Springer Proceedings in Mathematics & Statistics 305,
https://doi.org/10.1007/978-3-030-32808-5_4

# 1 Introduction

The initial idea of this paper was literally conceived at the very last stage of the preparation to the conference "*Symmetry versus Regularity: the first 50 years since Weisfeiler–Leman stabilization*" (briefly WL 2018). At that time the central object of our exposition, called by us the Paulus–Rozenfeld–Thompson graph (PRT-graph), denoted by $T$, was selected as the logo of WL 2018. Also, a relatively short note about $T$ was quickly prepared and posed on the home page of the planned event.

The automorphism group $G = \mathrm{Aut}(T)$ of $T$ has order 120 and is isomorphic to $A_5 \times \mathbb{Z}_2$, where $A_5$ is the smallest non-Abelian simple group, the alternating group of degree 5. Group $G$ has two orbits of length 20 and 6 in its action on the vertex set $V(T)$ of $T$, respectively. First action is faithful and literally coincides with the full symmetry group of the dodecahedron $\Delta$ (regarded as a Platonic solid or an antipodal distance regular graph of valency 3 and diameter 5). Second non-faithful action with kernel $\mathbb{Z}_2$ corresponds to the full automorphism group of the smallest non-trivial block design $\tilde{\mathcal{I}}$ with $v = 6$ points and $b = 10$ blocks. In this fashion, the famous Petersen graph $\Pi$ plays a kind of a key to comprehend unusual appearance of $A_5$ in two different faces: subgroup of index 2 of the full symmetry group of the quotient graph of the antipodal graph $\Delta$ and simultaneously the full automorphism group of the design $\tilde{\mathcal{I}}$ (sometimes called the hemi-icosahedron). This observation opened the box of Pandora. With each new section we were enjoying on the canvas of the graph $T$ unusual possibility to meet many attractive structures in combinatorics and geometry: embeddings of 5 tetrahedrons to $\Delta$, Schurian and non-Schurian association schemes, regular two-graphs, finite geometries, inversive planes among them, generalized quadrangles, etc. Thus, finally the current text absorbed all features of a quite detailed tutorial. Its goal is to travel with the reader on the crossroads of modern Algebraic Graph Theory (AGT) with a number of other areas of mathematics and computer algebra.

We stop here, very briefly announcing the entire content of the paper. Sections 2–5 play general introductory role. The group $A_5$, its transitive actions, the most significant action of degree 20 and association schemes corresponding to it, are subject of Sects. 6–9. After that, $A_5$ appears as the group of rotations of $\Delta$ together with master association scheme and its mergings, related to this group. Some nice combinatorial structures are treated in this context. In Sect. 16 a new start is given. We consider four regular two-graphs on 26 vertices and pay attention to our favourite two-graph $\mathcal{D}$ and its group. From $\mathcal{D}$ we naturally approach Steiner system $S(3, 6, 26)$ and association schemes, defined on its circles. Starting from Sect. 21, the master coherent configuration $W$ on 26 vertices of $T$ is evidently considered together with the place of $T$ inside of $W$. Special attention is paid to a few diagrams of $T$, which visually reflect some models, features and properties of $T$. Next Sects. 25 and 26, prepared jointly with Dennis Epple, briefly open promising new horizons towards links of our objects in AGT with structures in finite geometry, like inversive planes and generalized quadrangles. Two concluding sections are on purpose written in an expository style. Section 27 deals with a number of diverse deviations, not included into the main body of the text. The last Sect. 28 pursues more ambitious goals. This is a kind of an announcement of future tasks, plans and open problems, which naturally appear as corollaries of the selected style of presentation in suggested tutorial.

## 2 Basic Concepts

As was mentioned, the genre of this paper is a tutorial, which is prepared for quite an interdisciplinary audience, working on the crossroads of diverse parts of mathematics and computer algebra. While ideally it would be perfect to provide a detailed and self-contained account of all requested preliminaries, practically this is simply impossible. Indeed, such an account would be of a size, comparable with the size of the entire paper, and then we will be forced to approach a different style: small introductory course with a number of splashing examples, glued together by the mainstream topic.

To avoid such an unpleasant disproportion, we, after some hesitations, selected different structure and style of the current section on purpose. Namely, below we briefly discuss all required background concepts, coin terms (and abbreviations for some of them), to be used further, and briefly explain some most crucial links between them. Acting in this fashion, we expect that the reader will be forced to make some preparatory job before going ahead. The references, provided at the end of this section, aim to support the reader in such a self-preparation. On the other hand, we think that essential part of our audience may try to approach the paper quite immediately, postponing possible deviations to "educational" texts to rare special occasions, when they will become non-avoidable.

Essential feature of the selected style is to minimize the amount of formal definitions, appealing to the reader's ability to fill all details independently.

*2.1* Let $\Omega$ be a finite set of cardinality $n$, for example, $\Omega = [0, n-1]$, that is set of integers from 0 to $n-1$. We denote by $S(\Omega)$ the *symmetric group* $S_n$ of *degree* $n$, which consists of all permutations of $\Omega$, that is bijections of $\Omega$ onto $\Omega$. For a permutation $g \in S(\Omega)$ and $x \in \Omega$ the *image* of $g$ on $x$ is denoted by $x^g$. Group operation in $S(\Omega)$ is usual composition of bijections. A *permutation group* $(G, \Omega)$ is any subgroup of $S(\Omega)$; we say that $G$ *acts* (faithfully) on $\Omega$. We use such classical basic concepts as orbit of $(G, \Omega)$, transitivity and intransitivity, imprimitivity and primitivity, stabilizer $G_x$ of a point $x \in \Omega$ in $(G, \Omega)$, etc.

For a given permutation group $(G, \Omega)$ one considers its natural induced action $(G, \Omega^2)$ on the cartesian square $\Omega^2$: for $(a, b) \in \Omega^2$ and $g \in G$ we get $(a, b)^g = (a^g, b^g)$. The *rank* $r(G, \Omega)$ is the number of the *2-orbits* of $(G, \Omega)$, that is orbits of $(G, \Omega^2)$. A 2-orbit $R$ is *symmetric* if $R^T = R$, where $R^T = \{(b, a) \mid (a, b) \in R\}$. Primitive rank 3 permutation groups with two non-trivial (non-reflexive) symmetric 2-orbits are usually called *rank 3 groups*. Nowadays all rank 3 groups are known, it is one of the consequences of the classification of finite simple groups (CFSG).

Two permutation groups $(G, \Omega)$ and $(H, \Omega)$ are called *2-equivalent*, if they share the same set of 2-orbits. This is, indeed, an equivalence relation on the set of all permutation groups acting on the set $\Omega$. For each equivalence class, the *2-closure*

$(G^{(2)}, \Omega)$ contains all groups $(G, \Omega)$ in its class. Similar, more refined equivalence relation on the set of all permutation groups of *degree* $n$ (subgroups of $S(\Omega)$) may be defined for arbitrary positive integer $k$. As a result, any permutation group $(G, \Omega)$ is $k$-closed for a suitable $k$, $1 \le k \le n - 1$. For our text, 2-closed groups are in the centre of attention, as the automorphism groups of a suitable (directed) graph with coloured arcs.

Creation of modern permutation group theory goes back to H. Wielandt, though his classical texts, see [116], are nowadays a bit obsolete and difficult for general audience. We refer here to [21, 36, 68, 74, 81].

*2.2* The language of coherent algebras or coherent configurations (briefly CCs) was created about 50 years ago independently by B. Ju. Weisfeiler and A. A. Leman (Moscow) and D. G. Higman (USA). It can be developed in parallel terms of matrix algebras and binary relations. Typically, initial level of a minimal proficiency in this area means automatic skills to switch from one language to second one and vice versa. Some starting concepts here may be traced even to I. Schur, this explains the reasons of concepts, being introduced soon. So for arbitrary permutation group $(G, \Omega)$ of degree $n$ we consider its representation by group of corresponding permutation matrices of order $n$. Let $V(G, \Omega)$ be the *centralizer algebra* of $(G, \Omega)$ in the full matrix algebra of order $n$ over field $\mathbb{R}$ (or $\mathbb{C}$). It turns out that such algebras $V(G, \Omega)$ have some special additional properties (in comparison with all matrix algebras), which can be expressed in both algebraic and graph-theoretical terms. We formulate these properties as axioms and call all objects, which satisfy these properties, *coherent algebras*. Each coherent algebra $W$ of rank $r$ has a special standard basis consisting of (0, 1)-matrices. Interpreting these matrices as adjacency matrices of an $n$-vertex (directed) graph, we approach a concept of a *coherent configuration* (CC). Those CCs, which appear from the centralizer algebra of a suitable permutation group, are called *Schurian*, otherwise one talks of a *non-Schurian* CC. Investigation of Schurian CCs may and should be regarded as a chapter of modern permutation group theory (with or without the use of CFSG). Real challenge in combinatorics and graph theory (especially in AGT) is related to non-Schurian CCs.

*2.3* For an arbitrary graph $\Gamma = (\Omega, R)$ with vertex set $\Omega$ and arc set $R$ one may find (compute) its automorphism group $G = \mathrm{Aut}(\Gamma)$, regarded as permutation group of degree $|\Omega|$. After that, the set 2-orb$(G, \Omega)$ of the 2-orbits of $(G, \Omega)$ may be obtained and thus the standard basis of $V(G, \Omega)$ in other terms. While this procedure may be fulfilled quite efficiently with the aid of modern software, like COCO (see Sect. 3 for more details), still there is no proof that the result may be obtained in polynomial time of $|\Omega|$. In fact, this is one of the origins of the difficulties in the solution of the famous graph isomorphism problem. A great advantage of the pioneering paper [115], see [114] for more comprehensive introduction, is that there exists an efficient polynomial time procedure, which returns for an arbitrary graph $\Gamma$ its *Weisfeiler–Leman closure* $WL(\Gamma)$, that is the smallest coherent algebra containing the adjacency matrix $A(\Gamma)$. Implicitly, this procedure will be frequently regarded in this text, though we will not aim to face it literally on the level, which is friendly for the reader. Suitable texts, related to subsections 2.2 and 2.3, are [4, 21, 74, 79] with classical sources [58, 114, 115].

*2.4* We exploit a number of background concepts in AGT, more or less without special definitions and explanations. Considered graphs are mainly undirected, though some directed ones are also treated sometimes, typically, as basic graphs of a suitable CC. A very significant special case of CC is *association scheme* (AS); in our terms, this is a CC with one fibre. In other words, only one of basic graphs of an AS is reflexive. Note that some authors call such ASs as *homogeneous CCs*. For a *symmetric* AS, all basic graphs are undirected; for a *commutative* AS, its corresponding coherent algebra is commutative.

An undirected graph $\Gamma = (V, E)$ is called *distance regular* (DRG), if it is connected regular graph of valency $k$ and diameter $d$ and, in addition, the dimension of $WL(A(\Gamma))$ is equal to $d + 1$. Very significant metrical invariants of a DRG are described with the aid of its intersection array, or, more visually, intersection diagram. We will face special classes of DRGs, such as bipartite and antipodal, these concepts will be discussed in ad hoc manner. Each DRG $\Gamma$ of diameter $d$ defines canonically an AS, which is its WL-closure, such an AS is called *metric* AS with $d$ classes and of rank $d + 1$. A very special particular case of a DRG appears for $d = 2$, this is *strongly regular graph* (SRG). Typically, the intersection array of an SRG $\Gamma$ is described via its parameter set $(v, k, \lambda, \mu)$, which consists of the cardinality $|V(\Gamma)|$ of the vertex set and valencies of vertices, edges and non-edges, respectively. Schurian DRGs are called *distance-transitive graphs* (DTGs), Schurian SRGs are *rank 3 graphs*. In our considerations, we will be interested in coherent subalgebras of a given (computed) coherent algebra $W$, typically, those ones, which define an AS. In this case, we talk of mergings (fusions) of $W$. Sometimes the lattice of all fusions might be of a special interest. The main "hero" of this text, the PRT-graph $T$, will appear many times and in diverse contexts. Recall that it is non-Schurian, however, in many senses, the closest for being rank 3 graph, among all SRGs on 26 vertices. The book [16] is definitively the Bible of the theory of DRGs. The monograph [7] plays similar role for ASs, while [6] perfectly reflects origins of the symmetric ASs, which go back to the design of statistical experiments. Text [74] is written in a spirit, close enough to the current paper.

*2.5* Balanced incomplete block designs (BIBDs) with parameter set $(v, b, k, r, \lambda)$ will appear in this text. Some of these structures have additional property to be a Steiner 2-design or a Steiner 3-design. Steiner 3-design on $v = k^2 + 1$ points with blocks of size $k + 1$ is usually called an *inversive plane of order* $k$, denoted by $S(3, k + 1, k^2 + 1)$. It turns out that inversive planes of order 5 are closely related to our master graph $T$. All these objects will be introduced in ad hoc manner.

## 3 Computer-Aided Tools and Catalogues with Data

This project essentially depends on the use of computers. The way of our proceeding goes back, in particular, to a few research groups in former USSR. One of the groups, guided by I. A. Faradžev, was pioneering in the area what he called constructive enumeration of combinatorial objects. Essential achievements in this direction were presented in collection of papers [40]. On the other hand, M. Klin and a couple of his younger colleagues were using computers for enumeration of those association schemes which appear as a merging of a prescribed coherent configuration, in particular, association scheme. First efforts in this direction were developed by late V.A. Zaichenko at Moscow, common Ph.D. student of Faradžev and Klin. Basing on this experience, Faradžev created a very helpful computer package COCO for construction of the centralizer algebra of a given permutation group $(G, \Omega)$ and enumeration of its homogeneous coherent subalgebras together with their automorphism groups. (A few parallel languages are sometimes used in this area.) This, together with a number of examples of its applications was announced in [41]. With more details its algorithmic background and methodology is outlined in [42], see also [77] for a brief, though quite vivid outline. With the development of the computer algebra system GAP [46] together with its share package GRAPE (and installed to it nauty), see [91, 106], it became possible to think about creation of modern version of COCO as a kind of a share package of GAP. Some preliminary versions of such packages, created by Pech and Reichard and temporarily denoted by COCO-II P and COCO-II R, respectively, are available for the members of informal circle of researchers around Klin. Both these packages were used in the framework of the current project. The reader will realize soon that our results are initiated and inspired by some classical achievements in the constructive enumeration of small SRGs.

In our project, we are also essentially using some significant data, which are available via Internet from a few diverse sources. The most essential source was the website of Andries Brouwer [13]. In particular, we refer to the catalogue of all feasible parameter sets of SRGs (up to 1300 vertices) together with reasonably full information, regarding existence and properties of such graphs. This site [13] also contains many very attractive subsections; one of them is related to so-called nice graphs. There is no universal definition of such graphs. Nevertheless, it is clear that the language of AGT provides a helpful background for many common agreements, which classes and concrete examples of graphs should be regarded as nice. One of the foremost goals of the current paper is to attract the reader's attention to initial members of a few classes of such graphs and to share our pleasure from enjoying of striking examples of some very concrete and relatively small graphs, as well as, combinatorial and geometric structures related to them. Of course, we were using some other attractive sources of information, available on the Internet, a couple of them will be briefly mentioned in ad hoc manner. We also pay the reader's attention to a couple of special paradigms, which naturally appear at the process of human understanding of data, which were obtained with the aid of a computer, see again [77].

First related concept is *computer-free explanation*, under which we mean successful description of a combinatorial or geometric structure, originally constructed with the aid of a computer. Such an explanation should be reasonably human-friendly and should allow the reader to understand description (of say incidence structure $\gamma$), clearly observing all its ingredients, such as points, blocks and incidence relation. *Computer-free interpretation* is a much higher level of comprehension, on which the reader may enjoy clear computer-free proof of the existence of $\gamma$, as well as of its significant features. We expect that the reader will enjoy some examples of both explanation and interpretation, reading our quite long paper. Moreover, development of practical use of both these paradigms forms one of the foremost goals of the current project.

## 4 Brief Historical Overview

In this section, we provide a brief overview of activities, which resulted in the discovery of the main object of the current paper, the SRG $T$.

The central concept of our interest, an SRG, was subject of consideration at the down of the theory of association schemes, as one of the classes of an association schemes with two classes, see, e.g. [12]. However, mainly imprimitive association schemes, as well as two classical series of triangular graphs and graphs $L_2(n)$ were considered at that time. Moreover, association schemes were not investigated as an independent subject of interest, but in conjunction with partially balanced incomplete block designs (PBIBDs). The book [26] (first edition, 1954) still remains an interesting source of diverse information, regarding design of statistical experiments. Note that this catalogue in certain portion relies on the use of computer-aided information. Recall that the wording SRG is not used in [26]: it was coined by Bose in 1963, nine years later after first edition of [26]. One more source of the concept of an SRG, missed for a while at the area of design of statistical experiments is [18], second part in [17]. The concept of an SRG is visible already in [18], clearly no wording, however, here it is related strictly to finite geometries, namely, to the question of the existence of a finite affine plane of a prescribed order. The wording SRG was coined in the seminal paper [11] by R. C. Bose, which was submitted in 1962. SRGs here are treated mainly as point graphs of putative partial geometries. A number of clever and deep requirements for the geometricity of an SRG are coined in [11]. Still even a small improvement of the numerical requirements appears to be regarded as a significant event in AGT. Even before the coining of the concept of an SRG the problem of the enumeration (up to isomorphism) of SRGs with prescribed parameters has been attacked. At that time, an SRG was considered as one of the classes (usually, with smaller valency), of an association scheme with two classes. For a while, two main targets were the association schemes, algebraically isomorphic (in modern terminology) to triangular schemes $T(m)$ of order $\binom{m}{2}$ and lattice graph schemes $L_2(m)$ of order $m^2$, respectively. Here the classical results are that SRGs with parameters of $T(m)$ are unique, provided $m \neq 8$ and schemes of type $L_2(m)$, provided $m \neq 4$. The final efforts in such a classification were done by Chang [24] and Shrikhande [100] in 1959–1960, respectively.

Start of the use of computer era in the enumeration of SRGs and association schemes was going in conjunction with essential progress in theory, namely, by coining the concepts of cellular (coherent) algebra by Weisfeiler and Leman [115] and Higman [58], respectively. The collection of papers [114] already provided a quite detailed account of the first related results, both theoretical and computer aided. It seems that the paper [1] is a pioneering contribution, where a computer-aided result, related to enumeration of SRGs with the prescribed set of parameters, was reported. The main goal of the authors was to find an SRG $\Gamma$ such that $\mathrm{Aut}(\Gamma)$ acts intransitively on the vertex set $V(\Gamma)$. The proof of this property of $\mathrm{Aut}(\Gamma)$ was done implicitly via comparison of the subgraphs, induced by the neighbours of two vertices of the constructed graph with parameters (26, 10, 3, 4). The graph itself was not described explicitly at the paper. However, the comparison of the provided two neighbour subgraphs with such a collection for all SRGs of order 26, which appears in [114], allows to conclude that the authors of [1] simply rediscovered the block graph of the Steiner system on 13 vertices, which has group of order 39. (This group has two orbits on the point set.) Shrikhande and Bhat were reporting (see [101–103]) about their attempts to enumerate SRGs on 25 and 26 vertices a couple of years later on. Unfortunately, their results were incomplete and contained some inaccuracies.

First serious attempt to enumerate all SRGs in these two families was initiated by J. J. Seidel in the framework of his activities to create a network of researchers, working around the concept of a two-graph. At that time (cca. 1973), there were already at least two centres (Eindhoven and Moscow), which were equipped by necessary computer tools, as well as supported by strong experts in development of algorithms and further programming. For a couple of years, these centres were acting independently. Thus we start from a brief outline of the results, which were presented in technical report [94]. It consists of six chapters, including Introduction and theoretical background in Chaps. 2–4 (SRGs, backtracking and conference matrices of order 26). Chapter 5 describes all used Algol procedures, Chap. 6 contains tables of results. List of references includes nine items, two papers by Shrikhande and Bhat. Chapter 4, where the strategy of enumeration is described, is essential for us. Paulus starts from four evident geometric SRGs on 25 and 26 vertices (two from Latin squares of order 5 and two from STSs on 13 points). After that, four corresponding two-graphs are obtained and all related SRGs are enumerated. In this fashion, the author obtains 15 SRGs of order 25, and 10 SRGs of order 26. Remark on p. 27 of [94] plays an essential role (in the context of the current exposition). In particular, it claims: "*Up till now we do not know whether there exist other equivalence classes*". Further, he mentions efforts of F. C. Bussemaker to enumerate all SRGs on 25 vertices. Next essential reference in this story is paper [96] by M. Z. Rozenfeld (sometimes other spellings of her last name are used). This short note (published in a prestige Soviet journal) contains results of the arranged computer search in order to enumerate all SRGs on up to 28 vertices. References to [11, 24, 101–103] are mentioned as motivation of conducted research. A brief outline of the used algorithms, as well as auxiliary programs by other colleagues from the same team, are provided. The results of the fulfilled computer-aided enumeration of all existing SRGs are discussed. It is evident that the author reached full list of SRGs, well known nowadays. An interesting relation of some ad hoc local equivalence of graphs, as well as report about switching classes (in slightly obsolete terminology), is provided. A detailed report MR042 7130 to [96] by P. Hell rigorously claims that only algorithm by Rozenfeld evidently "*completed the search, thereby establishing that no other configurations exist*". Next significant text (before [114]) is preprint [2]. It extends a brief outline, presented in [96], providing also adjacency lists of all enumerated SRGs. In addition, partial results (26 discovered SRGs with 29 vertices) are presented. A helpful supplement contains a list of all feasible parameter sets for SRGs on up to 300 vertices. We mention also paper [30], which contains a lot of helpful information, regarding the SRGs discussed above, as well as corresponding 3-class association schemes of order 52. This paper, as well as its review by J. J. Seidel, evidently claim that M. Rozenfeld was the first in 1973, who showed that the list of known SRGs is full. We now turn to the results obtained by D. M. Thompson, presented in his Ph.D. thesis [112] and in paper [113]. The graph $T$ was rediscovered by Thompson in an interesting new context. Though Thompson refers to [114], his way of thinking of $T$ is fresh and contains promising potential for further generalizations. In a sense, Thompson's construction is a kind of nice amalgamation of arguments, borrowed from design and graph theory. In our paper, we follow spirit of Thompson, however not obligatorily his ways of reasonings. This justifies the selected name for the main structure $T$, considered in the current paper. The reader, who is interested to learn more about the history of AGT, is referred to sources in Sect. 10 of [79] and book [6].

## 5 Outline of the Main Goal

As it was already mentioned, this paper is written in the style of a *tutorial* on purpose (cf., e.g. [78]). Such a style combines a few diverse features of a scientific paper. Clearly, essential part of the text is formed by a number of mathematical claims and their justification (sometimes just an outline, though frequently a reasonably detailed proof). The central objects of interest are combinatorial structures, such as graphs and designs, originally obtained with the aid of a computer. This implies diversity in the ways of their presentation: from simple lists and sophisticated tables to diagrams and massives of data, the latter sometimes not strictly included in the text. Diagrams are of an independent significance: we are trying to make them reasonably nice, though, probably, some of them may appear more ugly in the eyes of the reader. The style of a tutorial also means that, while the goal is to introduce and to discuss diverse combinatorial and geometric structures on an axiomatical level, sometimes initial features of the considered classes of structures are better comprehended on concrete small examples.

Our central object is the *Paulus–Rozenfeld–Thompson* (briefly *PRT-*) *graph*, denoted by $T$. This is an SRG with the parameters $(v, k, \lambda, \mu) = (26, 10, 3, 4)$. As it was mentioned, there are exactly 10 such graphs up to isomorphism, known since 1973. In many diverse senses, the PRT-graph $T$ is the most symmetric among all (primitive) SRGs on 26 vertices. The group $\mathrm{Aut}(T)$ of order 120 is isomorphic to the direct product $A_5 \times \mathbb{Z}_2$, it has two orbits of length 20 and 6 on the vertex set $V(T)$ of $T$. Action on one of the orbits coincides with the action of the symmetry group of the dodecahedron $\Delta$ (regarded both as DRG and regular polytope). The group $\mathrm{Aut}(T)$ acts non-faithfully on the orbit of length 6, its action coincides with the automorphism group of the smallest non-trivial BIBD with the parameters

(6, 10, 3, 5, 2). Don Thompson was the first who observed this unique and essential property of $T$ and tried (quite successfully) to describe $T$ in terms of the growing language of coherent configurations at the time of publication of [113]. One more striking feature of the graph $T$ became clear to the authors in the framework of the recent conference WL 2018 at Pilsen, devoted to the 50th anniversary of the seminal paper [115]. The graph $T$ became the logo of the event, a short outline [53] of the current tutorial is available from the home page of this event. Immediately after this conference the authors learned from Peter J. Cameron (see his homepage [20]) an extraordinary nice feature of the graph $T$. It turns out that at the down of the AGT Norman Biggs, one of the creators of this part of mathematics, shared with his colleagues a kind of an informal hope. Why not to believe that each time, when there exist one (or a few) DRGs with the prescribed parameters, at least one of them has a vertex-transitive automorphism group. Clearly, this informal conjecture was rejected already in 1973. Nevertheless, we hope that the reader of the current paper will enjoy many properties of the graph $T$ and, in addition, will agree that from diverse point of view it behaves like "almost symmetric".

It would be not so easy to detect the next smallest size of parameters in the class of SRGs, where this informal Biggs' expectation is not fulfilled. This serves as an extra argument to agree that our main hero $T$ deserves a special attention, at least on the level of such a quite long tutorial. The curious reader may observe that the graph $T$ appears in relevance to many nice problems on the crossroads of AGT, design theory, finite geometries and permutation group theory. In particular, this is related to the fact that $26 = 2p$ and $26 = q^2 + 1$ for suitable prime numbers $p$ and $q$ (see, for example, Sect. 27). At this moment we just mention for the first time such structures as two-graphs, association schemes with three classes, double covers, inversive planes, which will become subject of our careful consideration. Also, a number of quite unusual non-Schurian association schemes will be naturally exposed in this text. It is also significant that many facts, which are valid for SRGs with $p^2 + 1$ vertices, $p = 5$, may be also observed for larger prime values of $p$. Such potential generalizations are subject of further work in progress, jointly with Dennis Epple. We hope that the beauty of the graph $T$ as one of the smallest and brightest examples of structures in the modern context "symmetry versus regularity" (cf. again with the agenda of the event WL 2018 at Pilsen) will be visible and enjoyed by the reader. This foremost goal was defining level of rigour at the presentation: from brief hints to a proof to quite detailed justification. Each time our decision was based on a desire to keep pace and clarity of a presentation pleasant and transparent. Once more, the graph $T$ is, in a sense, the main hero of this text. A number of planned follow-ups and diverse topics for extra discussions and deviations are briefly mentioned in the concluding sections.

# 6 Alternating Group $A_5$ and Its Subgroups

The alternating group $A_5$ of order 60 is our starting master group. It consists of all even permutations of the 5-element set $[0, 4] = \{0, 1, 2, 3, 4\}$ regarded as subgroup of index 2 in the symmetric group $S_5$ of degree 5, acting on the same set $[0, 4]$. Clearly, $A_5$ is the smallest non-Abelian simple group. In all forthcoming computations

$$A_5 = \langle (0, 1, 2, 3, 4), (2, 3, 4) \rangle,$$

that is group $A_5$ is generated by two elements of orders 5 and 3. The cycle index polynomial $C(A_5, [0, 4])$ is easily obtained by simple hand computations:

$$C(A_5, [0, 4]) = \frac{1}{60}\left(x_1^5 + 15x_1x_2^2 + 20x_1^2x_3 + 24x_5\right).$$

Note that elements of order 5 split into two conjugacy classes of size 12, while in other cases similarity and conjugacy coincide. List of subgroups of $A_5$ is also clear:

- 15 cyclic subgroups $\mathbb{Z}_2$;
- 10 cyclic subgroups $\mathbb{Z}_3$;
- 5 Klein subgroups $E_4$ of order 4;
- 6 cyclic subgroups $\mathbb{Z}_5$;
- 6 dihedral subgroups $D_5$ of order 10 and
- 5 alternating subgroups $A_4$ of order 12.

Recall that all permutation representations of a given abstract group are forming a structure of a ring, which is usually called *Burnside ring*, see [70]. The structure constants of the Burnside ring $B(G)$ of a group $G$ (*marks* of subgroups in $G$) are very helpful for the computations with centralizer algebras of permutation representations of $G$. Direct formulas for the counting of Burnside marks are known, e.g. [66]. The Burnside matrix for the symmetric group $S_5$ is presented in [70], p. 368. Its suitable subtable forms a table of Burnside marks of $A_5$. In principle, the latter table may be used for the computer-free justification of some computational results, to be considered below. Pros and contras of such different ways for computations are discussed in the concluding section.

# 7 Transitive Actions of $A_5$

Here we consider all transitive actions of $A_5$, up to conjugacy, and interpret them as symmetry group of suitable combinatorial structures. Special attention is paid each time to the *arity* of the appearing permutation group, that is the smallest value of $k$, for which the group is $k$-closed. Recall that there exists a bijection between different permutation representations of $A_5$, aka basic elements of $B(A_5)$, and conjugacy classes of subgroups of $A_5$. Because $A_5$ is simple, all its transitive actions, of degree larger than 1, are faithful.

## 7.1 Action of Degree 5

Here we consider natural action $(A_5, [0, 4])$. Recall that arity of $A_n$ (in natural action) is equal to $n - 1$, see [81]. We present relations of arity 4 and 5, $\Phi_{5,4}$ and $\Phi_{5,5}$, respectively, such that $\mathrm{Aut}(\Phi_{5,4}) = \mathrm{Aut}(\Phi_{5,5}) = A_5$. Consider paths of length 3: $P_1 = ([0, 4], \{\{0, 1\}, \{1, 2\}, \{2, 3\}\})$ and $P_2 = ([0, 4], \{\{0, 1\}, \{1, 2\}, \{2, 4\}\})$. Let $\Phi_{5,4} = P_1^{A_5}$ be the orbit of $P_1$ under the action of $A_5$.

**Proposition 1** *The following holds:*

*(a) There are exactly 60 different paths of length 3, regarded as subgraphs of $K_5$ with vertex set $[0, 4]$.*
*(b) The set $\Phi_{5,4}$ consists of 30 paths, $P_2 \notin \Phi_{5,4}$.*
*(c)* $\mathrm{Aut}(\Phi_{5,4}) = A_5$. □

The relation $\Phi_{5,5}$ is described with the aid of Fig. 1.

**Proposition 2** *The following holds:*

*(a) There are exactly 12 different pentagons regarded as cycles of length 5 in $K_5$.*
*(b) These pentagons split under the action of $(A_5, [0, 4])$ into two orbits of length 6.*
*(c) $\Phi_{5,5} = \{a, b, c, d, e, f\}$ forms orbit of pentagons under $(A_5, [0, 4])$.*
*(d)* $\mathrm{Aut}(\Phi_{5,5}) = A_5$.

***Proof*** Let $C_5$ be an arbitrary pentagon, $\mathrm{Aut}(C_5) \cong D_5$ is a subgroup of $A_5$, note that $|D_5| = 10$. Thus, orbit of length $\frac{5!}{10} = 12$ (under action of $S_5$) splits into two orbits of length 6 under the action of $A_5$. Routine inspection shows that $\Phi_{5,5}$ forms one of such orbits. $H = \mathrm{Aut}(\Phi_{5,5})$ acts transitively on 6 pentagons in $\Phi_{5,5}$, stabilizer of one pentagon coincides with $D_5$. Thus, $|H| = 6 \cdot 10 = 60$, $H \leq A_5$, therefore $H = A_5$. □

We will return to $\Phi_{5,5}$ below.

## 7.2 Action of Degree 6

This is exactly action $(A_5, \Phi_{5,5})$ on the set of six pentagons, depicted in Fig. 1. Recall an exceptional property of the group $A_6$: it contains two conjugacy classes of subgroups of index 6, both isomorphic to $A_5$. One of these classes corresponds to stabilizers of a point in natural action of $A_6$ of degree 6, another one is 2-transitive action of $A_5$ on 6-element set. A classical way to treat this action, due to isomorphism

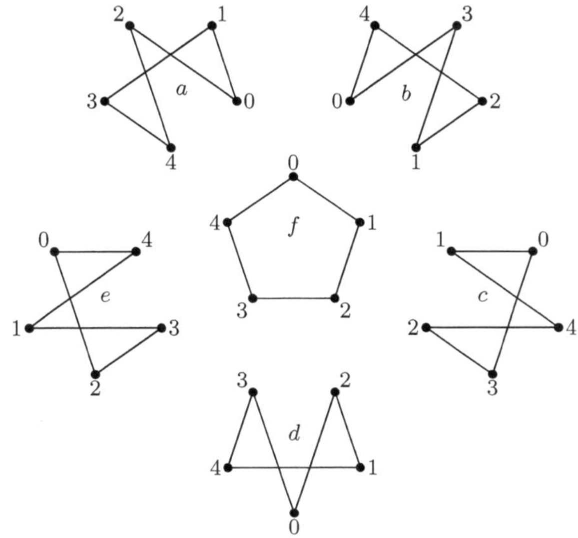


**Fig. 1** Six pentagons in $\Phi_{5,5}$

$PSL(2,5) \cong A_5$, is to interpret the 6-element set as projective line $PG(1,5)$, see, e.g. [59], Sect. 6.4 (iv). However, we will rely on considerations of combinatorial nature.

First way is to consider the *hemi-icosahedron*, regular non-orientable map of genus 1 with 6 vertices, 10 faces and 15 edges, see [31, 88]. Its rotational and full symmetry groups are isomorphic to $A_5$, and this is currently considered 2-transitive action of degree 6. Let us approach this object from purely combinatorial considerations. Thus, let $\mathcal{P} = \{a, b, c, d, e, f\}$ be the set of pentagons in $\Phi_{5,5}$, consider 10 edges of graph $K_5$ with vertex set $[0, 4]$. We describe incidence structure $\bar{\mathcal{I}} = (\mathcal{P}, \mathcal{B})$, in which blocks bijectively correspond to edges of $K_5$. Namely, each edge is incident to all pentagons, in which this edge appears. Clearly, $|\mathcal{B}| = 10$ and the resulted structure is depicted in Fig. 2: its triangles are blocks in $\bar{\mathcal{I}}$. Note that Fig. 2 is drawn on the projective plane. Thus, the antipodal vertices of the exterior hexagon are identified.

**Proposition 3** *The following hold:*

(a) *Ten faces of the hemi-icosahedron and its vertices form a BIBD $\bar{\mathcal{I}}$ with parameters* $v = 6$, $b = 10$, $k = 3$, $r = 5$, $\lambda = 2$.
(b) $\mathrm{Aut}(\bar{\mathcal{I}}) \cong A_5$, *where $A_5$ acts 2-transitively on the points and primitively on the blocks of $\bar{\mathcal{I}}$.*
(c) *The arity of $(A_5, \mathcal{P})$ is equal to 3.*
(d) $(A_5, \mathcal{P})$ *coincides with full group of symmetries of the hemi-icosahedron.*
(e) $(A_5, \mathcal{P}) = \langle (a, b, c, d, e, f), (a, c, f)(b, e, d) \rangle$, *here we show induced permutations from the starting presentation of $(A_5, [0, 4])$.*

***Proof*** Proof of (b) is the only ingredient, which requires a slight justification. Group $A_5$ acts transitively on the edges of $K_5$, which bijectively correspond to blocks of $\bar{\mathcal{I}}$.

**Fig. 2** Hemi-icosahedron

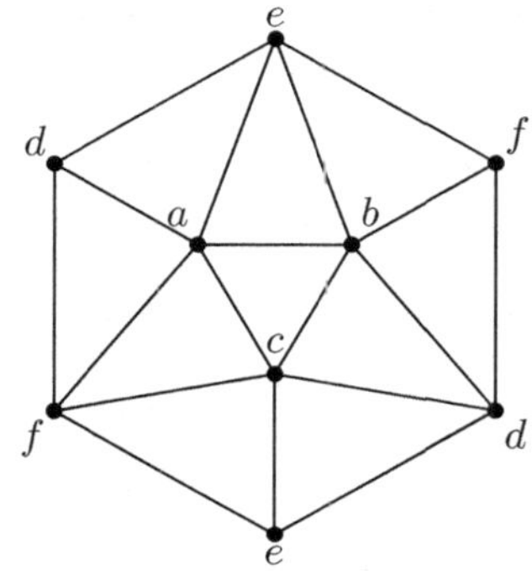


Let $H$ be the stabilizer of the block $\{a, b, c\}$ in $\text{Aut}(\bar{\mathcal{I}})$. It is transitive on the points of this block. Let $H_a$ be the stabilizer of $a$ in $H$. Then $H_a = \langle(b, c)(e, f)\rangle$, $|H_a| = 2$. Therefore, $|\text{Aut}(\bar{\mathcal{I}})| = 10 \cdot 3 \cdot 2 = 60$, and, because $A_5 \leq \text{Aut}(\bar{\mathcal{I}})$, $\text{Aut}(\bar{\mathcal{I}}) \cong A_5$. □

## 7.3 Action of Degree 10

We already understand this action in two ways: on the edges of $K_5$ and on the blocks of $\bar{\mathcal{I}}$. This action has rank 3 and is 2-equivalent to the induced action of $S_5$. The latter one is nothing else but the full automorphism group of the Petersen graph $\Pi$, probably the most famous object in graph theory, cf. [60].

We can consider a dual map to the hemi-icosahedron, which is called the *hemi-dodecahedron*. Then the action $\left(A_5, \left\{ {[0,4] \atop 2} \right\}\right)$ appears as rotational symmetry group as well as full symmetry group of the dual map. (Here $\{ {A \atop 2} \}$ denotes the set of all 2-subsets of the set $A$.) Taking into account that faces of dual map are pentagons, we obtain that action of $A_5$ of degree 10 is, at least, 5-closed. Taking into account that $\Pi$ is 3-arc transitive, see again [60], one easily gets that the current action is 4-closed. A stronger result was initially obtained with the aid of a computer.

**Proposition 4** *For the permutation group* $\left(A_5, \left\{ {[0,4] \atop 2} \right\}\right)$ *the followings hold:*

*(a) it is primitive rank 3 group with subdegrees 1, 3, 6;*
*(b) it is 3-closed;*
*(c) it is even 3-set closed.*

***Proof*** To prove (c), let us start with the group $S_5 \cong \text{Aut}(\Pi)$ and consider its orbit on the induced subgraphs of $\Pi$ of size 3, consisting of one edge. Clearly, such an orbit has length 60. Now, let us consider two subgraphs, $S'$ and $S''$ of $\Pi$, induced by $\{\{0, 1\}, \{0, 2\}, \{1, 3\}\}$ and $\{\{0, 1\}, \{0, 2\}, \{1, 4\}\}$, respectively. There are exactly two permutations in $S_5$, sending $S'$ to $S''$, namely, $(3, 4)$ and $(0, 1)(2, 4, 3)$, both are odd, thus not in $A_5$. This implies that orbit of length 60 splits into two orbits of length 30 under the action of $A_5$ with representatives $S'$ and $S''$. Automorphism group of any of these two relations regarded as totally symmetric relation of arity 3, clearly coincides with $\left(A_5, \left\{{[0,4] \atop 2}\right\}\right)$. □

## 7.4 Action of Degree 15

This transitive action of $A_5$ was considered with enough details in [78]. Thus, for the completeness, we simply repeat the results from there.

**Proposition 5** *Transitive action of $A_5$ of degree 15 has the following properties:*

*(a) It is similar to the action on the points of the line graph $L(\Pi)$, aka edges of $\Pi$.*
*(b) It is similar to the action of $(A_5, [0, 4])$ on the 1-factors of $K_5$ (with one isolated vertex).*
*(c) It has rank 6 with subdegrees 1, 1, 1, 4, 4, 4.*
*(d) Its centralizer algebra is non-commutative.*
*(e) It is 2-closed.*

## 7.5 Action of Degree 20

This permutation group is of central significance in this text. To obtain it, we consider induced action of $(A_5, [0, 4])$ on the set $\Omega$ of ordered pairs of distinct elements from $[0, 4]$, $|\Omega| = 20$. At the beginning, all properties of $(A_5, \Omega)$ were obtained with the aid of COCO; after that, many of them were justified without the use of a computer. It turns out that $(A_5, \Omega)$ is imprimitive rank 8 permutation group, its centralizer algebra is non-commutative. We will return to its consideration below.

## 7.6 Action of Degree 30

We investigated this action with the aid of COCO. The set $\Omega_{30}$ of cardinality 30, on which $A_5$ acts transitively, was identified with relational structure on $[0, 4]$ of a few kinds, in particular, with chains of length 3, regarded as disconnected graphs with one isolated point and three consecutive edges. Clearly, there are two orbits of $A_5$ on such structures, both of length 30. These two actions are similar as permutation groups. This action does not play a central role in our presentation; therefore, we discuss it informally. It turns out that $(A_5, \Omega_{30})$ has rank 16 with 14 2-orbits of valency 2 and one non-reflexive 2-orbit of valency 1. There are four pairs of antisymmetric 2-orbits. The action is 2-closed. COCO returns 29 non-trivial mergings, rank of which varies from 11 to 3. 22 of these mergings (with largest rank equal to 7) are symmetric. Exactly two of mergings are non-Schurian, first of rank 7 and group of order 120, the second symmetric merging of rank 6 and group of order 240. We will not discuss further this information, implicitly it will be relevant to some later considerations, while with full details will be subject of a planned separate publication.

### 7.7 Action of Degree 60

This is a regular action of the group $A_5$. Its centralizer algebra is isomorphic to the full group ring over $A_5$. Clearly, this action is 2-closed. All mergings of its centralizer algebra bijectively correspond to so-called *Schur rings* (briefly *S-rings*) over $A_5$. The constructive enumeration of all such S-rings for a while was a challenging task in AGT. It was fulfilled (clearly, with the aid of a computer) by the third author Matan Ziv-Av. First brief report about the achieved results appeared in [80]. Up to isomorphism, there are 161 non-trivial S-rings, rank of which varies from 3 to 33. There are 19 isomorphism classes of non-Schurian mergings (with rank from 4 to 14). General information about the obtained results, as well as some striking mergings, is considered in [80]. We believe that a more careful analysis of some results, reported in the current text, might be relevant for the desired deeper consideration of the S-rings over $A_5$. This is, surely, topic for a possible future research.

## 8 Master Association Scheme of Rank 8 on 20 Vertices

The rank 8 association scheme $\mathcal{M}$ on 20 vertices, considered below, plays the role of our master association scheme. It appears as the association scheme $2{-}orb(A_5, \Omega)$ for induced action of $A_5$ on the set $\Omega$ of all ordered pairs of distinct elements in $[0, 4]$.

**Proposition 6** *Let $\alpha = (0, 1) \in \Omega$, $(A_5)_\alpha$ stabilizer of $\alpha$ in $(A_5, \Omega)$. Then the following holds:*

*(a) $(A_5)_\alpha \cong \langle (2, 3, 4) \rangle$;*
*(b) $rank(A_5, \Omega) = 8$;*
*(c) the representatives of 2-orbits and corresponding subdegrees are given in Table 2.*

***Proof*** We describe the cycle index of $((A_5)_\alpha, [0, 4])$ as $\frac{1}{3}(x_1^5 + 2x_1^2x_3)$. Then, clearly, $rank(\mathcal{M}) = rank(A_5, \Omega) = \frac{1}{3}(20 + 2 \cdot 2) = 8$. The data in Table 2, originally obtained with the aid of COCO, is easily justified by simple hand computations. □

**Table 1** The bijection between labelling in COCO and the ordered pairs

| Label | Pair | Label | Pair | Label | Pair | Label | Pair | Label | Pair |
|---|---|---|---|---|---|---|---|---|---|
| 0 | (0, 1) | 1 | (1, 2) | 2 | (2, 3) | 3 | (1, 3) | 4 | (3, 4) |
| 5 | (2, 4) | 6 | (1, 4) | 7 | (4, 0) | 8 | (4, 2) | 9 | (3, 0) |
| 10 | (3, 2) | 11 | (2, 0) | 12 | (0, 3) | 13 | (4, 1) | 14 | (4, 3) |
| 15 | (3, 1) | 16 | (0, 4) | 17 | (0, 2) | 18 | (2, 1) | 19 | (1, 0) |

**Table 2** 2-orbits of group $(A_5, \Omega)$

| $R_i$ | Length | Representative | Mate |
|---|---|---|---|
| 0 | 1 | 0 | |
| 1 | 3 | 1 | 4 |
| 2 | 3 | 2 | |
| 3 | 3 | 5 | |
| 4 | 3 | 7 | 1 |
| 5 | 3 | 12 | |
| 6 | 3 | 13 | |
| 7 | 1 | 19 | |

Recall that in Table 2 $\alpha = 0$ denotes the pair (0, 1), labelling of all pairs created in COCO, is provided in Table 1. All 2-orbits besides $R_1$ and $R_4$ are symmetric. The table of structure constants is also provided, see Table 16 in the Supplement. They were calculated by COCO with the aid of command `inm`. The rectangle on the intersection of row $i$ and column $j$ contains numbers $p_{i,j}^k$, where $k$ runs from 0 to 7 (from top to bottom). In principle, all constants may be justified by routine simple hand computations. Fortunately, in our case this is not necessary, see discussion in forthcoming sections.

Note that $\mathcal{M}$ is not commutative, e.g. $p_{1,6}^1 \neq p_{6,1}^1$.

## 9 Mergings of $\mathcal{M}$ and Their Automorphism Groups

Using two next commands of COCO, `sub` and `aut`, we described all mergings of $\mathcal{M}$ and orders and ranks of their groups. The structure of small groups was clarified with the aid of GAP. The corresponding information is provided in Table 3. Here $\mathcal{M} = \mathcal{M}_0$ is initial association scheme, the trivial rank 2 merging is omitted. Essential fact is that COCO confirms that $\mathrm{Aut}(\mathcal{M}) = A_5$, that is $(A_5, \Omega)$ is 2-closed. GAP returns also that $\mathrm{Aut}(\mathcal{M}_1) \cong S_5$, $\mathrm{Aut}(\mathcal{M}_2) \cong A_5 \times \mathbb{Z}_2$, $\mathrm{Aut}(\mathcal{M}_3) \cong S_5 \times \mathbb{Z}_2$. Note that a generator of association scheme (in the sense of WL-stabilization) is easily deduced from the analysis of the Hasse diagram of mergings, typically, considered scheme has a few generators. A few generators of the initial scheme $\mathcal{M}$ will be considered below.

**Table 3** Mergings of the association scheme $\mathcal{M}$

| $\mathcal{M}_i$ | Rank | Merging | Commut. | Symm. | $\lvert \mathrm{Aut}(\mathcal{M}_i) \rvert$ | Generator |
|---|---|---|---|---|---|---|
| $\mathcal{M}_1$ | 7 | (2, 3) | No | No | 120 | $R_1$ |
| $\mathcal{M}_2$ | 6 | (1, 4) (5, 6) | | | 120 | $R_2$ |
| $\mathcal{M}_3$ | 5 | (1, 4) (2, 3) (5, 6) | | | 240 | $R_5 \cup R_6$ |
| $\mathcal{M}_4$ | 4 | (1, 4, 5, 6) (2, 3) | | | 122,880 | $R_2 \cup R_3$ |
| $\mathcal{M}_5$ | 3 | (1, 4, 2, 3, 6, 7) | | | 955,514,880 | $R_5$ |
| $\mathcal{M}_6$ | 3 | (1, 4, 2, 3, 5, 6) | | | 3715,891,200 | $R_7$ |
| $\mathcal{M}_7$ | 3 | (1, 4, 2, 3, 5, 7) | | | 955,514,880 | $R_6$ |

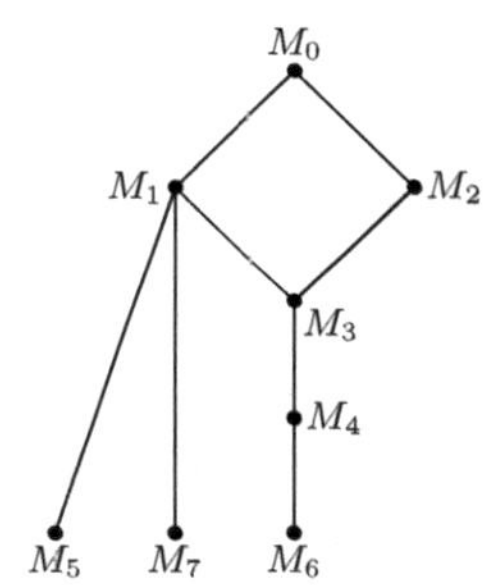


**Fig. 3** Hasse diagram for the mergings of the scheme $\mathcal{M}$

The last four "large" groups are identified by simple combinatorial reasonings. $\mathrm{Aut}(\mathcal{M}_4) \cong \widetilde{S_5^{(2)}} \wr S_2$, $\mathrm{Aut}(\mathcal{M}_5) \cong \mathrm{Aut}(\mathcal{M}_7) \cong S_5 \wr S_4$, $\mathrm{Aut}(\mathcal{M}_6) \cong S_{10} \wr S_2$. (Recall that we are using orthodoxal notation for the wreath product, like in [42].) Here the group $\widetilde{S_5^{(2)}}$ denotes the induced primitive rank 3 action of the group $S_5$ on the set $\left\{ {[0,4] \atop 2} \right\}$; in other words, the group $\mathrm{Aut}(\Pi)$ of the automorphisms of the Petersen graph $\Pi$. All mergings are Schurian: this is expected result, since, according to [55], all association schemes on 20 vertices are Schurian. We depict in Fig. 3 the Hasse diagram of all non-trivial mergings, for the readers convenience. Clearly, due to the natural Galois correspondence between Schurian association schemes and 2-closed transitive permutation groups, see [42], it is anti-isomorphic to the lattice of 2-closed overgroups of group $(A_5, \Omega)$ in the symmetric group $S(\Omega)$.

In what follows, we are giving a computer-free explanation (in the sense of [77]) for some results, presented in this section, step by step.

## 10 $A_5$ as the Group of Rotations of the Dodecahedron

Here we can think of three classical objects, the *dodecahedron* $\mathcal{D}$, one of the five Platonic solids, the *skeleton* $\Delta$ of $\mathcal{D}$, and the *map* $\mathcal{M}(\Delta)$, associated to $\Delta$.

Clearly, $\mathcal{D}$ consists of 20 vertices, 30 edges and 12 pentagonal faces, where vertices are regarded as suitable points of the three-dimensional Euclidean space. The skeleton $\Delta$ is a regular graph of valency 3 and girth 5. The map $\mathcal{M}(\Delta)$ may be defined in the framework of topological graph theory [52] as an embedding of a graph on surface. Then we are allowed to talk of the groups of rotations of map and full symmetry group of map, which sometimes may coincide. The *symmetry group* of a Platonic solid and its subgroup of rotations is defined as a suitable subgroup of isometries of the Euclidean 3D-space. Both groups induce subgroups of the automorphism group of the skeleton graph. The automorphism group of a map, also regarded as permutation group, acting on vertices, is defined as the group of all permutations, which preserve incidences between the three ingredients of map, see [33]. Group of *rotations* of map, in addition, is preserving *orientation* of each face. It was clear since the beginning of group theory, that the group of rotations of $\mathcal{D}$ is isomorphic to $A_5$, while the group of all symmetries to $A_5 \times \mathbb{Z}_2$. The text [32] is one of the classical sources from the twentieth century. Some modern variations of this claim are considered in [73].

To consider the rotation group of the map $\mathcal{M}(\Delta)$, we have to start from its planar depiction, like in Fig. 4. Let us virtually consider clockwise orientations of 11 pentagonal faces inside and anticlockwise orientation of the outer pentagon on the border. Clearly, this corresponds to the orientation of all faces of $\mathcal{M}(\Delta)$, embedded into sphere. In terms of our presentation, this is a 5-ary relation on the point set $\Omega$ of $\Delta$. Then the automorphism group of such a relation is isomorphic to $A_5$ and coincides with our copy of $(A_5, \Omega)$. A reasonably clear proof of such a claim can be found in [45], though without the use of language of 5-ary relations.

**Proposition 7** *The following holds:*

(*a*) *The group of space rotations of* $\mathcal{D}$*, as well as the group* $\mathrm{Aut}(\mathcal{M}(\Delta))$ *are isomorphic to* $A_5$.
(*b*) $\mathrm{Aut}(\Delta) \cong A_5 \times \mathbb{Z}_2$*, where* $A_5$ *coincides with* $(A_5, \Omega)$*,* $\mathbb{Z}_2 = \langle i \rangle$*, where*

$$i = (0, 19)(1, 18)(2, 10)(3, 15)(4, 14)(5, 8)(6, 13)(7, 16)(9, 12)(11, 17).$$

In what follows, we are interested in the 2-closure of rotation groups of the skeletons of Platonic solids. It seems that for the first time this question was treated in [81]. It turns out that the rotation groups of the tetrahedron, octahedron, cube and icosahedron are not 2-closed, while the group of the dodecahedron is. We give a computer-free proof.

**Proposition 8** *The group* $(A_5, \Omega)$*, as the rotation group of* $\Delta$*, is 2-closed.*

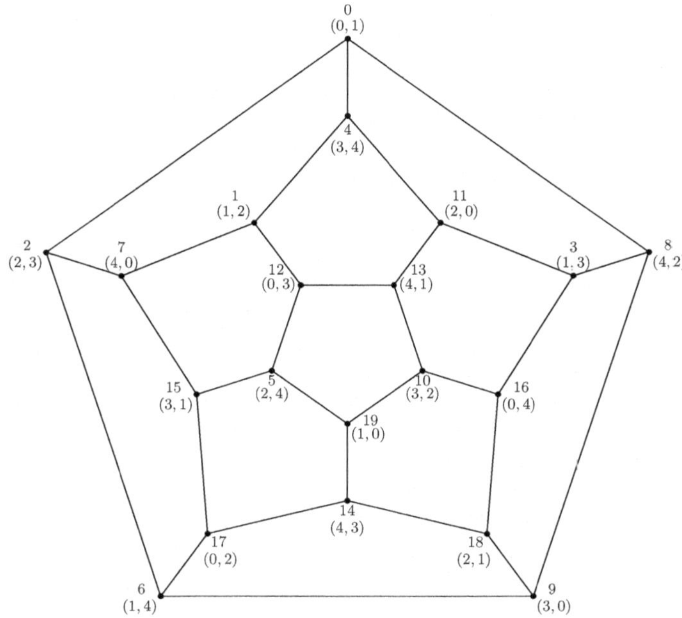


**Fig. 4** The dodecahedron $\Delta$ with two kinds of labels on its vertices

***Proof*** As we are aware from Sects. 8 and 9, $rank(A_5, \Omega) = 8$. The involution $i$ glues 2-orbits $R_1$ with $R_4$, and $R_5$ with $R_6$. Thus $\text{Aut}(\Delta) = \text{Aut}(\mathcal{M}_2) \cong A_5 \times \mathbb{Z}_2$, in particular, $A_5 \times \mathbb{Z}_2$ is 2-closed. Taking into account that $A_5^{(2)} \leq (A_5 \times \mathbb{Z}_2)^{(2)}$ and both groups keep the original ranks, we conclude that $A_5^{(2)} = A_5$. □

In what follows, under the word *dodecahedron* we mean a copy of graph $\Delta$ with 20 vertices from the set $\Omega$. Example of such a copy, namely, $\Delta = (\Omega, R_2)$, is depicted in Fig. 4, where vertices are labelled by elements from [0, 19], as well as by ordered pairs from [0, 4]. Note that the used bijection was listed in Table 1.

## 11 Some Diagrams of the Dodecahedron and Related Graphs

Our goal now is to provide a clear explanation of all basic graphs of $\mathcal{M}$. This will imply better understanding of the mergings of $\mathcal{M}$ and their groups. Recall that, the graph $\Delta$ is the unique distance-transitive graph (DTG) of valency 3 on 20 vertices, see [16]. Its intersection array has form (3, 2, 1, 1, 1; 1, 1, 1, 2, 3). $\Delta$ is an antipodal DTG of diameter 5. The intersection diagram, depicted in Fig. 5, is more suitable compact way for comprehension, cf. beautiful full Fig. 4.5. at [47].

**Fig. 5** Intersection diagram of $\Delta$

First, we interpret all basic graphs $\Delta_i$ of the metric association scheme $WL(\Delta) = \mathcal{M}_2$, having common vertex set $\Omega$. To each distance-$i$ graph $\Delta_i$ we associate its edge set.

**Proposition 9** *The following holds:*

*(a) The association scheme $\mathcal{M}_2$ is a metric association scheme, canonically generated by the graph $\Delta = \Delta_1$.*
*(b) The edge sets of the distance graphs $\Delta_i$ are defined by relations $R_2$, $R_1 \cup R_4$, $R_5 \cup R_6$, $R_3$, $R_7$, respectively.*
*(c) The graph $\Delta$ is a DTG.*
*(d) The graphs $\Delta$ and $\Delta_4$ provide two isomorphic copies of the dodecahedron graph, which share the same automorphism group.*
*(e) The distance-2 graph for $\Delta_4$ is equal to $\Delta_2$.*

***Proof*** We observe that the group $\mathrm{Aut}(\mathcal{M}_5) \cong S_5 \times S_2$ serves also as group $\mathrm{CAut}(\mathcal{M})$. Therefore $\mathrm{AAut}(\mathcal{M}) \cong E_4$, where $E_4 = \langle (2,3), (1,4)(5,6) \rangle$. This implies that graphs $\Delta_1$ and $\Delta_4$ are isomorphic. The claim (*e*) follows simply form the construction of distance-2 graph for $\Delta_4$. □

In what follows, we call $\Delta$ and $\Delta_4$ *opposite dodecahedrons* in $\mathcal{M}$.

Now, we wish to better understand the combinatorial sense of the isomorphism between $\Delta$ and $\Delta_4$. One of the possibilities would be to start from the planar Fig. 4 and to depict graph $\Delta_4$ on its canvas. Then we will get a non-planar diagram which corresponds to the well-known *stellated dodecahedron*, see [32].

We will exploit other geometric way to depict $\Delta$ and $\Delta_4$, which is related to the use of *Petrie polygons* in map $\mathcal{M}(\Delta)$. The recent paper [69] contains a helpful introduction to the related topics of the theory of regular maps.

Recall that Petrie polygon on a given map $\mathcal{M}$ appears via zigzag moving on $\mathcal{M}$. It turns out that $\mathcal{M}(\Delta)$ contains exactly 6 Petrie polygons, each of them is a 10-gon. All such six polygons form *Petrie dual* map of $\Delta$. The list of these polygons is as follows:

| | |
|---|---|
| (0, 2, 7, 15, 5, 19, 10, 16, 3, 8) | (0, 4, 1, 12, 5, 19, 14, 18, 9, 8) |
| (0, 2, 6, 17, 14, 19, 10, 13, 11, 4) | (1, 7, 2, 6, 9, 18, 16, 10, 13, 12) |
| (1, 4, 11, 3, 16, 18, 14, 17, 15, 7) | (3, 8, 9, 6, 17, 15, 5, 12, 13, 11) |

.

Similar list may be provided for $\Delta_4$. Two Petrie diagrams of $\Delta$ and $\Delta_4$ are simultaneously depicted in such a manner that the isomorphism of graphs is visible. Each diagram contains on the border a Petrie polygon. The remaining 10 vertices induce a subgraph isomorphic to $2 \circ C_5$. The stabilizer of the selected 10-gon is isomorphic to

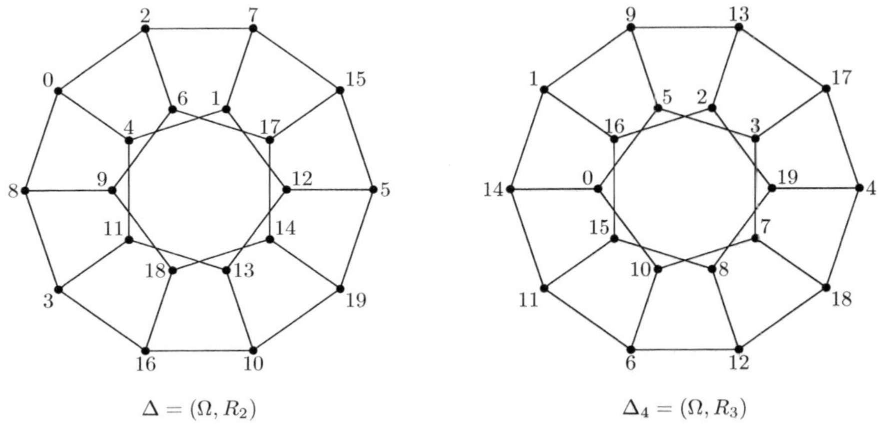


**Fig. 6** Two opposite dodecahedra with one evident Petrie polygon

**Fig. 7** The graph $\tilde{\Delta}$ depicted as $\Pi[E_2]$

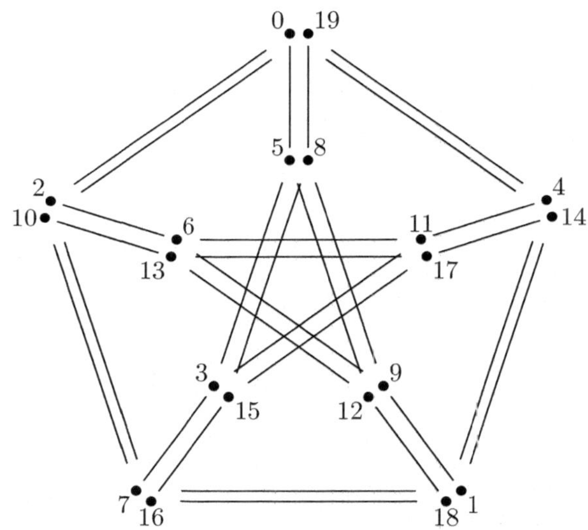


the dihedral group $D_5$ of order 10, which has two orbits of length 10 on the elements of $\Omega$ and orbits of length 1 and 5 on the set of Petrie polygons. A simple exercise allows to check that each of Petrie polygons appears, indeed, via zigzag moving on $\Delta$ (as well as on $\Delta_4$). Because $\Delta$ is an antipodal graph, its distance-5 graph is isomorphic to $10 \circ K_2$, each antipodal block includes vertices $(a, b)$ and $(b, a)$. The graph $\Delta_3$ will be considered in the next section.

First, let us consider graph $(\Omega, R_2 \cup R_3)$, which appears as disjoint union of edges from $\Delta$ and $\Delta_4$. As we are aware, it generates the association scheme $\mathcal{M}_4$. We denote this graph by $\tilde{\Delta}$, its diagram appears in Fig. 7, consisting of 10 "hypervertices" and

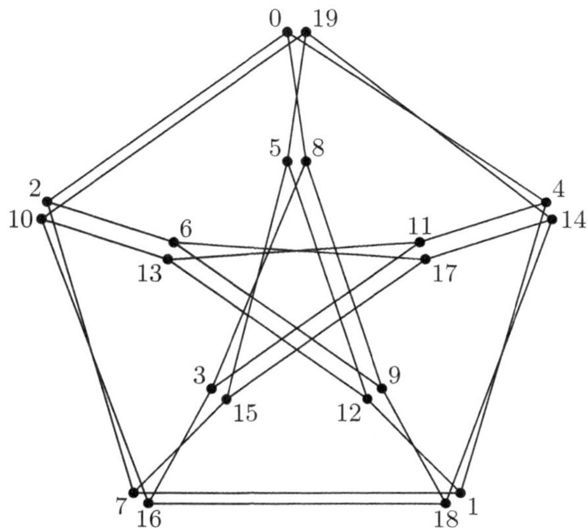


**Fig. 8** The graph $\Delta$, depicted as a half of $\tilde{\Delta}$

15 "hyperedges", shown by thick lines. Each hypervertex corresponds to induced empty subgraph $E_2$, while the hyperedge between $\{a, b\}$ and $\{c, d\}$ substitutes four edges $\{a, c\}$, $\{a, d\}$, $\{b, c\}$ and $\{b, d\}$. The reader is welcome to observe that $\tilde{\Delta}$ is isomorphic to the *composition* $\Pi[E_2]$ (see, e.g. [56]) of the Petersen graph with $E_2$. This observation immediately implies a computer-free proof of some claims.

**Proposition 10** *The following holds:*

*(a) The graph $\tilde{\Delta}$ is isomorphic to $\Pi[E_2]$.*

*(b) The* $\mathrm{Aut}(\tilde{\Delta}) = \mathrm{Aut}(WL(\tilde{\Delta})) = \mathrm{Aut}(\mathcal{M}_4)$ *is the rank 4 group* $\widetilde{S_5^{(2)}} \wr S_2$ *of order* $5! \cdot 2^5$.

*(c) The Petersen graph is a homomorphic image of the graph $\Delta$ with respect to antipodal imprimitivity system.* □

Note that part $(c)$ explains notation $\tilde{\Delta}$, selected for the considered graph.

Let us now concentrate on the graph $\Delta_2 = (\Omega, R_1 \cup R_4)$.

First, we return once more to the graph $\Delta$ and depict it in Fig. 8 as a "half" of $\tilde{\Delta}$. In other words, spanning subgraph of $\tilde{\Delta}$. Now for any vertex of $\Delta$ consider the set $N(x)$ of its neighbours in $\Delta$. We get a collection of 20 3-element subsets (see Table 4). Regard each subset $N(x)$ as a clique of the new graph $X$. Get the entire graph $X$ and check its properties.

**Proposition 11** *The following holds:*

*(a) The obtained graph $X$ coincides with the distance-2 graph $\Delta_2$ of $\Delta$.*

*(b) The graph $\Delta_2$ appears as a point graph of PBIBD $C(\Delta_2)$ with the parameters $v = b = 20$, $k = r = 3$, $\lambda_1 = 1$, $\lambda_2 = \lambda_3 = \lambda_4 = 0$.*

**Table 4** 3-element subsets as cliques in $X$

| | | | | |
|---|---|---|---|---|
| $\{0, 1, 11\}$ | $\{0, 3, 9\}$ | $\{0, 6, 7\}$ | $\{1, 5, 13\}$ | $\{1, 2, 15\}$ |
| $\{3, 4, 13\}$ | $\{5, 7, 17\}$ | $\{3, 10, 18\}$ | $\{4, 7, 12\}$ | $\{6, 14, 15\}$ |
| $\{2, 9, 17\}$ | $\{6, 8, 18\}$ | $\{9, 14, 16\}$ | $\{13, 16, 19\}$ | $\{2, 4, 8\}$ |
| $\{10, 11, 12\}$ | $\{8, 11, 16\}$ | $\{12, 15, 19\}$ | $\{5, 10, 14\}$ | $\{17, 18, 19\}$ |

*(c) The association scheme, related to* $C(\Delta_2)$*, is* $\mathcal{M}_3$ *with four classes.*
*(d)* $\mathrm{Aut}(\Delta_2) = \mathrm{Aut}(C(\Delta_2)) \cong S_5 \times S_2$.

***Proof*** Again, essentially benefits from the consideration of Fig. 8 and knowledge of the group $\mathrm{Aut}(\Pi)$. After conclusion $\mathrm{Aut}(\Delta_2) \geq S_5 \times S_2$, it is necessary to find base of $\mathrm{Aut}(\Delta_2)$ and to prove that $|\mathrm{Aut}(\Delta_2)| = 240$. □

We refer to [6] for the consideration of partially balanced incomplete block designs.

## 12 More About the Basic Graphs of the Master Association Scheme $\mathcal{M}$

Below, we add more information about the basic graphs of $\mathcal{M}$. Recall that it contains three imprimitivity systems, defined by the 2-orbits $R_5$, $R_6$, and $R_7$ of $(A_5, \Omega)$. While $R_7$ remains visible in terms of $\Delta$, both relations $R_5$ and $R_6$ are merged in $\mathcal{M}_2 = WL(\Delta)$. The reason of the existence of two disconnected graphs, isomorphic to $5 \circ K_4$ and defined by $R_5$ and $R_6$ is pretty clear. Indeed, in first case we get blocks of form $\{(a, b) \mid b \neq a\}$ for $a \in [0, 4]$, while in second case $\{(b, a) \mid b \neq a\}$ for $a \in [0, 4]$. These two imprimitivity systems are *orthogonal*, in the sense that two blocks from different systems intersect in at most one element of $\Omega$. This allows us to create one more drawing of $\Delta$, see Fig. 9.

Here relation $R_5$ corresponds to vertical, while $R_6$ to horizontal lines in depicted square (with empty places on the main diagonal). The antipodal relation corresponds to pairs of elements, transposed with respect to the main diagonal. A copy of $\Delta$ is also presented in this figure. Note that each vertex $x$ defines two rows and two columns, where $x$ and its antipodal mate do appear. Each of the remaining three rows and three columns contains exactly one element from the set $N(x)$ of the neighbours of $x$ in $\Delta$. The opposite dodecahedron to $\Delta$ can be similarly depicted on the same canvas. Thus we get once more previous claim, though in slightly different framework.

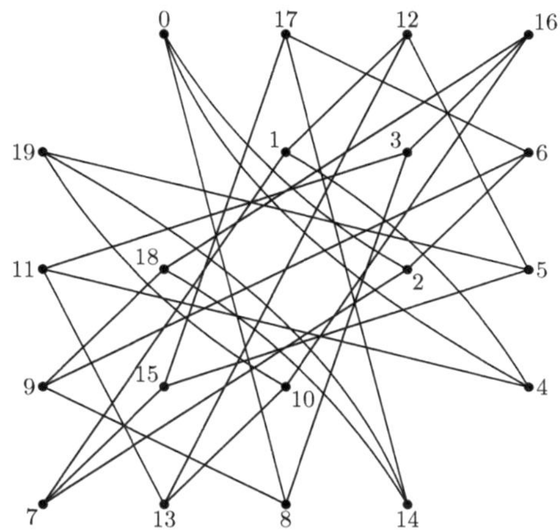


**Fig. 9** The dodecahedron graph $\Delta$ on the canvas of the imprimitivity systems of $\mathcal{M}$

**Proposition 12** *The following holds:*

*(a) The master association scheme appears as*

$$WL(R_2 \cup R_5) = WL(R_2 \cup R_6) = WL(R_3 \cup R_5) = WL(R_3 \cup R_6).$$

$$\mathrm{Aut}(R_2 \cup R_5) = \mathrm{Aut}(R_2) \cap \mathrm{Aut}(R_5) = (A_5, \Omega),$$

$$\mathrm{Aut}(R_2 \cup R_6) = \mathrm{Aut}(R_2) \cap \mathrm{Aut}(R_6) = (A_5, \Omega),$$

$$\mathrm{Aut}(R_3 \cup R_5) = \mathrm{Aut}(R_3) \cap \mathrm{Aut}(R_5) = (A_5, \Omega),$$

$$\mathrm{Aut}(R_3 \cup R_6) = \mathrm{Aut}(R_3) \cap \mathrm{Aut}(R_6) = (A_5, \Omega).$$

□

There is one more way to imagine one representative of two systems of blocks, that is clique of size 4 in the graph $\Delta_3$. For this purpose, we depict below one more diagram of $\Delta$, now with the aid of a Hamiltonian cycle. One of such cycles is depicted in Fig. 10.

Note that consideration of such cycles in dodecahedron has long history since Hamilton himself (1856) and Kirkman (1855). Papers [9, 10] by N. Biggs, one of the leaders of both AGT and history of graph theory, provide a nice survey of relevant activities and further developments. The topic was also reflected in many textbooks on combinatorics, see, e.g. [50].

**Proposition 13** *The following holds:*

*(a) There exist exactly 30 different Hamiltonian cycles in the given copy $\Delta$ of the dodecahedron graph.*
*(b) All these cycles form one orbit under the action of* $\mathrm{Aut}(\Delta)$.
*(c) The stabilizer $K$ of one cycle in* $\mathrm{Aut}(\Delta)$ *is isomorphic to the group $E_4$ of order 4.*

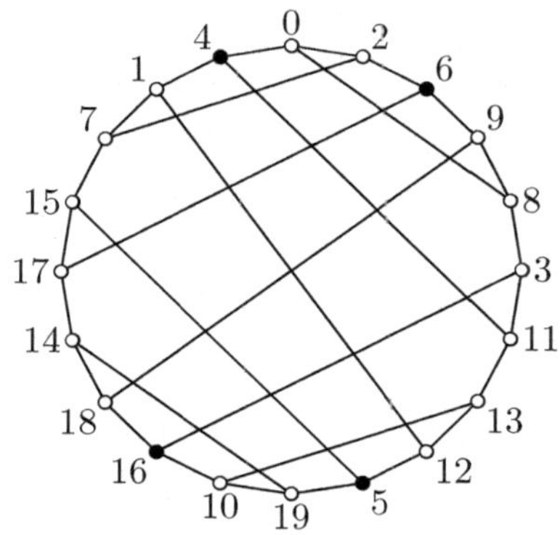


**Fig. 10** A Hamiltonian cycle in $\Delta$

***Proof*** We suggest to start from Fig. 10, on which one concrete cycle is depicted. Evident geometrical reflections of this diagram generate subgroup $E_4$ of $\mathrm{Aut}(\Delta)$, which stabilizes subset $\{4, 5, 6, 16\}$ of vertices, shown by black colour. Thus, the group $K$ acts transitively on this subset. Requiring that permutation $k \in K$, fixes, say 4, we obtain that $k = e$, the identity permutation. Thus, $K = E_4$, therefore the orbit of depicted cycle contains exactly $120/4 = 30$ cycles. Since Hamilton, all Hamilton cycles are described in terms of a planar diagram of $\Delta$, see [62] for modern outline: there are exactly 30 of them. □

Note that the selected clique in graph $\Delta_3$ coincides with the right vertical line in Fig. 9, while (horizontal) clique $\{7, 8, 13, 14\}$ is also visible from Fig. 10. A patient reader is welcome to check that these two cliques are transposed by antipodal involution $i$, while belong to different orbits of action of $(A_5, \Omega)$ on the 4-cliques of $\Delta_3$.

We also notice that existence of two systems of tetrahedrons, which are embedded into $\mathcal{D}$, regarded as a Platonic solid, is a favourite issue in geometry, cf. [32], see also a nice picture on the home page of J. Baez [5].

We converge with two antisymmetric 2-orbits $R_1$ and $R_4 = R_1^T$ of $(A_5, \Omega)$. To describe them, let us consider again graph $\Delta_2 = (\Omega, R_1 \cup R_4)$, the edge set of which splits into twenty 3-cycles, as it was described in the previous section. We have to assign a directed cycle $\mathbf{C_3}$ to each clique simultaneously. It turns out that it is enough to do this for one clique and then to get image of the obtained $\mathbf{C_3}$ applying $(A_5, \Omega)$. The list of 20 such images is given in Table 5.

**Table 5** The list of images for directed 3-cycles

| | | | | |
|---|---|---|---|---|
| (2, 4, 8) | (4, 7, 12) | (0, 6, 7) | (8, 11, 16) | (0, 1, 11) |
| (12, 15, 19) | (2, 9, 17) | (1, 2, 15) | (0, 3, 9) | (6, 8, 18) |
| (13, 19, 16) | (3, 4, 13) | (1, 5, 13) | (10, 11, 12) | (17, 18, 19) |
| (5, 7, 17) | (3, 10, 18) | (6, 14, 15) | (9, 16, 14) | (5, 14, 10) |

In this fashion, we get the list of all 60 arcs in $R_1$. Reversing these arcs leads to $R_4$. There is a clear way to explain the obtained directed triangles. Consider $\mathcal{D}$ on surface, fix orientation. Consider zigzag way of length 2 of the prescribed kind, fulfilled three times starting from vertex $x$: straight and right, back and right, again back and right. Get $\mathbf{C_3}$ on the set $N(x)$.

**Proposition 14** *The following equivalent procedures describe 20 directed cycles* $\mathbf{C_3}$*, which are forming relation* $R_1$*:*

*(a) Construct orbit of the directed cycle* $(2, 4, 8)$ *with respect to* $(A_5, \Omega)$*, induced on the set* $N(0)$*.*
*(b) Construct 20 cycles with the aid of zigzag procedure, as it was explained above.*
*(c) Interpret vertices of* $\Delta$ *as elements of* $\Omega$*; consider* $x_1 = (a, b)$*,* $x_2 = (b, c)$*,* $x_3 = (c, a)$*,* $a, b, c \in [0, 4]$*, then the cycle* $(x_1, x_2, x_3)$ *belongs to the considered orbit, defining* $R_1$*.* □

*Remark 1* For comparison, the relation $R_4 = R_1^T$ is formed by cycles of form $((c, a), (b, c), (a, b))$.

## 13 Directed Strongly Regular Graphs on 20 Vertices

In this section, we treat some well-known DSRGs of valency 4 and 7 on 20 vertices, interpret them and their group in the framework of current considerations. For this purpose, we exploit the association scheme $\mathcal{M}_1$. The following standard approach goes back to [76], see also [43, 82].

Consider BIBD $\gamma = (\mathcal{P}, \mathcal{B})$ with set $\mathcal{P}$ of points and set $\mathcal{B}$ blocks, having parameters $(v, b, k, r, \lambda)$. Let us restrict our consideration for $\lambda = 1$. Denote by $F$ the set $\{(x, b) \mid b \in \mathcal{B}, x \in b\}$ of incident point-block pairs. Usually, such pairs are called *flags* of $\gamma$. Clearly, $|F| = vr = bk$. It is possible to define natural binary relations on the set $F$, which together with reflexive relation form a rank 7 non-commutative association scheme. The adjacency algebra of this association scheme is usually called the *flag algebra* of $\gamma$. The valencies $k_i$ of basic relations, in suitable labelling, are $k_0 = 1$, $k_1 = k - 1$, $k_2 = r - 1$, $k_3 = k_4 = (k - 1)(r - 1)$, $k_5 = (k - 1)^2(r - 1)$, $k_6 = (k - 1)(r - 1)(r - k)$. In this labelling, $R_3$ and $R_4$ form an antisymmetric pair of relations, while all other relations are symmetric. The structure constants of the appearing rank 7 association schemes are provided, e.g. in [82]. The Proposition 6.1. in [82] classifies all DSRGs, which appear as suitable mergings of classes of the flag algebra of $\gamma$.

Here, we consider trivial case of $\gamma$ with parameters (5, 10, 2, 4, 1), which consists of elements of $[0, 4]$, regarded as points, and elements of $\left\{ {[0,4] \atop 2} \right\}$ as blocks. Then, as it is mentioned in the above-cited paper, we get two DSRGs with parameters (20, 4, 1, 0, 1) and (20, 7, 4, 3, 2), respectively. We also mention that, according to [43], both these DSRGs have automorphism group $S_5$, the first DSRG was already known to Duval [37], while the second was mentioned by him as an open case.

**Proposition 15** *The following holds:*

(*a*) *The graph* $\Gamma_1 = (\Omega, R_1 \cup R_7)$ *is a DSRG with parameters* $(20, 4, 1, 0, 1)$.
(*b*) *The graph* $\Gamma_2 = (\Omega, R_1 \cup R_5 \cup R_7)$ *is a DSRG with parameters* $(20, 7, 4, 3, 2)$.
(*c*) $WL(\Gamma_1) = WL(\Gamma_2) = \mathcal{M}_1$.
(*d*) $\mathrm{Aut}(\Gamma_1) = \mathrm{Aut}(\Gamma_2) = \mathrm{Aut}(\mathcal{M}_1) \cong S_5$.

***Proof*** Consider the trivial BIBD $\gamma = \left([0, 4], \left\{\binom{[0,4]}{2}\right\}\right)$. Establish a bijection between its flags $(a, \{a, b\})$ and ordered pairs $(a, b)$. Clearly, $\mathrm{Aut}(\gamma) = S_5$, where $S_5$ acts naturally on $[0, 4]$. Thus action $(S_5, F)$ on flags of $\gamma$ may be identified with the induced action of $S_5$ on $\Omega$. Now, proof of parts $(a)$ and $(b)$ follows from the results about flag algebras, which were discussed above. In principle, they may be confirmed directly using tensor of structure constants in Table 16, see the Supplement.

According to our interpretation of all basic graphs of $\mathcal{M}$, the relations $R_2$ and $R_3$ are merged by group $S_5$, while all other relations are preserved. Now, using Proposition 11, we conclude proof with the understanding that $\mathrm{Aut}(\mathcal{M}_1)$ is a proper subgroup of $\mathrm{Aut}(\mathcal{M}_3) \cong S_5 \times S_2$. Thus, the only choice is $\mathrm{Aut}(\mathcal{M}_1) \cong S_5$. □

A brief discussion of DSRGs $\Gamma_1$ and $\Gamma_2$ will be resumed in Sect. 27.

## 14 Brouwer's (0, 2)-Graph *N*6.3

A $(0, 2)$-graph is a connected graph, such that any two vertices have either 0 or 2 common neighbours. According to M. Mulder [92], $(0, 2)$-graphs are regular. This implies that $(0, 2)$-graphs form a particular case of Deza graphs, see [39]. There is a number of standard constructions for $(0, 2)$-graphs. This motivates special interest to "sporadic", in some sense, cases of these graphs. The results of computer-aided enumeration are reported in [14], see also home page [13] of A. E. Brouwer.

A quite unusual $(0, 2)$-graph, related to dodecahedron, is presented in Sect. 12 of [14]. On Brouwer's home page this is graph denoted by $N6.3$ and discussed with enough attention. (Here $N$ stands for the case of non-bipartite graphs.) It turns out that $N6.3$ "lives" inside of our master association scheme $\mathcal{M}$. More exactly, there are four isomorphic copies of such graphs, which can be merged from basic relations of $\mathcal{M}$. One copy $\Gamma_3'$ of $N6.3$ is depicted (conditionally) in Fig. 11. We have regular graph of valency 6 with vertex set $\Omega$. Its edge set consists of edges of initial dodecahedron $\Delta$, and in addition, of the edges of graph $5 \circ K_4$, corresponding to horizontal lines, as in the previous section. Here vertices from the same cliques are depicted by five kinds of symbols.

**Proposition 16** *Let us consider the graph* $\Gamma_3' = (\Omega, R_2 \cup R_5)$. *Then:*

(*a*) *The graph* $\Gamma_3'$ *is a* $(0, 2)$-*graph of valency 6.*
(*b*) *The graph* $\Gamma_3'$ *has diameter 3 with local distribution* $1 + 6 + 12 + 1$ *around each vertex.*

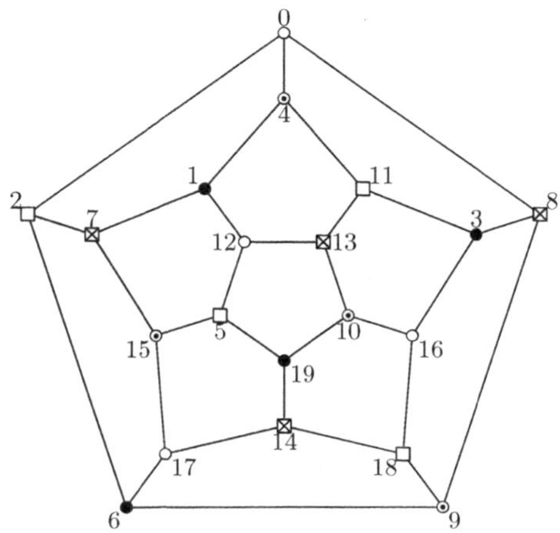


**Fig. 11** A conditional diagram of (0, 2)-graph $N6.3$

*(c) The subgraph induced by the neighbours of any* $x \in \Omega$ *is isomorphic to* $K_3 + E_3$.
*(d)* $WL(\Gamma_3') = \mathcal{M}$.
*(e)* $\mathrm{Aut}(\Gamma_3') = (A_5, \Omega)$.
*(f) The graph* $\Gamma_3'$ *is the unique vertex-transitive non-bipartite* $(0, 2)$*-graph of valency 6 on 20 vertices, up to isomorphism.*

***Proof*** Part $(f)$ was proved in [14] with the aid of a computer. Part $(a)$ can be proved with the aid of a diagram. It is enough to take representatives of six symmetrized 2-orbits of $(A_5, \Omega)$ and to check that elements of each pair have exactly 0 or 2 common neighbours in $\Gamma_3'$. Alternatively, one can count $(A_2 + A_5)^2$, using the tensor of structure constants of $\mathcal{M}$. Proof of $(b)$ and $(c)$ require simple inspections, for example, for $0 \in \Omega$. Parts $(d)$ and $(e)$ are corollaries of Galois correspondence between mergings of $\mathcal{M}$ and overgroups of $(A_5, \Omega)$, which was many times exploited in this paper. □

**Proposition 17** *Similar statement is valid for graphs* $\Gamma_3'' = (\Omega, R_2 \cup R_6)$, $\Gamma_3''' = (\Omega, R_3 \cup R_5)$ *and* $\Gamma_3^{iv} = (\Omega, R_3 \cup R_6)$. □

## 15 Master Scheme $\mathcal{M}$ as Infima of Rank 7 and Rank 6 Schemes

Let $\mathcal{N}_1 = \{S_0, S_1', \dots, S_l'\}$ and $\mathcal{N}_2 = \{S_0, S_1'', \dots, S_m''\}$ be two association schemes, defined on the same set $\Omega$. Let $\mathcal{N}_1 \wedge \mathcal{N}_2 = WL(S_1', \dots, S_l', S_1'', \dots, S_m'')$. Then $\mathcal{N}_1 \wedge \mathcal{N}_2$ is called the *infima* of $\mathcal{N}_1$ and $\mathcal{N}_2$, the term goes back to [6]. A significant message is that infima of two association schemes is not obligatory an association scheme, since it might be a coherent configuration with more than one fibre. Note also that while in [6] each association scheme is commutative by default, in our text this is not required.

Looking again on the lattice of mergings of our master association scheme $\mathcal{M}$ we easily observe that $\mathcal{M} = \mathcal{M}_1 \wedge \mathcal{M}_2$. The advantage of this observation is that both $\mathcal{M}_1$ and $\mathcal{M}_2$ are "standard" ones, in one or other sense. Indeed, $\mathcal{M}_1$ is rank 7 non-commutative association scheme defined on flags of a BIBD. As we discussed above, we are aware of its structure constants on theoretical level.

The scheme $\mathcal{M}_2$ is metric one, canonically generated by DTG $\Delta$. Because we know the intersection array of $\Delta$, the structure constants of $\mathcal{M}_2$ can be obtained from the intersection array by calculation, see [7]. This information provides a nice challenge to obtain the tensor of structure constants of $\mathcal{M}$ (which is already given in Table 16, see the Supplement) on a purely theoretical level, without use of any computer. We face this challenge conceptually, not trying to fulfil literally all necessary activities.

**Proposition 18** *The following holds.*

*(a) Tensors of structure constants of $\mathcal{M}_1$ and $\mathcal{M}_2$ are known on theoretical level.*
*(b) $\mathcal{M} = \mathcal{M}_1 \wedge \mathcal{M}_2$.*
*(c) The tensor of structure constants of $\mathcal{M}$ is uniquely recovered from the structure constants of $\mathcal{M}_1$ and $\mathcal{M}_2$.*

***Proof*** Each of pairs $R_2$ and $R_3$, $R_1$ and $R_4$, $R_5$ and $R_6$ of basic relations remains non-merged in $\mathcal{M}_1$ or $\mathcal{M}_2$. This allows to reconstruct those structure constants, which do not appear evidently in $\mathcal{M}_1$ or $\mathcal{M}_2$. For this purpose, we join still unknown structure constants to suitable arrays, write appearing systems of linear equations and solve them in unique way. □

It is clear that practical use of the above proposition still relies on quite routine computations. Nevertheless, we think that it provides some theoretical curiosity.

## 16 Four Regular Two-Graphs on 26 Vertices and Beyond

Recall that a *two-graph* $\Phi = (V, T)$ consists of set $V$ of points, also called *vertices*, and a collection $T$ of 3-subsets, which are called the *triples* (sometimes the *odd triples*), such that each 4-subset of $V$ contains, as subsets, even number of elements of $T$. Any two-graph is a particular case of incidence structure. Thus, isomorphism of two-graphs, automorphism group of a two-graph are defined naturally, as it appears in design theory. A two-graph $(V, T)$ is called *regular*, if it forms a BIBD. The concept goes back to G. Higman and J. Seidel, we refer to [22, 47, 98, 107] among many other nice sources, related to this part of AGT.

Given a simple graph $\Gamma = (V, E)$ the set $T$ of 3-subsets of $V$, whose induced subgraph in $\Gamma$ contains an odd number of edges gives rise to a two-graph $(V, T)$. This is a standard source of appearance of two-graphs, it also explains wording

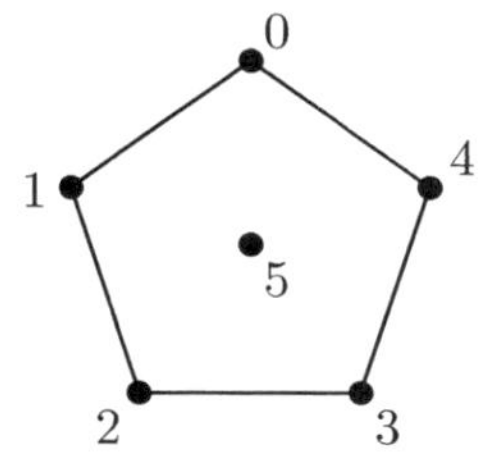


**Fig. 12** A 5-cycle with an isolated point

odd triples. In fact, the same two-graph $(V, T)$ can be obtained, as above, from exactly $2^{n-1}$ different graphs with $n$-vertex set $V$. The collection of these graphs forms a *switching class* on $V$, bijectively associated with $(V, T)$. For each graph $\Gamma = (V, E)$ the group $\mathrm{Aut}(\Gamma)$ is a subgroup of $\mathrm{Aut}(\Phi)$, where $\Phi = (V, T)$ is the two-graph obtained from $\Gamma$. A *clique* of a two-graph $(V, T)$ is a subset $C$ of $V$ such that every 3-subset of $C$ is in $T$. For a given two-graph $\Phi = (V, T)$ and $x \in V$ define graph $\Phi_x$ on the same set of vertices $V$ such that $\{a, b\}$ is an edge of $\Phi_x$ if and only if $\{a, b, x\} \in T$. Clearly, $x$ is an *isolated point* in $\Phi_x$. The subgraph of $\Phi_x$, induced by $V \setminus \{x\}$, is sometimes called a *child* of $\Phi$ in the point $x$. The following criterion is very helpful, see, e.g. [47].

**Proposition 19** *Let $\Phi = (V, T)$ be a two-graph on $n + 1$ points. Then the followings are equivalent:*

*(a) $\Phi$ is a regular two-graph;*
*(b) for $x \in V$ all the children $\Phi_x$ are regular graphs;*
*(c) all the children $\Phi_x$ are $(n, k, \lambda, \mu)$-SRGs with $k = 2\mu$;*
*(d) one child $\Phi_x$ of $\Phi$ is an $(n, k, \lambda, \mu)$-SRG with $k = 2\mu$.* □

**Example 1** Let us consider the pentagon $C_5$ with an isolated point, which is depicted in Fig. 12. Then the regular two-graph $\Phi = ([0, 5], T)$, to the switching class of which the depicted graph belongs to, contains the following set $T$ of triples:

$$T = \{\{0, 1, 3\}, \{0, 1, 5\}, \{0, 2, 3\}, \{0, 2, 4\}, \{0, 4, 5\},$$

$$\{1, 2, 4\}, \{1, 2, 5\}, \{1, 3, 4\}, \{2, 3, 5\}, \{3, 4, 5\}\}.$$

Indeed, the graph $C_5$ is triangle-free, thus the only odd triples are induced subgraphs with one edge. Note that $C_5$ is (5, 2, 0, 1)-SRG, thus, according to the above proposition, we get a regular two-graph $\Phi$ on 6 vertices. The group $\mathrm{Aut}(\Phi)$ contains $\mathrm{Aut}(C_5) = D_5$ and acts transitively on six children isomorphic to $C_5$. This easily implies that $\mathrm{Aut}(\Phi) \cong A_5$.

The reader is welcome to check that the list of six-point children of $\Phi$ literally coincides with the one, depicted in Fig. 1, while the design $\tilde{I}$, considered in Proposition 3, is equal to $\Phi$.

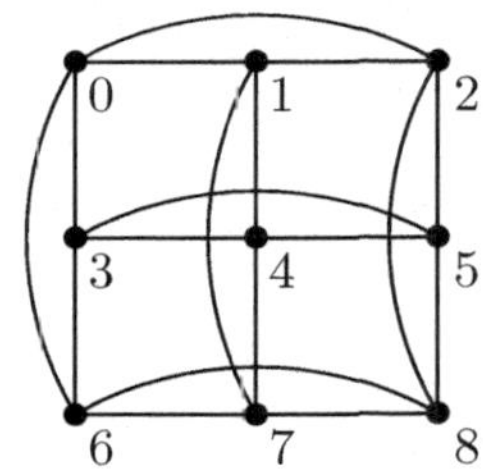


**Fig. 13** The lattice square graph $L_2(3)$

Recall that if $\Gamma$ is a graph, $A(\Gamma)$ its adjacency matrix, then the matrix $S = S(\Gamma) = J - I - A(\Gamma)$ is called the *Seidel matrix* of $\Gamma$. Clearly, eigenvalues of $A(\Gamma)$ and $S(\Gamma)$ may be expressed one through other. The great advantage of the Seidel matrix is that for all graphs $X$ in the same switching class the spectrum of $S(X)$ is the same, thus, we are allowed to talk of the *spectrum* of a two-graph. Moreover, a two-graph is regular if and only if it has just two distinct eigenvalues.

In principle, a given two-graph $\Phi$ can contain different kinds of SRGs (anyway, they are called *children* of $\Phi$), namely, those with and without isolated points. Here, the following criterion is crucially significant.

**Proposition 20** *Let $\Gamma = (V, E)$ be a regular graph, which yields the two-graph $\Phi = (V, T)$. Then $\Phi$ is a regular two-graph if and only if $\Gamma$ is an SRG with eigenvalues $k, r, s$, satisfying $n = 2(k - r)$ or $n = 2(k - s)$.* □

**Example 2** This time we start from the graph $L_2(3)$, which is depicted in Fig. 13. It is easy to check that $L_2(3)$ is isomorphic to the Paley graph $Pal(9)$ of order 9. Clearly, $L_2(3)$ is an SRG with parameters $(9, 4, 1, 2)$, thus condition $k = 2\mu$ is fulfilled. This implies that $L_2(3)$ is a child of a regular two-graph $\Phi_{10}$ on 10 vertices. Recall that the full parameter set of the Petersen graph $\Pi$ is $(10, 3, 0, 1, 1, -2)$. Here $n = 10 = 2(k - s) = 2 \cdot (3 - (-2))$. This implies that the switching class of $\Phi_{10}$ also contains one more SRG, namely, $\Pi$. Finally, we obtain that $\text{Aut}(L_2(3)) \cong S_2 \wr S_3$ and $\text{Aut}(\Pi) \cong S_5$ are subgroups of $\text{Aut}(\Phi_{10})$. Note that $\Phi_{10}$ is an *extension* of $L_2(3)$, where it is regarded in classical terms of design theory. This follows that $|\text{Aut}(\Phi_{10})| = 10 \cdot |\text{Aut}(L_2(3))| = 10 \cdot 72 = 720$ and, moreover, $\text{Aut}(\Phi_{10}) \cong S_6$. Here $S_6$ may be identified with the induced action on the set of partitions to two 3-element subsets of the 6-element set.

Note also that if we consider the graph $L_2(3)$ in Fig. 13 together with an isolated point 9, then $\{0, 1, 2, 9\}$ is a clique in $\Phi_{10}$. It turns out that $\Phi_{10}$ contains exactly 30 such cliques, which define the block set of the unique (up to isomorphism) Steiner system $S(3, 4, 10)$.

Now we are prepared to consider an interesting case of regular two-graphs on 26 points. The history of the investigations of these objects was discussed in Sect. 4. Below, we provide a formal outline of related results, using the same notation, like in the article "Paulus graphs" on home page [13] of A. Brouwer. We use the concept of *geometric SRG* exactly, like it was defined in the seminal paper [11], by R. C. Bose.

**Proposition 21** *The following holds:*

(*a*) *Up to isomorphism, geometric SRGs with parameters* (25, 12, 5, 6) *correspond to the main classes of Latin squares of order 5.*
(*b*) *There are exactly two such main classes, one of them implies* $Pal(25)$*, another one geometric SRG* $LS(5)$.
(*c*) $\mathrm{Aut}(Pal(25)) \cong (E_{25} : \mathbb{Z}_{12}) : \mathbb{Z}_2$, $\mathrm{Aut}(LS(5)) \cong (S_4 + S_3)^{pos}$, *regarded as maximal subgroup of* $A_7$.

***Proof*** Part $(a)$ is a particular case of a general claim, due to Bose. Parts $(b)$ and $(c)$ are considered in a reasonably friendly style and with enough details in [57]. □

**Proposition 22** *The following holds:*

(*a*) *Up to isomorphism, geometric SRGs with parameters* (26, 10, 3, 4) *bijectively correspond to Steiner systems* $S(2, 3, 13)$.
(*b*) *There exist exactly two non-isomorphic classes of* $S(2, 3, 13)$.
(*c*) *The automorphism groups of two geometric SRGs on 26 vertices P26.05 and P26.09 are isomorphic to* $S_3$ *and* $\mathbb{Z}_{13} : \mathbb{Z}_3$*, respectively.*

***Proof*** Again, first, we rely on Bose. After that credits go to K. Zulauf (1897) and their followers, see monograph [27] for exhaustive consideration of STSs and exact references. □

**Proposition 23** *The following holds:*

(*a*) *There are exactly four regular two-graphs on 26 vertices, up to isomorphism.*
(*b*) *Each of these regular two-graphs is uniquely determined by a geometric SRG on 25 or 26 vertices.*
(*c*) *The basic information about the four two-graphs is reflected in Table 6.*
(*d*) *Altogether there are 15 SRGs with parameters* (25, 12, 5, 6) *and 10 with parameters* (26, 10, 3, 4)*, up to isomorphism.*

***Proof*** As we discussed in Sect. 4, part $(a)$ was proved with the aid of a computer and a computer-free proof still was not given. Though proofs of parts $(c)$ and $(d)$ also rely on the use of a computer, here, in principle, the results may be justified by hand using case analysis. □

Note that in Table 6 for a two-graph $\Phi$ group size is $|\mathrm{Aut}(\Phi)|$, while last two columns provide number of children of $\Phi$ on 25 and 26 points. (Children on 25 points are also called *descendants* of $\Phi$.)

## 17 Regular Two-Graph $D$ and Its Group

Let $\Delta_1 = Pal(25)$ be the Paley graph with 25 vertices. The graph $Pal(25) = (\mathbb{F}_{25}, E)$ has as the vertex set the set of elements of the finite field $\mathbb{F}_{25}$, while for

**Table 6** Regular two-graphs on 26 vertices

| Name | Geometric SRG | Group size | 25 vertices | 26 vertices |
|---|---|---|---|---|
| A | STS(13):P26.05 | 6 | 8 | 5 |
| B | LS(5) | 72 | 4 | 2 |
| C | STS(13):P26.09 | 39 | 2 | 2 |
| D | Pal(25) | 15,600 | 1 | 1 |

$x, y \in \mathbb{F}_{25}$, pair $\{x, y\} \in E \iff x - y$ is a non-zero square in $\mathbb{F}_{25}$. This definition implies that $Pal(25)$ is a Cayley graph over the group $E_{25}$ with the connection set consisting of all square elements in the multiplicative group $\mathbb{F}_{25}^*$, which is isomorphic to the cyclic group $\mathbb{Z}_{24}$. It is well known that $Pal(25)$ is an SRG with parameters $(25, 12, 5, 6)$, see [22]. The automorphism group of $Pal(q)$, $q$ arbitrary prime-power, was considered in many publications, see discussion in survey [67]. Thus, we get that $\mathrm{Aut}(Pal(25)) \cong E_{25} : \mathbb{Z}_{12} : \mathbb{Z}_2$ (where $\mathbb{Z}_2 = \mathrm{Aut}(\mathbb{F}_{25})$), is a group of order 600.

Our second model $\Delta_2$ appears as a Latin square graph $SRG(\mathbb{Z}_5)$ over the cyclic group $\mathbb{Z}_5$. This is also a classical class of SRGs. From the general theory of such graphs, see again survey [57], it follows that $\mathrm{Aut}(\Delta_2) \cong (\mathbb{Z}_5^2 : \mathrm{Aut}(\mathbb{Z}_5)).S_3 \cong (\mathbb{Z}_5^2 : \mathbb{Z}_4).S_3$. We get the same group of order 600, though it is described in a slightly different form.

The third model $\Delta_3$ appears in terms of *amorphic association schemes*. The concept goes back to [51], though the presentation there is quite obsolete. Here we follow the recent paper [85]. Thus let $\mathcal{A}_5$ be the full classical amorphic association scheme of rank 7 over the field $\mathbb{F}_5$. Its non-reflexive basic graphs bijectively correspond to the parallel classes of the Desarguesian affine plane of order 5. Thus every basic graph is isomorphic to $5 \circ K_5$ and is a disconnected graph of valency 4. The group $\mathrm{Aut}(\mathcal{A}_5)$ is isomorphic to the group $\mathbb{Z}_5^2 : \mathbb{Z}_4$ of order 100. The group $\mathrm{CAut}(\mathcal{A}_5)$ is isomorphic to $Aff(2, p) = \mathbb{Z}_p^2 : GL(2, p)$. This implies that $\mathrm{CAut}(\mathcal{A}_5)/\mathrm{Aut}(\mathcal{A}_5)$ is isomorphic to $PGL(1, 5)$ in its action of degree 6 on the projective line $PG(1, 5)$, consisting of 6 projective points. Finally, the group $\mathrm{AAut}(\mathcal{A}_5)$ is isomorphic to $S_6$.

It is significant to mention that $\mathrm{CAut}(\mathcal{A}_5)/\mathrm{Aut}(\mathcal{A}_5) \cong PGL(1, 5)$ is also isomorphic to $S_5$ (in its 3-transitive sporadic action of degree 6). This implies that all rank 3 mergings of classes of $\mathcal{A}_5$ with valencies 1, 12, 12 are isomorphic. In other words, our model $\Delta_3$ for the rank 3 graph on 25 vertices of valency 12 appears as a gluing of three arbitrary parallel classes of the affine plane of order 5.

So let us regard $\Delta_3$ as $\Delta_3 = (\mathbb{Z}_5^2, E)$, where $\Delta_3$ is a Cayley graph $\mathrm{Cay}(\mathbb{Z}_5^2, X)$ defined by connection set

$$X = \bigcup_{x \in \mathbb{Z}_5^*} \{(x, 0), (0, x), (x, x)\}.$$

**Proposition 24** *Let $\Delta_1$, $\Delta_2$ and $\Delta_3$ be three graphs with 25 vertices of valency 12, as described above. Then the followings hold:*

**Table 7** The Kramer–Mesner matrix for $D$

| Pairs\Triples | 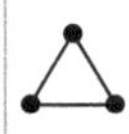 | 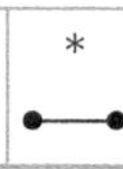 | 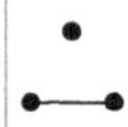 | $\tilde{\lambda}$ |
|---|---|---|---|---|
| • * | 0 | 12 | 0 | 12 |
| •—• | 5 | 1 | 6 | 12 |
| • • | 0 | 0 | 12 | 12 |

*(a)* *$\Delta \cong \Delta_1 \cong \Delta_2 \cong \Delta_3 \cong Pal(25)$ is an SRG with parameters* $(25, 12, 5, 6)$*;*
*(b)* *the automorphism group* $\mathrm{Aut}(Pal(25))$ *is a rank 3 group of order 600;*
*(c)* *the graph $\Delta$ is self-complementary, i.e. $\bar{\Delta} \cong \Delta$;*
*(d)* *the graph $\Delta$ has exactly 15 maximal cliques of size 5, which bijectively correspond to the cliques in parallel classes of the three basic graphs $5 \circ K_5$, merged into $\Delta_3$.*

***Proof*** The isomorphism of models $\Delta_2$ and $\Delta_3$ is evident. To prove the isomorphism with $\Delta_1$ it is enough to construct a copy of the field $\mathbb{F}_{25}$ and a copy of $Pal(25)$, and to compare it with other graphs. Each Paley graph is self-complementary: interchange squares and non-squares in the role of connection set. Clearly, graph $\Delta_3$ has at least 15 cliques of size 5. Extra inspection reveals that these are all the cliques. □

At this stage, we join to the graph $\Delta_3$ an extra vertex $\infty$ and construct, starting from the obtained disconnected graph $\tilde{\Delta}_3$ on 26 vertices, the two-graph $D$ corresponding to it. From the general theory of two-graphs the resulted two-graph has to be regular, since for $\Delta_3$ we have $k = 2\mu$. Let us justify this fact from the scratch and investigate the group $G = \mathrm{Aut}(D)$.

We are using naive version of the procedure of combinatorial extension (with the aid of Kramer–Mesner matrices), see, e.g. [84]. The result is presented in Table 7. Its rows and columns correspond to the orbits of $\mathrm{Aut}(\tilde{\Delta}_3) \cong \mathrm{Aut}(\Delta_3)$ on induced subgraphs of size 2 and 3, respectively. The entry in each item is a coefficient of the projection in the sense of [84]. In this fashion, starting from the graph $\tilde{\Delta}_3$ with 26 vertices and 150 edges, we obtain a two-graph $D$, which turns out to be a BIBD with $v = 26$ vertices, $b = 1300$ blocks, $k = 3$, $r = 150$ and $\tilde{\lambda} = 12$. To describe the automorphism group $G = \mathrm{Aut}(D)$ we briefly discuss three possible ways below.

First, due to E. Shult, is considered on very detailed level in [104], where certain procedure of the extension of the automorphism group of a rank 3 graph with $k = 2\mu$ is presented. It can be successfully applied to all of the Paley graphs. The reader who is interested (following the spirit of the current text) to justify "honestly" the fact that the group $G$ acts 2-transitively on the set $\mathbb{F}_{25} \cup \{\infty\} = PG(1, 25)$, may rely on the approach due to T. Holyoke [61], see also [54]. Namely, we select arbitrary element from $\mathbb{F}_{25}$, say 0, and construct an involution $i$, which transposes 0 and $\infty$. In advance, all triples from $D$ are classified into four classes, depending on whether they include only 0, or only $\infty$, or both 0 and $\infty$, or none of them. Finally, suitable candidates for action of $i$ on $PG(1, 25) \setminus \{0, \infty\}$ are selected and, at least one, is approved to belong to $G$.

However, we refer to the classical survey [99], which strongly influenced the development of the theory of two-graphs. Namely, Theorem 13.5 in [99] claims (for the case considered by us) that the group $\mathrm{Aut}(D)$ contains the group $PSL(2, 25)$ acting 2-transitively on $PG(1, 25)$. Following this paper, the considered two-graph $D$ should be called the *Paley two-graph* on 26 points.

Finally, it is easy to understand the structure of maximal cliques in $D$. Each such a clique, containing $\infty$, has size 6 and includes, in addition, a 5-clique from $\Delta$. Due to transitivity of $G$, the same is correct for any element of $PG(1, 25)$. Therefore, there are exactly $26 \cdot 15/6 = 65$ maximal 6-cliques in $D$.

**Proposition 25** *The Paley two-graph D has the following properties:*

*(a) It is a BIBD with parameters* $(26, 1300, 3, 150, 12)$.
*(b) It has 26 descendants with one isolated point, each isomorphic to* $Pal(25)$.
*(c) D has exactly 65 maximal cliques of size 6.*
*(d)* $\mathrm{Aut}(D) \cong PSL(2, 25) : \mathbb{Z}_2$ *is group of order 15600, which acts 2-transitively on the set* $PG(1, 25)$.
*(e) The complement* $\bar{D}$ *is isomorphic to D.*

***Proof*** To prove $(d)$ consider the stabilizer $G_\infty$, which is isomorphic to the group $\mathrm{Aut}(\Delta)$. In formulation of $(e)$ the complement means all triples of elements in $PG(1, 25)$, which do not appear in $D$. To prove $(e)$ construct the two-graph, starting from $\bar{\Delta}$. Since $\Delta \cong \bar{\Delta}$, the obtained two-graphs are isomorphic. From the above procedure of construction of $D$, it is clear that the two-graph of $\bar{\Delta}$ coincides with $\bar{D}$. □

## 18 From the Two-Graph $D$ to the Steiner System $S(3, 6, 26)$

Now we start from the pair of complementary two-graphs $\{D, \bar{D}\}$. According to Proposition 25, each of them has 65 6-cliques. Let us form an incidence structure $\gamma = (V(T), \mathcal{B})$, where $\mathcal{B}$ is a union of all 6-cliques appearing in $D$ or $\bar{D}$.

**Proposition 26** *The followings hold:*

*(a) The incidence structure* $\gamma$ *is a Steiner design* $S(3, 6, 26)$.
*(b) The group* $\mathrm{Aut}(\gamma)$ *contains a subgroup* $PSL(2, 25) : \mathbb{Z}_2$ *acting transitively on the point set of* $\gamma$.

***Proof*** By definition, $\mathcal{B} = \mathcal{B}_1 \cup \mathcal{B}_2$, where $\mathcal{B}_1$ and $\mathcal{B}_2$ are the sets of 6-cliques in $D$ and $\bar{D}$, respectively. According to Proposition 25, the group $PSL(2, 25) : \mathbb{Z}_2$ preserves both $D$ and $\bar{D}$ and therefore also both $\mathcal{B}_1$ and $\mathcal{B}_2$. This implies $(b)$. To prove $(a)$, due to transitivity of $\mathrm{Aut}(\gamma)$ it is enough to consider triples $\{\infty, x, y\}$, where $x, y$

are vertices of $Pal(25)$. We are aware that $Pal(25)$ contains exactly 15 5-cliques. Each such a 5-clique together with point $\infty$, clearly, forms a 6-clique in $D$ and thus a block in $\mathcal{B}_1$. Similarly, we get 15 blocks in $\mathcal{B}_2$, also containing point $\infty$. Let $\mathcal{B}_\infty$ be the set of all blocks in $\mathcal{B}$, containing $\infty$, $|\mathcal{B}_\infty| = 30$. Altogether there are $25 \cdot 24/2 = 300$ triples of form $\{\infty, x, y\}$. The revealed 30 blocks in $\mathcal{B}_\infty$ contain exactly $30 \cdot \frac{5 \cdot 4}{2} = 300$ pairwise different triples of the kind $\{\infty, x, y\}$. Thus, finally, we obtain that $\gamma$ is $S(3, 6, 26)$. □

**Corollary 1** *The incidence structure* $\gamma$ *is a 2-design with parameters* $v = 26$, $b = 130$, $k = 6$, $r = 30$, $\lambda_2 = 6$.

***Proof*** We count parameter $b$ as

$$b = \frac{\binom{26}{3}}{\binom{6}{3}} = \frac{26 \cdot 25 \cdot 24}{6 \cdot 5 \cdot 4} = 130, \quad \text{after that} \quad \lambda_2 = \frac{130 \cdot \binom{6}{2}}{\binom{26}{2}} = 6.$$

□

At this stage, we have clear understanding of only 30 blocks in $\gamma$ out from 130, those which are in $\mathcal{B}_\infty$. Of course, full list of blocks may be (and in fact was) obtained with the aid of a computer. It turns out that the incidence structure $\gamma$, up to isomorphism, is a well-known object.

A Steiner design $S(3, n+1, n^2+1)$ with $n \geq 2$ is called an *inversive plane* (or sometimes *Möbius plane*) of order $n$, see [34, 63]. Usually, blocks of inversive planes are called *circles*. In fact, many authors are relying on initial axioms of purely geometrical nature, which are valid also for infinite planes. However, these axioms in finite case imply the definition above, given in purely combinatorial terms. Each inversive plane of order $n$ appears as a combinatorial extension of an affine plane of order $n$ with one extra point $\infty$ added to the point set of the starting affine plane. The object of our interest in this paper, namely, the inversive plane of order 5, is unique up to isomorphism, see [25, 35]. One of the classical models for finite inversive planes goes back to B. Segre. An *oval* in an affine plane of order $n$ is a set of $n+1$ points, with no three collinear ones. In this direction circles of an inversive plane, which do not contain $\infty$, can be interpreted as ovals in the corresponding (residue) affine plane. If we start with classical affine plane $AG(2, q)$, in our case $q = 5$, then for the construction of ovals in $AG(2, q)$ a suitable conic should be revealed, which is described with the aid of a polynomial of degree 2 with two variables. Another classical way to get a model of inversive plane seems to be more convenient for our purposes. The model $\mathcal{M}(5)$ of $S(3, 6, 26)$, which is obtained on this way, is usually called a *Miquelian plane* (of order 5). We start from the group $PGL(2, 25)$ of order $26 \cdot 25 \cdot 24$, which acts sharply 3-transitively on projective line $PG(1, 25)$. If the line $PG(1, 25)$ is obtained via field $\mathbb{F}_{25}$, then $\mathbb{F}_{25}$ contains a unique copy of the field $\mathbb{F}_5$. This implies that a suitable subset $X$ of projective points in $PG(1, 25)$ forms a projective line $PG(1, 5)$. Moreover, the setwise stabilizer of $X$ in $PG(1, 25)$ is isomorphic to $PGL(2, 5)$. Now consider the incidence structure $\mathcal{M}(5) = (PG(1,25), X^{PGL(2,25)})$, where $X^{PGL(2,25)}$ is the orbit of $X$ under the induced action of $(PGL(2,25), PG(1,25))$ on the set of images of $X$.

**Proposition 27** *The following holds:*

*(a) The incidence structure $\mathcal{M}(5)$ is an $S(3,6,26)$.*
*(b) The Steiner system $\gamma$ and $\mathcal{M}(5)$ are isomorphic.*

***Proof*** See result 4.1 in [63], which justifies $(a)$ for arbitrary prime-power $q$. The proof of $(b)$ follows from $(a)$ and the above-mentioned uniqueness of $S(3,6,26)$. □

**Corollary 2** *The group* Aut$(\gamma)$ *is a 3-transitive group of order 31200 and is isomorphic to $P\Gamma L(2,25)$.*

***Proof*** Since the incidence structures with block sets $\mathcal{B}_1$ and $\mathcal{B}_2$ are isomorphic, the group Aut$(\gamma)$ is at least twice larger than Aut$(D)$. The fact of 3-transitivity of Aut$(\gamma)$ follows from the transitivity of Aut$(D)$ on $\mathcal{B}_1$ and $\mathcal{B}_2$. There are three ways to finish the proof and to justify desired order and structure of Aut$(\gamma)$:

- to trust GAP;
- to construct the external design $\gamma^\infty$ and to describe its automorphism group, see definitions in [63];
- to use the known catalogue of primitive groups of degree 26 (it is available in many sources, in particular in GAP library) and to conclude that the group is maximal subgroup of $S_{26}$ and has order twice larger than Aut$(D)$. □

We still wish to get understanding of the remaining 100 circles in $\gamma$, those which do not contain $\infty$, aka blocks of external structure $\gamma^\infty$. Of course, one way is purely algebraic via construction of the field $\mathbb{F}_{25}$ and getting list of blocks of $\mathcal{M}(25)$, literally. However, we prefer a more combinatorial approach.

Below, we denote by $\Sigma$ arbitrary (isomorphic) copy of the Paley graph $Pal(25)$. For a vertex $x$ of $\Sigma$, as usually, $\Sigma_1(x)$ and $\Sigma_2(x)$ denote the subgraphs of $\Sigma$ induced by the neighbours and non-neighbours of $x$, respectively.

**Lemma 1** *The followings hold:*

*(a) The graph $\Sigma_1(x)$ is a regular graph of valency 5 with 12 vertices.*
*(b) The graph $\Sigma_2(x)$ is a regular graph of valency 6 with 12 vertices.*
*(c) The graph $\Sigma_2(x)$ is isomorphic to the complement $\overline{\Sigma_1(x)}$ of $\Sigma_1(x)$.*
*(d) $K =$ Aut$(\Sigma_1(x)) \cong$ Aut$(\Sigma_2(x))$ is a transitive group of order 48, which is isomorphic to $S_3 \times D_4$.*
*(e) The group $K$ contains a cyclic regular subgroup $\mathbb{Z}_{12}$ of order 12.*
*(f) $\Sigma_1(x) \cong Cay(\mathbb{Z}_{12}, \{2,3,6,9,10\})$.*
*(g) $\Sigma_2(x) \cong Cay(\mathbb{Z}_{12}, \{1,4,5,7,8,11\})$.*
*(h) The graph $\Sigma_2(x)$ contains exactly 24 induced pentagons.*
*(i) The subgraph of $\Sigma$, induced by $\Sigma_1(x) \cup \{x\}$, contains exactly 24 induced subgraphs isomorphic to the wheel graph $W_5$. (Here $W_5$ is a graph with 6 vertices and 10 edges, the complement to which is isomorphic to pentagon together with an isolated point.)*

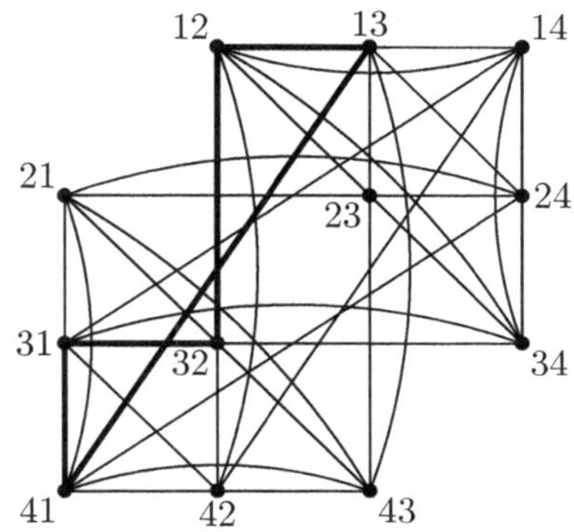


**Fig. 14** Graph $N'$ with one induced pentagon

***Proof*** Parts $(a)$ and $(b)$ are consequences of standard use of the parameter set $(v, k, \lambda, \mu)$ of an SRG, which for $\Sigma$ is $(25, 12, 5, 6)$. Isomorphism of $P(25)$ with its complement $\overline{P(25)}$ implies $(c)$.

Let us now construct the graph $\Sigma_2(x)$. First way is to use model of $P(25)$ as union of 3 classes in classical amorphic scheme on 25 points. Then we get the diagram of graph $N'$, which is depicted in Fig. 14. It is the subgraph induced on the non-neighbours of the vertex 00. Note 4 horizontal, 4 vertical and 4 diagonal triangles, which are partitioning the edge set of $N'$.

The graph $N''$ in Fig. 15 is a Cayley graph $Cay(\mathbb{Z}_{12}, \{1, 4, 5, 7, 8, 11\})$. An example of isomorphism

$$f = \begin{pmatrix} 12 & 13 & 14 & 21 & 23 & 24 & 31 & 32 & 34 & 41 & 42 & 43 \\ 0 & 5 & 4 & 2 & 1 & 9 & 3 & 7 & 8 & 10 & 11 & 6 \end{pmatrix}$$

was found with the aid of a computer. Thus $N' \cong N''$. To describe the group $K = \text{Aut}(N'')$, we recall that $Hol(\mathbb{Z}_{12}) \cong \mathbb{Z}_{12} : \{1, 5, 7, 11\}$, where the group $\text{Aut}(\mathbb{Z}_{12}) \cong E_4$ is acting on $\mathbb{Z}_{12}$ by multiplication modulo 12 in $\mathbb{Z}_{12}$ of elements from $\{1, 5, 7, 11\}$. The circulant graph $N''$ is a *rational circulant* (in the sense of [75]). Therefore, the group $K$ contains group $Hol(\mathbb{Z}_{12})$ and has order at least 48. The group $Hol(\mathbb{Z}_{12})$ is isomorphic to direct product of group $Hol(\mathbb{Z}_4)$ and $Hol(\mathbb{Z}_3)$, which, in turn, are isomorphic to $D_4$ and $S_3$, respectively. Both $D_4$ and $S_3$ are 2-closed, thus $Hol(\mathbb{Z}_{12})$ is also 2-closed and coincides with the direct product of $D_4$ and $S_3$, which has order 48. Note that as a permutation group of degree 12, this is group #28 in the catalogue of transitive permutation groups, presented in [29].

A copy of a pentagon in graph $N'$ is shown (black edges) in Fig. 14, where it looks quite artificially. Much more naturally the induced pentagons are visible in Fig. 15. Indeed, graph $N''$ has two "natural" Hamiltonian cycles, namely, $Cay(\mathbb{Z}_{12}, \{1, 11\})$ and $Cay(\mathbb{Z}_{12}, \{5, 7\})$. These two cycles are interchanged by the group $K$. One induced pentagon has vertices in the set $\{10, 11, 0, 1, 2\}$. There are 12 such pentagons, formed by 4 edges from the first Hamiltonian cycle and one edge from $Cay(\mathbb{Z}_{12}, \{4, 8\}) \cong 4 \circ C_3$. Clearly, we get 12 more pentagons with the aid of the second Hamiltonian cycle. Easy inspection reveals that we get all pentagons.

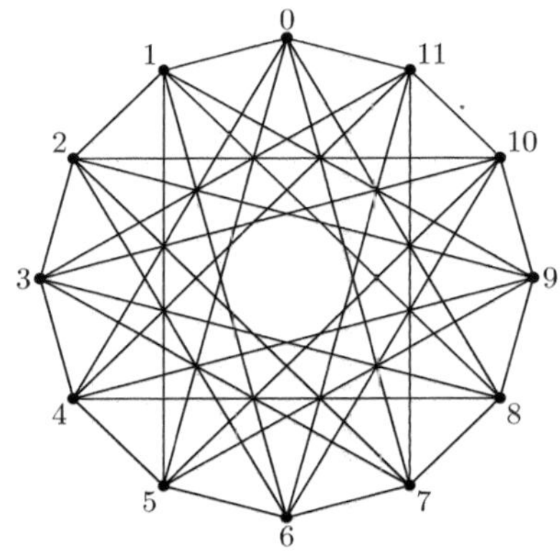


**Fig. 15** Cayley graph $N'' = Cay(\mathbb{Z}_{12}, \{1, 4, 5, 7, 8, 11\})$

To prove $(i)$ consider in $\Sigma_1(x)\{x\}$ the subgraph induced by vertices from pentagon $\Sigma_2(x)$ and $x$. Take into account that $C_5 \cong \overline{C_5}$. □

**Proposition 28** *Let $\gamma = S(3, 6, 26)$ be the inversive plane of order 5. The group $\mathcal{G} = \mathrm{Aut}(\gamma)$ acts on the set $PG(1, 25)$ of the points of $\gamma$. Let $\mathcal{G}_\infty$ be the point stabilizer of $\infty$ in $\mathcal{G}$. Then the followings hold:*

*(a)* $|\mathcal{G}_\infty| = 1200$.
*(b) Under the action of $\mathcal{G}_\infty$ the 130 circles from $\gamma$ form two orbits: one of length 30 and second of length 100.*
*(c) The first orbit of length 30 is formed by $\infty$ together with 5-cliques in six graphs $5 \circ K_5$, which appear in classical full amorphic association scheme $\mathcal{M}_{25}$ on 25 points.*
*(d) The second orbit with 100 circles can be obtained by the following procedure: consider the graph $\Sigma$ formed by union of three classes in $\mathcal{M}_{25}$, fix any point $x$ and add to it all vertices of any induced pentagon in the subgraph $\Sigma_2(x)$ in the considered copy $\Sigma$ of the Paley graph $P(25)$.*

***Proof*** The presented claims follow from the previous considerations in this section. It remains to justify the number of circles in $(d)$ with respect to the described procedure. Indeed, for each vertex $x$ of $\Sigma$ we get 24 circles. Concrete circle, regarded (without vertex $x$) as complete graph $K_5$, has exactly 12 subgraphs isomorphic to a pentagon. Denote by $\tilde{b}$ the number of different circles, which may be obtained from the selected copy $\Sigma$ of $P(25)$, we obtained that $25 \cdot 24 = \tilde{b} \cdot 12$, and thus $\tilde{b} = 50$. Note that the remaining 50 circles may be obtained applying the same procedure to the complement $\bar{\Sigma}$. However, we can remain in the framework of the same copy $\Sigma$. Then extra 50 circles are obtained from unions of $x$ with pentagon in $\Sigma_1(x)$. □

**Fig. 16** Intersection diagram of $\mathcal{Z}$

## 19 Association Schemes on Circles of $S(3, 6, 26)$

Let $S(3, k, v)$ be an arbitrary Steiner 3-design with the block set $\mathcal{B}$. We define relations $R_i$, $i \in \{1, 2\}$, on the set $\mathcal{B}$, such that for $x, y \in \mathcal{B}$, $(x, y) \in R_i$ if and only if $|x \cap y| = i$. The relation $R_3$ includes pairs $(x, y)$, $x, y \in \mathcal{B}$, such that $|x \cap y| = 0$. Consider colour graph $\mathcal{M}(\mathcal{B}) = (\mathcal{B}, \{R_0, R_1, R_2, R_3\})$, here, as usually, $R_0$ is the diagonal relation on $\mathcal{B}$. According to [23], the Steiner design $S(3, k, v)$ is called *block-schematic* if $\mathcal{M}(\mathcal{B})$ is a rank 4 association scheme. (In fact, definition in [23] is given in a more general setting.) Cameron claims [23] (without proof) that the designs $S(3, n+1, n^2+1)$, for $n > 2$ even, are block-schematic. In this case, the valencies of basic graphs are $1, n^2-1, n^2(n+1)/2, n(n-1)(n-2)/2$. The intersection numbers of $\mathcal{M}(\mathcal{B})$ for $n$ even are given in [44]. Note that when $n > 2$, $n$ even, the scheme $\mathcal{M}(\mathcal{B})$ is primitive.

One can regard case $n = 2$ as degenerate one: here valency $v_3 = 0$, while $v_1 = 3$, $v_2 = 6$. Then we get rank 3 primitive association scheme instead of rank 4 primitive association scheme. The reader is welcome to check that $S(3, 3, 5)$ trivially consists of all 3-subsets of 5-set of projective line $PG(1, 4)$. The relation $R_1$ defines the Petersen graph $\Pi$, while $R_2$ corresponds to the complement $\bar{\Pi}$. Clearly, there is no pair of disjoint 3-subsets of $PG(1, 4)$.

The case $n = 4$ is the first non-trivial one for even $n$. Looking on the intersection matrix $P_3$, given in [44], we observe that here it is a 3-diagonal one. According to the general theory of association schemes, see, e.g. [7], this implies that the graph $\mathcal{Z} = (\mathcal{B}, R_3)$ is a primitive DRG of diameter 3 and valency 12, with the intersection diagram as on the Fig. 16.

This graph $\mathcal{Z}$, commonly called the *Doro graph*, was constructed in 1981 with the aid of a computer by V.A. Zaichenko. We refer to the concluding section for extra discussion of this and other graphs.

It was already known to Cameron that the Miquelian plane $M(p)$ for $p$ odd, is not block-schematic. Let us, however, return to the consideration of the "half" of the Miquelian plane $HM(p)$. For $p = 5$, using notation of Sect. 18, it may be defined as $HM(p) = (PG(1, 25), \mathcal{B}_1)$, where $\mathcal{B}_1$ is the set of 6-cliques of the two-graph $\mathcal{D}$, coming from the Paley graph $Pal(25)$. Clearly, a similar structure may be introduced for the Miquelian plane $M(q)$ for arbitrary odd prime-power $q$.

For the case $p = 5$ the structure $HM(5)$ is a 2-design with the parameters $v = 26$, $b = 65, k = 6, r = 15, \lambda = 3$. Good news are now coming from the same paper [44].

**Proposition 29** *The half-Miquelian plane $HM(p)$ is block-schematic, the association scheme, corresponding to it for $p > 3$, is primitive rank 4 association scheme with valencies $v_0 = 1$, $v_1 = p^2 - 1$, $v_2 = p(p^2-1)/4$, $v_3 = p(p-1)(p-3)/4$.*

***Proof*** Again, the intersection matrices for the association scheme $\mathcal{M}(\mathcal{B}_1)$, related to $HM(p)$, are given in [44]. As the authors claim, one needs to make direct calculations to get them. □

From the point of view of Proposition 29, the case $p = 3$ can be again regarded as degenerate, however, by no means trivial one. Let us face it in order to help the reader to get a new visual insight.

**Example 3** *Half of $\mathcal{M}(3)$ with 15 blocks.* Let us start from $Pal(9) = L_2(3)$ together with an isolated vertex $\infty$, as in Fig. 17 (left). We switch this graph with respect to coclique $\{0, 4, 8\}$ and get the graph depicted in the middle of Fig. 17. After that, this graph is redrawn on the right-hand side of Fig. 17 in a nicer manner, hopefully convincing the reader that we, indeed, approach the Petersen graph $\Pi$. Recall that the latter, less famous diagram, aims to explain that $\Pi$ is hypo-Hamiltonian (see [60], p. 6.). Now, we have to express 15 blocks in $\mathcal{B}_1$ in terms of $Pal(9) \cup K_1$, as well as of $\Pi$. For this purpose, it is helpful to interpret both graphs in the language of line graphs. Indeed, $Pal(9) \cong L(K_{3,3})$, while $\Pi = \overline{T_5}$, where $T_5 = L(K_5)$ is the triangular graph on 10 vertices. Therefore it is convenient to exploit the methodology of graphical designs, as it is developed in [8]. In fact, $M(3)$ appears there as design $\mathcal{D}(3, 4)$. Moreover, $HM(3)$ is also visible from this text: it is formed by 15 quadrangles, regarded as subgraphs of $K_5$. On the other hand, in terms of $K_{3,3}$, we, in addition to 6 cliques of size 4 depicted above (formed by $\infty$ and subgraphs $K_{1,3}$ in $K_{3,3}$), need 9 more substructures of $K_{3,3}$. Each such a substructure is a union of two edge-disjoint paths of length 2. Finally, $\mathrm{Aut}(HM(3))$, contains, according to our construction, group $H_1 = \mathrm{Aut}(K_{3,3}) \cong S_2 \wr S_3$ of order 72, as well as $H_2 = \mathrm{Aut}(K_5) \cong S_5$ of order 120. Using a very naive version of the method of group amalgams, we get one more justification of the fact that $\mathrm{Aut}(HM(3)) \cong S_6$. This may also serve as another proof of the classical exceptional isomorphism between $P\Gamma SL(9)$ and $S_6$, cf. [59], p. 149. Finally, we recall that for $p = 3$ the relation $R_3$ for $\mathcal{M}(\mathcal{B}_1)$ is empty. Thus, we get a rank 3 primitive association scheme, whose basic graphs are the triangular graph $T_6$ and its complement $\overline{T_6}$. We refer to [38] for more detailed consideration of the case $p = 3$.

Now, we return to the central case for this paper $p = 5$. Considering the corresponding rank 4 association scheme on 65 cliques of size 6 in the two-graph $\mathcal{D}$ and looking on matrix $P_3$ for it, presented in [44], we observe that two values in it, equal to $(p + 1)(p - 5)/2$, are zero for $p = 5$. In other words, $P_3$ is again a 3-diagonal matrix, thus the graph $I = (\mathcal{B}_3, R_3)$ is a primitive DRG of diameter 3 with intersection diagram as on the Fig. 18.

The graph $I$ was constructed by A.A. Ivanov in 1980 with the aid of a computer, using the procedure of transitive extension, see [65]. Nowadays, this procedure was repeated with the aid of COCO. For this purpose, we considered induced action of $S_5$ or 4 sets consisting of $\emptyset$, $\{a, b\}$, ordered 4-cycles with vertices $a, b, c, d$; ordered pentagons with vertices $a, b, c, d, e$. Here distinct letters denote distinct elements of $[0, 4]$. Like it was reported in [65], we get four rank 3 mergings in the obtained coherent configuration of rank 49. All mergings define isomorphic copies of the

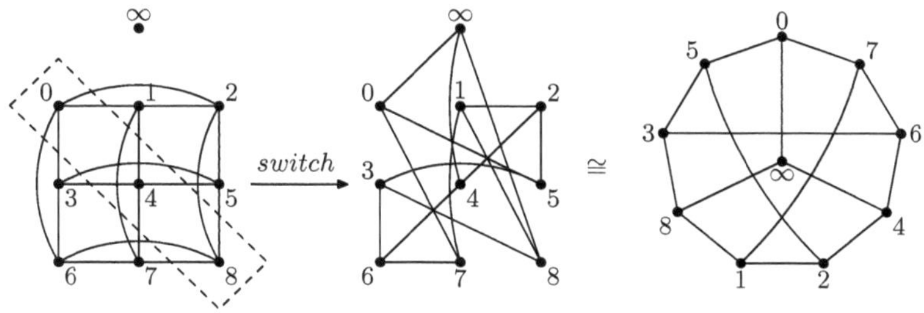


**Fig. 17** Switching $Pal(9) \cup \{\infty\}$ to Petersen graph

**Fig. 18** Intersection diagram of $I$

graph $I$. Note that here $\mathrm{Aut}(I) \cong \mathrm{Aut}(HM(5)) \cong PGL(2, 25)$, while the stabilizer of a point in $\mathrm{Aut}(I)$ is isomorphic to $S_5 \times \mathbb{Z}_2$.

Everything, presented in this section till this moment, was touching quite known facts, though, hopefully, from fresh viewpoint. Now we pursue new research tasks. Thus, we construct two-graphs $\mathcal{D}$ and $\overline{\mathcal{D}}$ from the pair of complementary graphs $Pal(25)$ and $\overline{Pal(25)}$, find common group $\mathrm{Aut}(\mathcal{D})$ and $\mathrm{Aut}(\overline{\mathcal{D}})$, and detect in it subgroup $PSL(2, 25) = \langle g_1, g_2 \rangle$ in its 2-transitive action on $PG(1, 25)$. Here

$$g_1 = (2, 24, 22, 5, 19, 17, 4, 14, 12, 3, 9, 7)(6, 25, 23, 21, 20, 18, 16, 15, 13, 11, 10, 8),$$

$$g_2 = (0, 1, 4)(2, 3, 5)(6, 17, 13)(7, 21, 8)(9, 16, 11)(10, 24, 23)(14, 25, 18)(15, 22, 20).$$

Using COCO, we induce an action of $(PSL(2, 25), PG(1, 25))$ on the orbits of two 6-cliques, one in $\mathcal{D}$, one in $\overline{\mathcal{D}}$, namely, $\{0, 1, 2, 3, 4, 5\}$ and $\{0, 1, 6, 11, 16, 21\}$. We obtain induced action $(PSL(2, 25), \mathcal{B})$, where $\mathcal{B} = \mathcal{B}_1 \cup \mathcal{B}_2$ is a set of 130 circles of $\mathcal{M}(5)$. The list of these circles is presented in Supplement, Table 15. After that, we construct the coherent configuration $\mathcal{M}(\mathcal{B})$, which this time has two fibres and rank 14. Note that the 2-closure of $(PSL(2, 25), PG(1, 25))$ is twice larger, has order 15600 and is isomorphic to the group $P\Gamma SL(2, 25)$.

The restrictions $\mathcal{M}(\mathcal{B}_1)$ and $\mathcal{M}(\mathcal{B}_2)$ on the fibres $\mathcal{B}_1$ and $\mathcal{B}_2$ are isomorphic primitive rank 4 association schemes with valencies 1, 10, 24, 30. Up to the used notation, the intersection numbers of both these association schemes coincide with those, presented in [44], provided $p = 5$. There are three pairs of antisymmetric 2-orbits between fibres $\mathcal{B}_1$ and $\mathcal{B}_2$. According to the COCO notation, these are $R_4$ and $R_7 = R_4^T$ of valency 30, $R_5$ and $R_8 = R_5^T$ of valency 15 and $R_6$ and $R_9 = R_6^T$ of valency 20. It turns out that relations $R_6$ and $R_9$ may be explained in a block-schematic manner. Namely, they correspond to the pairs of circles from $\mathcal{B}_1$ and $\mathcal{B}_2$, which have empty intersection. Remaining four relations between $\mathcal{B}_1$ and $\mathcal{B}_2$ include pairs of circles which have exactly two joint projective points. They can be distinguished in a bit more sophisticated manner, using methodology of block intersections.

Following [44], for $\gamma = S(3, 6, 26)$ we introduce a graph $\Gamma(\gamma)$ with vertex set $\mathcal{B}$ such that two circles in $\mathcal{B}$ are adjacent if and only if they have exactly one common point. According to [44] for $p$ odd, and, in particular, for $p = 5$, the graph $\Gamma(\gamma)$ is disconnected and has two isomorphic connectivity components. For $p = 5$, this fact is independently confirmed by COCO, the vertex sets of the two components are exactly $\mathcal{B}_1$ and $\mathcal{B}_2$. Now, we describe a procedure of distinguishing of relations $R_4$, $R_5$ from $\mathcal{B}_1$ to $\mathcal{B}_2$ (and of relations $R_7$, $R_8$ from $\mathcal{B}_2$ to $\mathcal{B}_1$).

- We select a reference circle $c_0 = \{0, 1, 2, 3, 4, 5\}$ from $\mathcal{B}_1$.
- We find the set $X$ of the neighbours of $c_0$ in the graph $(\mathcal{B}_1, R_3)$ (disjoint circles in $\mathcal{B}_1$), $|X| = 10$.
- We select reference circles $c_{65} = \{0, 1, 6, 11, 16, 21\}$, $c_{68} = \{0, 1, 8, 13, 18, 23\}$, such that $(c_0, c_{65}) \in R_4$, $(c_0, c_{68}) \in R_5$. (We are using labelling of circles as in S.2.)
- We find subsets $X_4$ and $X_5$ of circles in $X$, which have empty intersection with $c_{65}$ and $c_{68}$, respectively. Here

$$X_4 = \{\{7, 12, 13, 14, 23, 24\}, \{8, 12, 17, 18, 19, 24\},$$

$$\{7, 8, 9, 14, 18, 22\}, \{9, 13, 17, 22, 23, 24\}\},$$

$$X_5 = \{\{6, 10, 12, 19, 21, 25\}, \{7, 14, 16, 20, 21, 25\},$$

$$\{9, 11, 15, 16, 20, 22\}, \{6, 10, 11, 15, 17, 24\}\}.$$

- We consider subgraphs (of size 4), which are induced by $X_4$ and $X_5$ in graph $(\mathcal{B}_1, R_3)$.
- We get empty graph $E_4$ and 1-factor $2 \circ K_2$, respectively.
- We claim that relations $R_4$, $R_5$ are distinguished by intersections.

COCO returns all non-trivial association scheme mergings of coherent configuration $\mathcal{M}(\mathcal{B})$. There are four such mergings: $\mathcal{M}_1 - \mathcal{M}_4$. Merging $\mathcal{M}_1$ is obtained by gluing isomorphic relations on $\mathcal{B}_1$ and $\mathcal{B}_2$ and symmetrization of relations between $\mathcal{B}_1$ and $\mathcal{B}_2$. The association scheme $\mathcal{M}_1$ is a rank 7 Schurian association scheme with valencies 1, 30, 24, 10, 30, 15, 20, $\mathrm{Aut}(\mathcal{M}_1) = \mathrm{Aut}(\gamma) = P\Gamma L(2, 25)$. Mergings $\mathcal{M}_3$ and $\mathcal{M}_4$ are quite predictable. They have rank 5 and 3 and groups $S_2 \wr P\Gamma SL(2, 25)$ and $S_2 \wr S_{65}$, respectively. The most interesting is the non-Schurian rank 6 merging with group $P\Gamma L(2, 25)$ and valencies 1, 30, 24, 10, 45, 20. Its basic relations have the following natural explanation:

- three block-schematic classes of valency 30, 24, 10, which appear via union of isomorphic relations on $\mathcal{B}_1$ and $\mathcal{B}_2$;

– two bipartite relations defined for circles between $\mathcal{B}_1$ and $\mathcal{B}_2$ via size of intersection of the circles.

Let us call inversive plane $\gamma = (\mathcal{P}, \mathcal{B})$ *bipartite-schematic*, if the following conditions are satisfied:

– graph $\Gamma(\gamma)$ is disconnected with two algebraically isomorphic components, induced by partition $\mathcal{B} = \mathcal{B}_1 \cup \mathcal{B}_2$;
– restrictions of $\mathcal{M}(\mathcal{B})$ on $\mathcal{B}_1$ and $\mathcal{B}_2$ are algebraically isomorphic primitive rank 4 association schemes;
– two remaining classes define bipartite graphs with bipartite components $\mathcal{B}_1$ and $\mathcal{B}_2$ and edges, corresponding to the size of intersection of circles (0 or 2).

Clearly, according to our definition for a bipartite-schematic inversive plane $\gamma$ its WL-closure $WL(\mathcal{B})$ has rank 6.

**Theorem 1** *The followings hold:*

*(i) The unique Miquelian plane $\mathcal{M}(5)$ is bipartite-schematic.*
*(ii) The schematic closure $WL(\mathcal{B})$ of $\mathcal{M}(5)$ is a non-Schurian rank 6 association scheme with valencies 1, 30, 24, 10, 45, 20.*
*(iii) $WL(\mathcal{B})$ is the WL-closure of any of two bipartite graphs (of valency 45 or 20), which are determined by cardinality of intersection of circles (2 or 0, respectively) between sets $\mathcal{B}_1$ and $\mathcal{B}_2$.*

***Proof*** Initially, proof was obtained with the aid of a computer, mainly via the use of COCO. Later on, some ingredients of justification were reconstructed in a more human-friendly manner. □

We refer to the concluding sections to additional discussion of the introduced concept of a bipartite-schematic plane. One more association scheme, related to $\mathcal{M}(5)$, of order 260, will be considered later on.

## 20 Some Association Schemes on 52 Points and Beyond

We start from the consideration of the *canonical double cover* of a given graph $\Gamma = (V, E)$, frequently denoted by $CDC(\Gamma)$, see, e.g. [86]. In fact, CDC coincides with the *direct product* $\Gamma \times K_2$. In other words, $V(CDC(\Gamma)) = V \times \{0, 1\}$, while unordered pair $\{(x, 0), (y, 1)\} \in E(CDC(\Gamma)) \iff \{x, y\} \in E(\Gamma)$. It is clear that $\mathrm{Aut}(\Gamma \times K_2)$ contains group $\mathrm{Aut}(\Gamma) \times \mathbb{Z}_2$. A graph $\Gamma$ is called *unstable* if $|\mathrm{Aut}(\Gamma \times K_2)| > 2|\mathrm{Aut}(\Gamma)|$. On the other hand, $\Gamma$ is called *stable* if $|\mathrm{Aut}(\Gamma \times K_2)| = 2|\mathrm{Aut}(\Gamma)|$. Clearly, $\Gamma \times K_2$ is a bipartite graph, sometimes it is helpful to denote parts $V \times \{0\}$ and $V \times \{1\}$ by $V$ and $V'$, respectively. To avoid triviality for consideration of efficient criteria of stability, usually it is assumed that the initial graph $\Gamma$ is connected, non-bipartite and does not have a pair of distinct vertices with the same neighbourhood. The following proposition (proved in [110]) has special significance for us.

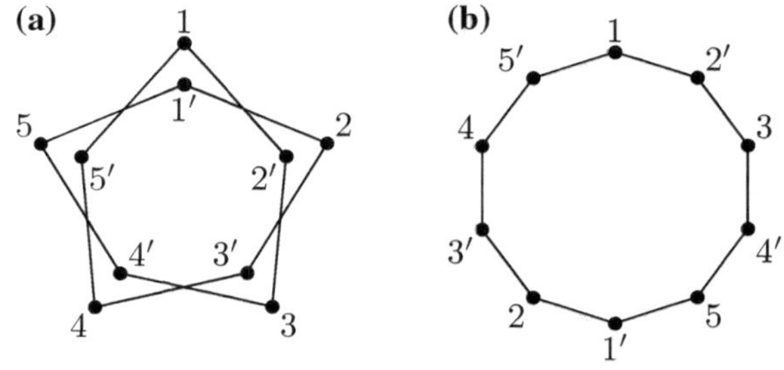


**Fig. 19** The canonical double cover of a pentagon

**Proposition 30** *Let $\Gamma$ be an SRG with parameter set $(v, k, \lambda, \mu)$ such that $k > \mu \neq \lambda \geq 1$. Then $\Gamma$ is stable.*

The proof is given in [110]. □

**Example 4** Graph $L_2(3) = Pal(9)$ is an SRG with parameters (9, 4, 1, 2). According to the previous proposition $L_2(3)$ is stable. The automorphism group of $CDC(L_2(3))$ is of order 144. Using COCO we checked that the WL-closure of $CDC(L_2(3))$ is bipartite, Schurian rank 6 association scheme with automorphism group of order 144.

**Example 5** The pentagon $C_5$ is an SRG with parameters (5, 2, 0, 1). Since here $\lambda = 0$, the sufficient condition in Proposition 30 does not hold. However, it is easy to construct $CDC(C_5)$, see Fig. 19. Its part (a) shows $C_5 \times K_2$, while the same graph appears in part (b) as $C_{10}$. Clearly, $\mathrm{Aut}(C_{10}) = D_{10}$ is the dihedral group of order 20, which is twice larger than $\mathrm{Aut}(C_5) = D_5$. Thus, the pentagon is also a stable graph.

**Example 6** The graph $L_2(4)$ is an SRG with parameters (16, 6, 2, 2). Here $\lambda = \mu$, thus investigation of the $CDC(L_2(4))$ is of a definite interest. We will approach this graph by three different ways.

*Way 1.* $\mathrm{Aut}(L_2(4)) \cong S_4 \uparrow S_2$ is a primitive rank 3 permutation group of degree 16. We construct group $(S_4 \uparrow S_2) \times \mathbb{Z}_2$ of order $2 \cdot 2 \cdot (4!)^2 = 2304$ and investigate it with the aid of COCO. As it is expected, COCO replies that we have 2-closed rank 6 permutation group with subdegrees 1, 1, 6, 6, 9, 9. The corresponding association scheme has 7 mergings (all Schurian), five of rank 4 and two of rank 3. Two rank 4 mergings have group of the same order 23040, however with subdegrees 1, 15, 15, 1 and 1, 6, 9, 16, respectively. Analysing the Hasse diagram of these mergings, we recognize that the second merging coincides with the WL-closure of $CDC(L_2(4))$, and thus in this case, the graph $L_2(4)$ is not stable: the automorphism group of its double cover is 10-times larger than the order of $\mathrm{Aut}(L_2(4)) \times \mathbb{Z}_2$. Using GAP we identify this larger group as $E_{16} : S_6$. Moreover, we conclude that $CDC(L_2(4))$ is a bipartite DRG (even a DTG) of valency 6 and diameter 3 with intersection diagram on the Fig. 20.

*Way 2.* Here we consult [16] and recall that the approached graph is the incidence graph of the *nicest biplane* of order 4, that is symmetric BIBD with parameters

**Fig. 20** Intersection diagram of $CDC(L_2(4))$

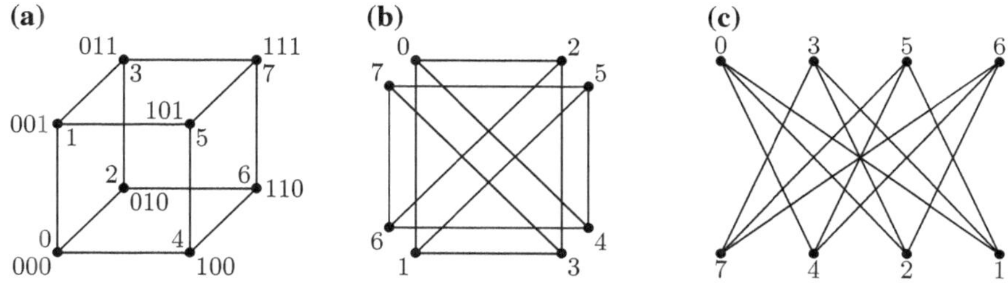


**Fig. 21** Three drawings of the three-dimensional cube $Q_3$

(16, 6, 2). In fact, there are exactly three (up to isomorphism) such designs, see [63]. Note that all biplanes of order 4 were already known to Hussain [64]. From [16] we may obtain one more name for $CDC(L_2(4))$: it is isomorphic to the *folded 6-cube* $\square_6$. This implies, in particular, that $\mathrm{Aut}(\square_6) \cong E_{32} : S_6$ (the structure of the group, which was already communicated to us in *Way 1*, when we were using GAP in addition to COCO), see also home page of A. Brouwer [13].

*Way 3.* This way is also mentioned in [16], it goes back to [97], see also [110]. Namely, to each SRG $X = (V, E)$ with parameters $(v, k, \lambda, \mu)$ where $\lambda = \mu$, we may associate an incidence structure $R(X) = (V, \mathcal{B})$, where the set $\mathcal{B}$ coincides with the subsets $X(v)$, $v \in V$ of all neighbours of vertices in $X$. It is an easy exercise to prove that $R(X)$ is a symmetric BIBD with parameters $(v, k, \lambda)$. As a result, we conclude that the incidence graph of $R(X)$ is a bipartite DRG of diameter 3 with intersection array $(k, k-1, k-\lambda; 1, \lambda, k)$ Clearly, the automorphism group of the incidence graph of $R(X)$ may be much larger in comparison with $\mathrm{Aut}(X) \times \mathbb{Z}_2$, thus, implying non-stability of $X$. Exactly this happens for our case $X = L_2(4)$.

The reader is already familiar with the concept of an *antipodal DRG*, the dodecahedron provides a nice example of such a graph. Antipodal graphs of diameter 3 are of a special interest, because they serve as examples of antipodal covers of complete graphs. Here, we are especially interested in the *twofold antipodal covers of complete graphs* $K_m$: these are DRGs with intersection array $(m-1, c_2, 1; 1, c_2, m-1)$. The vertex set of such a graph $X$ allows partition $\pi$ with cells of size 2. The quotient graph $X/\pi$ of a DRG $X$ with respect to $\pi$ is the complete graph $K_m$.

**Example 7** (*The three-dimensional cube* $Q_3$) This graph is simultaneously twofold antipodal cover of $K_4$ and bipartite, see Fig. 21. Therefore, it does not belong to the class of graphs which are in the middle of our interest in this section. The goal of its presentation is to stress its special additional features, which will be significant later on. In fact, $Q_3$ has intersection array (3, 2, 1; 1, 2, 3). Note that the vertices of $Q_3$

are labelled both by binary strings of length 3 and corresponding numbers in $[0, 7]$, in part (a) of Fig. 21.

Now, we introduce the central construction in this section. We follow the notation of [110]. Let us start with a graph $\Omega$ and let $\Omega'$ be an isomorphic copy of $\Omega$ with isomorphism $\varphi$, defined as $v^{\varphi} = v'$, $v \in V(\Omega)$. Let us add two extra vertices $\infty$ and $\infty'$, which do not belong to $V(\Omega)$ and $V(\Omega')$. Create the graph $\Omega^* = SDC(\Omega)$, which is called the *Shult double cover* of $\Omega$. The vertex set of the graph $\Omega^*$ is $V(\Omega^*) = \{\infty\} \cup V(\Omega) \cup V(\Omega') \cup \{\infty'\}$ and adjacencies are defined as follows:

- (i) $\infty$ is adjacent with every vertex of $\Omega$, while $\infty'$ is adjacent with every vertex in $\Omega'$.
- (ii) $\Omega$ and $\Omega'$ are subgraphs of $\Omega^*$.
- (iii) If $v_1 \in V(\Omega)$ and $v_2' \in V(\Omega')$, then $\{v_1, v_2'\} \in E(\Omega^*)$ if and only if $v_1 \neq v_2$ and $\{v_1, v_2\} \notin E(\Omega)$.

The following theorem goes back to E. Shult [105], though he was not using its current formulation.

**Theorem 2** *Let $\Omega$ be an SRG with parameters $(v, k, \lambda, \mu)$ such that $k = 2\mu$. Then the followings hold:*

- *(i) The Shult double cover $SDC(\Omega) = \Omega^*$ is a twofold antipodal cover of the complete graph $K_{v+1}$.*
- *(ii) The graph $\Omega^*$ is an antipodal DRG of diameter 3 with intersection array $(v, v - \lambda - 1, 1; 1, v - \lambda - 1, v)$.*
- *(iii) The WL-closure of $\Omega^*$ is a rank 4 antipodal association scheme with valencies $1, v, v, 1$.*

***Proof*** The results in (i) and (ii) are classical background statements on association schemes with three classes, see, e.g. [9, 90]. The last item is a trivial reformulation of the fact that $\Omega^*$ is a DRG. □

**Example 8** ($SDC(C_5)$ *as the icosahedron*) We start from the pentagon $C_5$, that is the smallest non-trivial Paley graph $Pal(5)$. Recall that it is an SRG with parameters $(5, 2, 0, 1)$. Thus the Theorem 2 holds for it. Again, all arguments are visible, see Fig. 22. Note that both diagrams are planar, first one literally explains the skeleton $J$ of the icosahedron as $SDC(C_5)$, while the second diagram appeals to traditional depiction of $J$. Good news from these diagrams are that they make visual subgroups $D_5$ and $S_3$ of $\mathrm{Aut}(J)$. Using involution $\varphi$, as it is defined above, an interested reader can reconstruct the full group $\mathrm{Aut}(J) \cong A_5 \times \mathbb{Z}_2$ from this information.

The considered example is a particular case of the general situation.

**Proposition 31** *The automorphism group* $\mathrm{Aut}(SDC(Pal(q)))$ *is isomorphic to the group* $\mathbb{Z}_2 \times P\Sigma L_2(q)$.

***Proof*** This result goes back to the early beginning of the theory of two-graphs. An accurate detailed proof is presented in [111], we use the notation from this paper, where $q$ is an arbitrary odd prime-power. □

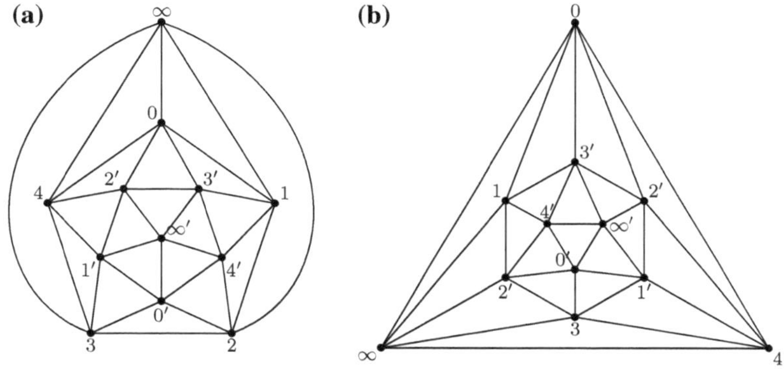


**Fig. 22** The Shult double cover of a pentagon

*Remark 2* When $q = 5$, we get the group $P\Sigma L_2(5) = PSL(2, 5)$, which is isomorphic to $A_5$, this being in agreement with Example 8.

**Example 9** (*Shult double cover of* $Pal(25)$) Here we were using COCO. The starting group was $H = \mathrm{Aut}(Pal(25))$ of degree 25 and order 600. We induced an intransitive action of $H$ of degree 52 with orbits of lengths 1,1,25,25. Clearly, we obtained a Schurian coherent configuration with 4 fibres and rank 24. COCO returns 7 merging association schemes, among them three imprimitive rank 3 mergings, generated by $25 \circ K_2$ and $2 \circ K_{26}$ (twice). There are also two rank 4 mergings, isomorphic to the direct product of rank 2 association schemes on 26 and 2 vertices (with valencies 1,1,25,25). The only interesting, though predicted result (due to elements of Shult's graph extension theory presented above), is the existence of two isomorphic imprimitive rank 4 mergings, generated by antipodal DRG of valency 25 and diameter 4. The automorphism group of both schemes has order 31200, it is isomorphic to the group $P\Gamma SL(2, 25) \times \mathbb{Z}_2$. The stabilizer of a point in this group is isomorphic to the starting group $H$.

*Remark 3* The appearance of two isomorphic antipodal DRGs in the obtained result is pretty clear. Indeed, there is a small degree of freedom in the description of $SDC(\Omega)$: an extra vertex $\infty$ can be joined either with all vertices of $\Omega$, or with all vertices of $\Omega'$. Up to isomorphism, the result will be the same. In algebraic terms, the group $\mathbb{Z}_2 = \langle\varphi\rangle$, appears as a direct multiplier in the total group $\mathrm{Aut}(\Omega^*)$.

A slight deviation, concluding this section, sheds extra light on some links between two considered covering constructions. It is proved in [110] that the Shult double cover of Paley graph is unstable. This motivates the interest to consider the skeleton $J$ of the icosahedron, again.

**Example 10** (*Double cover of the icosahedron*) We have arranged two similar experiments with the aid of COCO. First time, starting from $\mathrm{Aut}(J)$, a coherent configuration with two fibres of size 12 was created, on which $\mathrm{Aut}(J)$ acts diagonally.

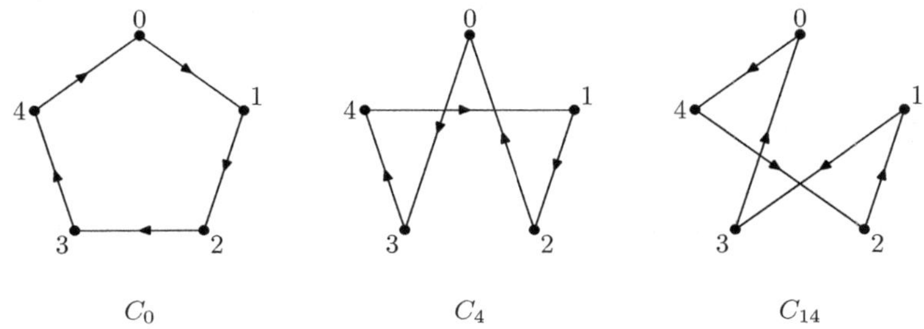


**Fig. 23** Three directed pentagons

The obtained coherent configuration $W_1$ has rank 16, $\mathrm{Aut}(W_1) \cong A_5 \times \mathbb{Z}_2$. COCO returns 32 mergings, 26 of them symmetric ones. Among these mergings, there are four isomorphic rank 6 Schurian mergings with valencies 1, 1, 5, 5, 10, 2. Each of these mergings is the WL-closure of a suitable copy of $CDC(J)$, with automorphism group of order 480, isomorphic to $E_4 \times S_5$. This confirms Theorem 1.5 in [111] for $q = 9$, which claims that $\Omega^* = SDC(Pal(q))$ is not stable and describes the full group $\mathrm{Aut}(CDC(\Omega^*))$ for arbitrary value of $q$.

There is also a pair of non-Schurian rank 5 mergings in $W_1$. To explain it better, one more experiment was arranged. This time we started from the induced action of $S_5$ on the set of all directed pentagons, sharing common vertex set $[0, 4]$. We get a transitive rank 8 action of degree 24 and subdegrees $1^4, 5^4$. Clearly, its 2-closure is twice larger: consider an involution which transposes each directed pentagon with its opposite pentagon. The corresponding coherent configuration $W_2$ has 13 mergings, 11 of them are symmetric. Still we face two isomorphic rank 5 symmetric non-Schurian association schemes with valencies 1, 1, 2, 10, 10. Such a merging, say, first, is the WL-closure of a graph $\Sigma$, which is implicitly introduced in Fig. 23. Here, graph $\Sigma$ on the set of 24 directed pentagons, is obtained via merging of two 2-orbits of the introduced action of $S_5 \times \mathbb{Z}_2$, with representatives $(C_0, C_4)$ and $(C_0, C_{14})$, (we use notation of COCO), see Fig. 23.

An enough challenged reader is welcome to prove independently that the introduced graph $\Sigma$ has the prescribed properties. In fact, this concrete non-Schurian association scheme is a particular case of a quite general class of (typically) non-Schurian association schemes, which are invariant with respect to some imprimitive actions of the groups $PSL(2, q)$ on the so-called quasi-projective points (in terms of S. Reichard), see Sect. 28 for extra discussion.

## 21 Master Coherent Configuration $W$ on 26 Vertices

Now, in continuation of Sect. 8, we describe our master coherent configuration of rank 14 on 26 vertices. Let $V = \Omega \cup \mathcal{P}$ be the set of cardinality 26, consisting of the

**Table 8** Pentagons forming orbit $\mathcal{P}$

| Pentagons | a | b | c | d | e | f |
|---|---|---|---|---|---|---|
| COCO label | 21 | 22 | 23 | 25 | 24 | 20 |

**Table 9** Remaining relations of $W$

| $R_i$ | Length | Representative | Mate |
|---|---|---|---|
| 8 | 3 | 20 | 10 |
| 9 | 3 | 22 | 11 |
| 10 | 10 | 0 | 8 |
| 11 | 10 | 3 | 9 |
| 12 | 1 | 20 | |
| 13 | 5 | 21 | |

set $\Omega$ of ordered pairs of elements from $[0, 4]$, like in Sect. 8, and the set $\mathcal{P}$ of six pentagons, forming relation $\Phi_{5,5}$, like in Sect. 7.2. Clearly, sets $\Omega$ and $\mathcal{P}$ are disjoint. Thus, we consider induced action $(A_5, V)$. This is faithful intransitive action with two orbits of degree 20 and 6. First, this action was constructed and investigated with the aid of COCO. The labelling of the elements in $\Omega$, as given in Table 1, is preserved. Elements of $\mathcal{P}$ are labelled as in Table 8. Note that this labelling was created by COCO, automatically. Here we use notation for pentagons, which was presented in Fig. 1.

COCO returns that the obtained rank 14 coherent configuration $W$ has two fibres with reference representatives 0 and 20. The description of first eight basic relations coincides with one, presented in Table 2. The remaining six relations of $W$ are reflected in Table 9.

Note that reference point for $R_8$ and $R_9$ is still 0, while for further relations it is 20. Clearly, $\text{Aut}(W) = A_5$, though this was confirmed again by COCO. Using GAP, we described two other groups, related to $W$. Namely, $\text{CAut}(W) \cong \mathbb{Z}_2 \times S_5$, while $\text{AAut}(W) \cong E_8$. The list of generators of $\text{AAut}(W)$ in action on relations of $W$ is as follows: $\text{AAut}(W) = \langle g_1, g_2, g_3 \rangle$, where $g_1 = (2, 3)$, $g_2 = (1, 4)(5, 6)$, $g_3 = (8, 9)(10, 11)$. The quotient group $\text{CAut}(W)/\text{Aut}(W)$ is an index 2 subgroup of $\text{AAut}(W)$. It consists of the following involutions, acting on relations (together with the identity):

$$\text{CAut}(W)/\text{AAut}(W) = \{id, g_2, g_1g_3, g_1g_2g_3\}.$$

To confirm all this new data without the use of a computer, first note that $A_5$ acts 2-transitively on the set $\mathcal{P}$. Now consider pair $(a, b) \in \Omega$ and pentagon $p \in \mathcal{P}$. Then $((a, b), p)$ belongs to $R_8$ and $R_9$ if and only if pentagon $p$ contains, or does not contain edge $\{a, b\}$, respectively. Understanding of the colour group $\text{CAut}(W)$ follows from the analysis of mergings of $\mathcal{M}$, which was fulfilled in Sect. 9. Indeed, both overgroups $S_5$ and $S_5 \times \mathbb{Z}_2$ of $(A_5, \Omega)$ naturally appear there through the scope of Galois correspondence.

**Proposition 32** *The following four permutations from* $\mathrm{AAut}(W)$ *in action on the relations of* $W$ *form a coset in* $\mathrm{AAut}(W)$*, which is complementary to subgroup* $\mathrm{CAut}(W)/\mathrm{Aut}(W)$ *of* $\mathrm{AAut}(W)$*:*

$$t_1 = (2,3), \quad t_2 = (1,4)(2,3)(5,6), \quad t_3 = (8,9)(10,11),$$

$$t_4 = (1,4)(5,6)(8,9)(10,11).$$

***Proof*** Proof is based on the analysis of the data above, which were obtained with the aid of COCO and GAP. As it was mentioned a few times before, the proof can be also repeated by hand computations, provided such a training task will develop suitable skills of the reader. □

## 22 The Graph *T* as Merging of *W*

COCO returns that $W$ has two association scheme mergings, each of them defines an SRG.

**Proposition 33** *Let* $W = (\Omega \cup \mathcal{P}, 2\text{--}orb(A_5, \Omega \cup \mathcal{P}))$ *be a Schurian rank 14 coherent configuration, which is defined by 2-orbits of the induced intransitive action of* $A_5$ *on the sets* $\Omega$ *and* $\mathcal{P}$*.*

*(i)* *The coherent configuration* $W$ *has two rank 3 mergings* $W_1$ *and* $W_2$*, described in terms of relations of* $W$ *as* $(0,12)(1,4,13,9,11,2,3)(5,6,8,10,7)$ *and* $(0,12)(1,4,13,8,10,2,3)(5,6,9,11,7)$.
*(ii)* *The graphs* $\Gamma_1$ *and* $\Gamma_2$ *defined as* $(\Omega \cup \mathcal{P}, R_5 \cup R_6 \cup R_7 \cup R_8 \cup R_{10})$ *and* $(\Omega \cup \mathcal{P}, R_5 \cup R_6 \cup R_7 \cup R_9 \cup R_{11})$ *are SRGs with parameters* $(26, 10, 3, 4)$.
*(iii)* $W_1$ *and* $W_2$ *appear also as mergings of an algebraic fusion of* $W$*, which corresponds to* $t_2 \in \mathrm{AAut}(W)$*,* $t_2 = (1,4)(2,3)(5,6)$.
*(iv)* $\mathrm{Aut}(\Gamma_1) = \mathrm{Aut}(\Gamma_2) = A_5 \times \mathbb{Z}_2$.
*(v)* *The graphs* $\Gamma_1$ *and* $\Gamma_2$ *are isomorphic copies of the PRT-graph* $T$*.*
*(vi)* *The graphs* $\Gamma_1$ *and* $\Gamma_2$ *are switching equivalent.*

***Proof*** Parts (i), (ii), (iv) are results of COCO; part (iii) follows from Proposition 32. The isomorphism between $\Gamma_1$ and $\Gamma_2$ is justified by the permutation $(2,3)(8,9)(10,11) \in \mathrm{CAut}(W)/\mathrm{Aut}(W)$. According to the list of Paulus graphs in [13], there is a unique SRG on 26 vertices with largest possible group of order 120, this is graph $T$. Thus, clearly, both $\Gamma_1$ and $\Gamma_2$ are isomorphic to $T$. Permutation $t_3 = (8,9)(10,11)$ on relations of $W$ fulfils the role of switching between $\Gamma_1$ and $\Gamma_2$. □

In what follows, we will regard graph $\Gamma_1$ as our master copy of the graph $T$, as it appears in the framework of $W$. Note that $W$ has 9 symmetrized 2-orbits, up to permutation $t_2$, only 7 of them are distinguishable. This means that computer-free justification of the fact that $\Gamma_1$ is, indeed, an SRG, requires 7 easy inspections. Pick up a representative $\{x, y\}$ of each of these 2-orbits, and find common neighbours of $x$, $y$ in $\Gamma_1$, thus confirming that their total amount is equal to 3 or 4, respectively. We leave this exercise to an interested reader.

## 23 Some Diagrams of the Graph $T$

According to the spirit of this paper, visual images are essential, and frequently crucially inalienable parts of the presentation. Thus, in this section we present a few reasonably nice diagrams of the graph $T$. Each diagram reflects some part of symmetry of $T$ (more rigorously, is invariant with respect to the suitable subgroup of the group $\mathrm{Aut}(T)$), as well as, may reflect a suitable substructure of $T$. We are trying to accompany diagrams by some, hopefully helpful, hints.

Diagram 1 appears on the canvas of the graph $\Delta$, see Fig. 4 in Sect. 10. Recall that the induced subgraph $(\Omega, R_5 \cup R_6 \cup R_7)$ of $T$ on $V(\Delta)$ corresponds to distance 3

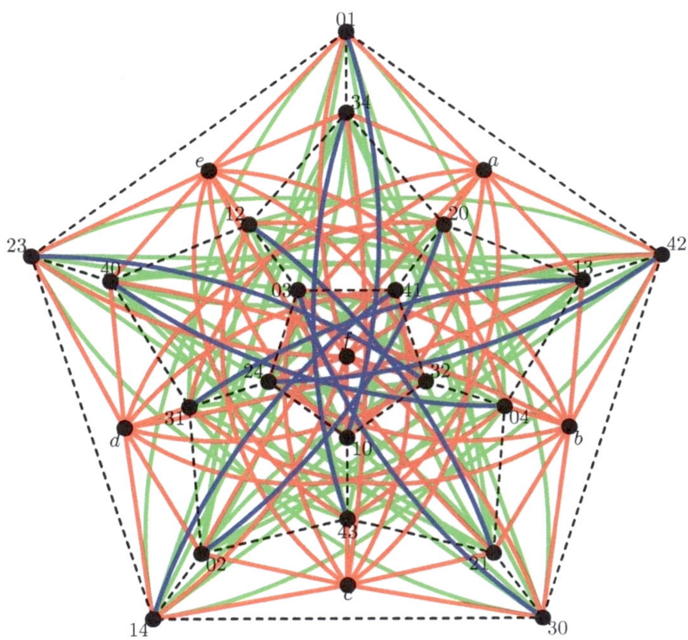


**Fig. 24** Diagram 1 of the graph $T$

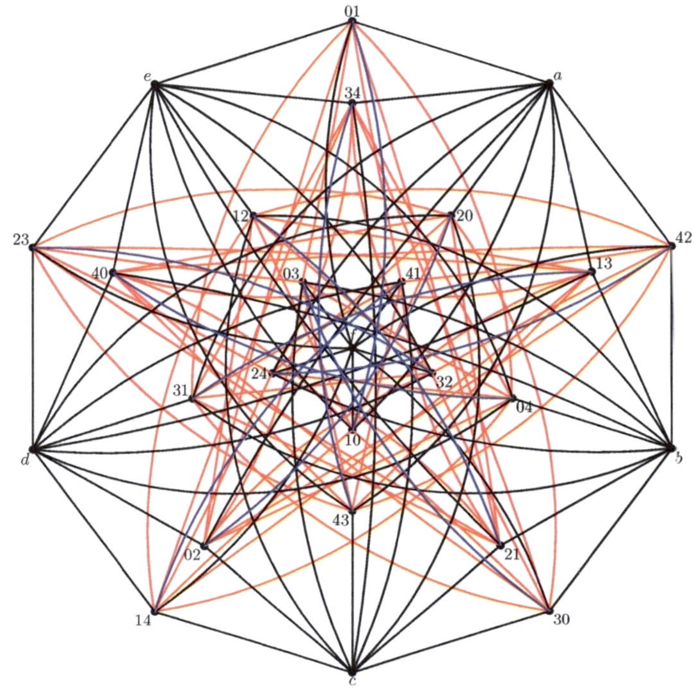


**Fig. 25** Diagram 2 of the graph $T$

and 5 graphs of $\Delta$, while the subset $\mathcal{P}$ induces a 6-coclique of $T$. This is the main message of Diagram 1, which is depicted in Fig. 24. In this fashion, it reflects full symmetry of $T$.

Diagram 2 is of a similar nature. However, here, instead of classical planar depiction of the dodecahedron, we rely on its Petrie diagram, as it is given in Fig. 6. The corresponding Diagram 2 appears in Fig. 25. One may claim that here the dihedral group $D_{10} \cong D_5 \times \mathbb{Z}_2$ of order 20 is posed "at the centre of the universe".

Diagram 3, which appears in Fig. 26, on purpose is adjusted to the induced subgraph $(\Omega, T(\Omega))$ of $T$ on the 20-vertex $\Omega$. Here we start from canvas of $\Delta$, as it is given in Fig. 9. The advantage of Fig. 26 is that quite dense induced subgraph $T(\Omega)$ of valency 7 appears in clear symbolic terms: two vertices are adjacent if they are in the same line, column (both bold), or are joined by a (thin) skew line.

The overgraph $T$ requires stronger imagination from the reader. Here six new vertices (corresponding to pentagons $(0, 1, 2, 3, 4)$, $(0, 1, 3, 4, 2)$, $(0, 3, 1, 2, 4)$, $(0, 1, 4, 2, 3)$, $(0, 2, 1, 4, 3)$ and $(0, 2, 3, 1, 4)$ from $\mathcal{P}$) are added. Each edge of each pentagon implies two edges in $T$: if $\{a, b\}$ is an edge in pentagon, then the pairs $(a, b)$ and $(b, a)$ are adjacent to the vertex corresponding to that pentagon. Thus, altogether we get 10 neighbours of vertex $p$ in $\mathcal{P}$ at the set $\Omega$.

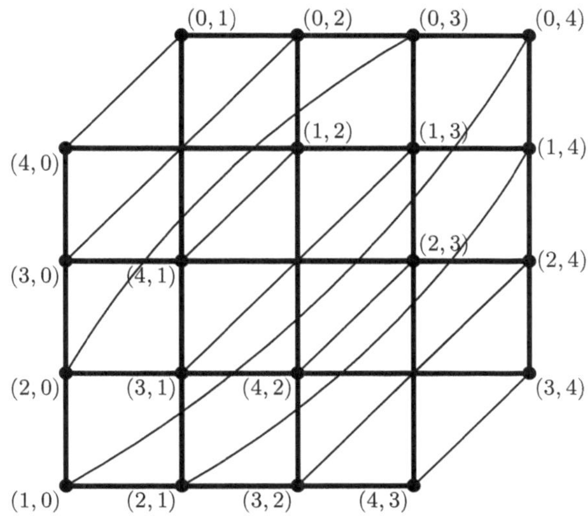


**Fig. 26** The subgraph of $T$ induced on the set $\Omega$

We refer to the logo of a recent conference WL2018, see [28], where first two diagrams are depicted in colours, while for Diagram 1 even an animation is provided. Finally, it should be mentioned that art of depiction of helpful pictures is becoming more popular in the area of Discrete Mathematics, see discussion in the concluding section.

## 24 Interactions Between Some Diagrams of $T$, Its Models and Properties

Here we present a few models of the graph $T$. The created common roof makes possible to discuss some properties of $T$ explicitly, and its links to other combinatorial structures, which were visible in less formal manner before.

**Model 1 (Master model)** Vertices of $T$ are elements of $V = \Omega \cup \mathcal{P}$. Pair

$$\{(x_1, y_1), (x_2, y_2)\} \in E(T) \iff ((x_1 = x_2) \vee (y_1 = y_2) \vee (x_1 = y_2 \wedge y_1 = x_2)).$$

Pair $\{(a, b), p\}$, $p \in \mathcal{P}$ is an edge in $T$ if and only if $\{a, b\} \in p$. Graph $T$ is invariant with respect to $A_5$, which naturally acts on $V(T)$. Group $\mathbb{Z}_2$ is generated by the involution, which transposes $(a, b)$ and $(b, a)$ from $\Omega$ and leaves each $p \in \mathcal{P}$ on the place. Thus, $\mathrm{Aut}(T) \geq A_5 \times \mathbb{Z}_2$. The ordered triple $x = ((0, 1), (1, 2), (2, 3))$ provides an example of a *base* for $\mathrm{Aut}(T)$. In other words, if $g \in \mathrm{Aut}(T)$ leaves all components of $x$ on the place, then $g = id$, that is the identity permutation.

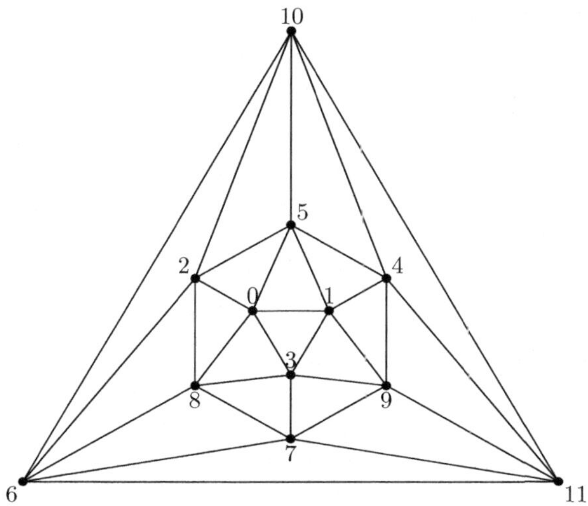


**Fig. 27** The icosahedron graph

Now, counting all possible images of $x$ with respect to $\mathrm{Aut}(T)$, we justify that their total amount is at most 120. This finally implies that $\mathrm{Aut}(T) = A_5 \times \mathbb{Z}_2$ with actions of $A_5$ and $\mathbb{Z}_2$, explained above.

**Model 2 (Dodecahedral model)** Here we start from the graph $\Delta$, skeleton of the dodecahedron, not relying on coordinatization of its vertices by pairs from $[0, 4]$. Let $V' = V(\Delta)$ be the vertex set of $\Delta$, $V''$ set of six Petrie polygons in $\Delta$, cf. Sect. 11. Put $V = V' \cup V''$.

Two vertices from $V'$ are adjacent in the constructed $T''$, if they are at distance 3 or 5 in $\Delta$. Petrie polygon from $V''$ is adjacent to vertex in $V'$, if it contains this vertex. To prove isomorphism between $T' = T$ (from Model 1) with $T''$, it is enough to establish a bijection between six Petrie polygons and six pentagons as in Fig. 1. For this purpose, we recall that subgraph of $\Delta$, induced by the complement to Petrie polygon, is isomorphic to $2 \circ C_5$.

**Model 3 (Icosahedral model)** We start from explicit diagram of the graph $Ic$, the skeleton of the icosahedron, see Fig. 27. Recall that the alternating group $A_5$ acts as subgroup of index 2 in $\mathrm{Aut}(Ic)$, it is generated by permutations $h_1 = (1, 2, 3, 5, 8)(4, 6, 9, 10, 11)$, $h_2 = (0, 1, 3)(2, 4, 7)(5, 9, 8)(6, 10, 11)$. Permutation $h_3 = (0, 11)(1, 6)(2, 9)(3, 10)(4, 8)(5, 7)$ is generating subgroup $\mathbb{Z}_2$ in presentation of $\mathrm{Aut}(Ic) = A_5 \times \mathbb{Z}_2$, namely, it transposes antipodal vertices of $Ic$. The vertex set $V$ of the coming model $T'''$ again splits to subsets $V'$ and $V''$. $V'$ is set of triangles of $Ic$ (thus dual to vertex set of dodecahedron), while $V''$ consists of six antipodal pairs of $Ic$, visible from $h_3$ (again, this set is dual to the set of Petrie polygons in previous model). Clearly, $V''$ is a coclique of size 6 in $T'''$, two triangles from $V'$ are adjacent if they do not have common vertices, antipodal pair is adjacent to a triangle, if they intersect in one vertex of triangle. For the readers convenience,

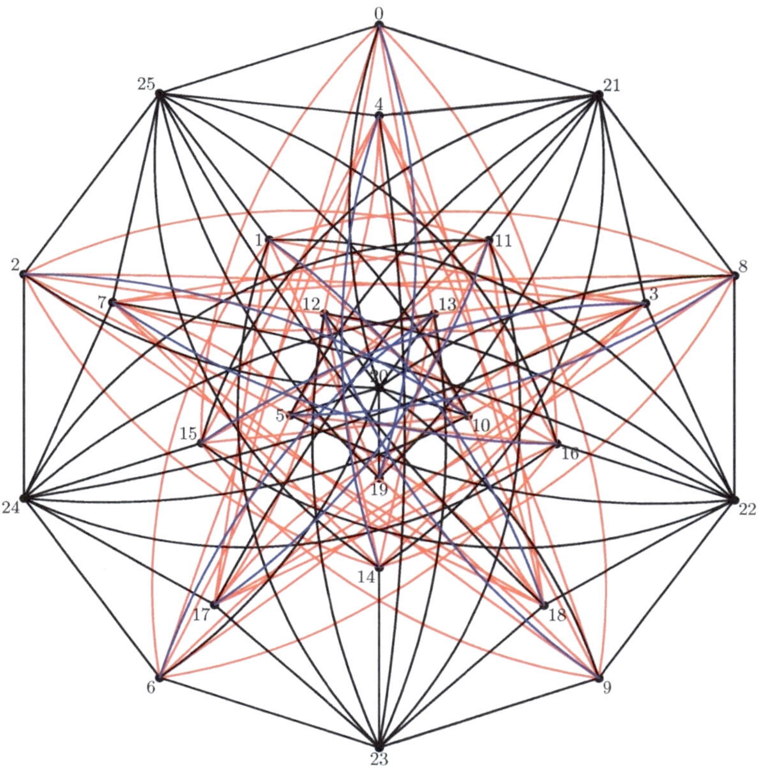


**Fig. 28** The PRT-graph with COCO labelling

triangle {0, 1, 3} is adjacent to {0, 11}, {2, 5, 10}, {4, 5, 10}, {6, 10, 11} (we show just representatives of 2-orbits).

Isomorphism of models $T'''$ and $T''$ follows immediately from the classical duality between $\Delta$ and $Ic$.

**Model 4** Here we create a diagram of our master copy of the graph $T$ for the first time. It will be denoted by $T_0$ a few times below. It is obtained from any of the previous descriptions of $T$, using labelling of its vertex set $V(T)$ by the elements of set [0, 25]. The used labelling was presented earlier in Table 1 and Table 8. The resulted diagram appears below in Fig. 28. It is the same diagram like Diagram 2 in Fig. 25. However, this time we use uniform notation for $V(T)$ by elements of [0, 25]. To this notation we will refer in what follows.

**Model 5 (Switching from Paley graph** $Pal(25)$**)** In fact, we are giving two models $5'$ and $5''$. In both cases, we first proceed backwards. We start from the master copy $T_0$ of $T$, as it is given in the last diagram. We select vertex $x \in V(T_0)$, consider partition of $V(T_0)$ into the sets $\{x\} \cup T_2(x)$ and $T_1(x)$. Recall that here $T_1(x)$ and $T_2(x)$ are the sets of neighbours and non-neighbours of $x$ in $T = T_0$, respectively. After switching with respect to this partition we get a graph, which consists of an isolated vertex $x$ and regular graph on 25 vertices of valency 12. According to the classical Seidel's theory of two-graphs [99], the latter graph is an SRG with parameters of $Pal(25)$. In fact, this graph is isomorphic to $Pal(25)$, due to the properties of the two-graph $D$. It remains to mention that we were working with vertices $x = 0$ and $x = 25$, getting Paley graphs $P'$ and $P''$, respectively. A helpful feature of $Pal(25)$ is that it contains exactly 15 cliques of size 5, which form decomposition of the edge set to three spanning subgraphs of the form $5 \circ K_5$. This is why both graphs $P'$ and $P''$ are presented by lists of corresponding 5-cliques.

First, we present the results for $x = 0$. We list the partition of $P' = Pal(25)$ with isolated vertex 0 into 15 cliques of size 5, see Table 10. After that, we present four diagrams with decomposition of the canonical copy $T_0$ of graph $T$ into parts, which are relevant to switching between $P'$ and $T_0$. See Figs. 29, 30, 31 and 32. In similar manner, we present the results for $x = 25$. We get partition into 5-cliques as in Table 11. (Again, each column presents one of the spanning subgraphs $5 \circ K_5$.) After that, again we are able to present four diagrams, related to the decomposition of $T_0$ with respect to vertex 25, see Figs. 33, 34, 35 and 36.

**Table 10** The list of 5-cliques in $P'$

| | | |
|---|---|---|
| {1, 3, 13, 20, 22} | {1, 6, 15, 21, 23} | {1, 8, 10, 12, 16} |
| {2, 10, 19, 21, 24} | {2, 3, 14, 16, 17} | {2, 5, 11, 13, 15} |
| {4, 5, 6, 12, 17} | {4, 9, 10, 13, 18} | {3, 6, 18, 24, 25} |
| {7, 8, 14, 15, 18} | {5, 8, 19, 22, 25} | {4, 14, 19, 20, 23} |
| {9, 11, 16, 23, 25} | {7, 11, 12, 20, 24} | {7, 9, 17, 21, 22} |

**Fig. 29** Ten edges incident to 0

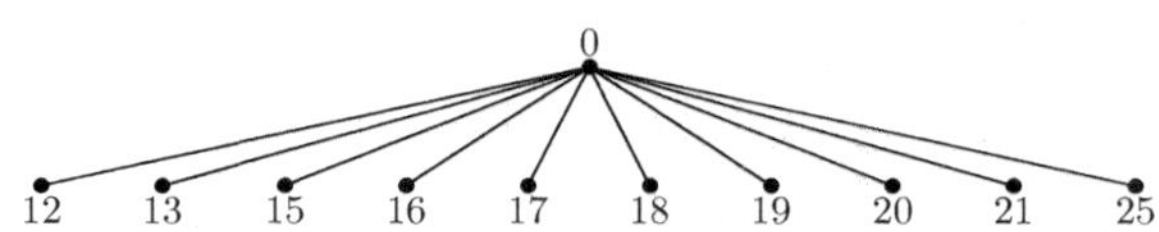


**Fig. 30** Fifteen edges between the neighbours of 0

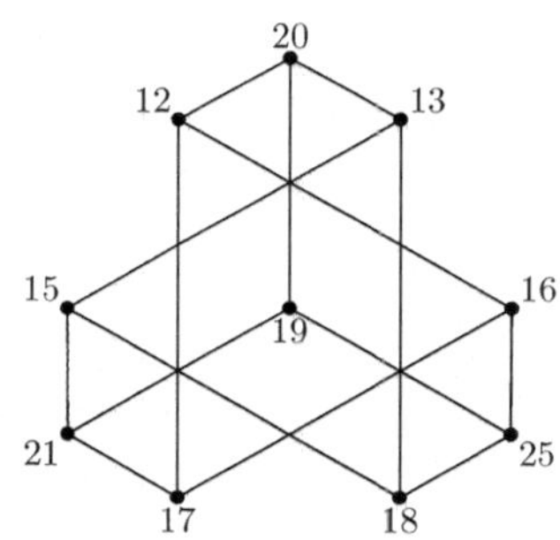

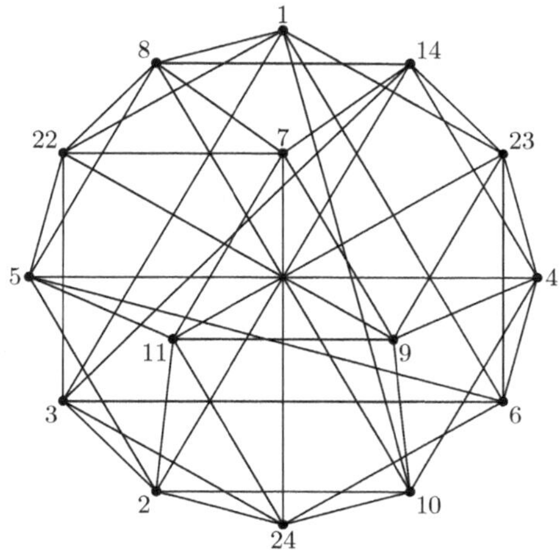


**Fig. 31** Forty-five edges between the non-neighbours of 0

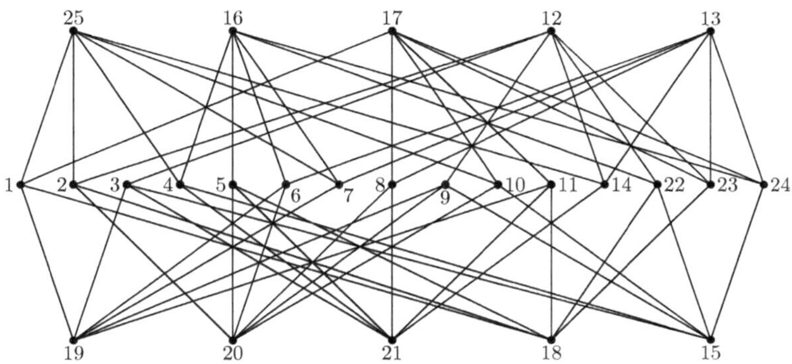


**Fig. 32** Sixty edges between the neighbours and non-neighbours of 0

**Table 11** The list of 5-cliques in $P''$

| | | |
|---|---|---|
| {0, 3, 6, 18, 24} | {0, 5, 8, 19, 22} | {0, 9, 11, 16, 23} |
| {1, 5, 10, 11, 21} | {1, 9, 12, 18, 20} | {1, 13, 15, 19, 24} |
| {2, 9, 14, 15, 22} | {2, 6, 10, 13, 23} | {2, 8, 17, 18, 21} |
| {4, 8, 13, 16, 20} | {3, 7, 15, 16, 21} | {3, 4, 10, 12, 22} |
| {7, 12, 17, 19, 23} | {4, 11, 14, 17, 24} | {5, 6, 7, 14, 20} |

**Fig. 33** Ten edges incident to 25

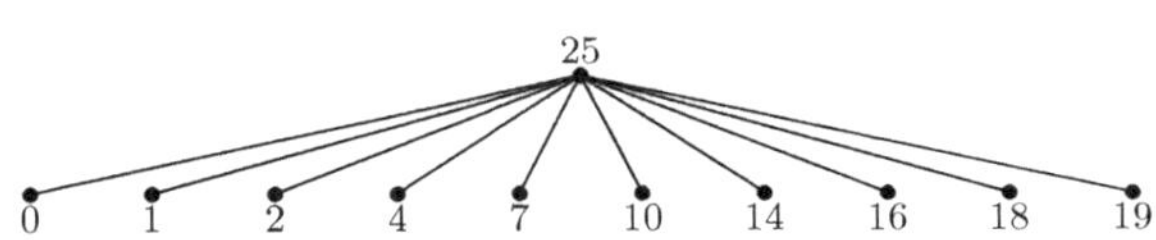

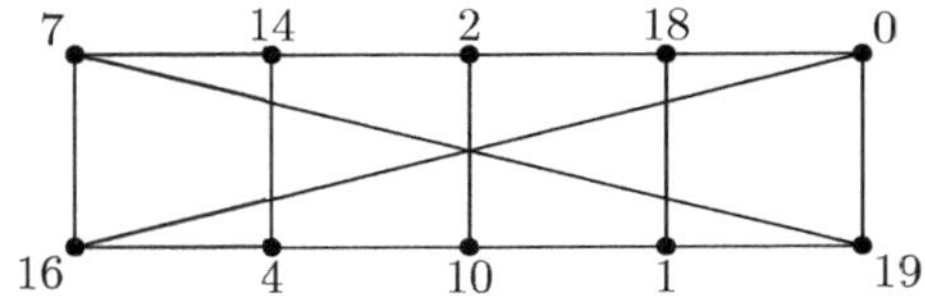


**Fig. 34** Fifteen edges between the neighbours of 25

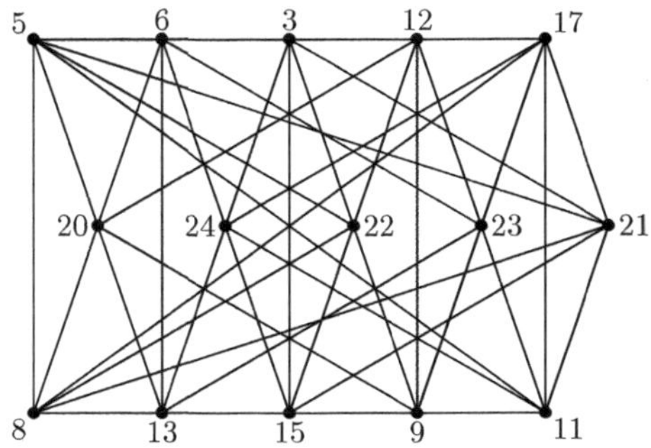


**Fig. 35** Forty-five edges between the non-neighbours of 25

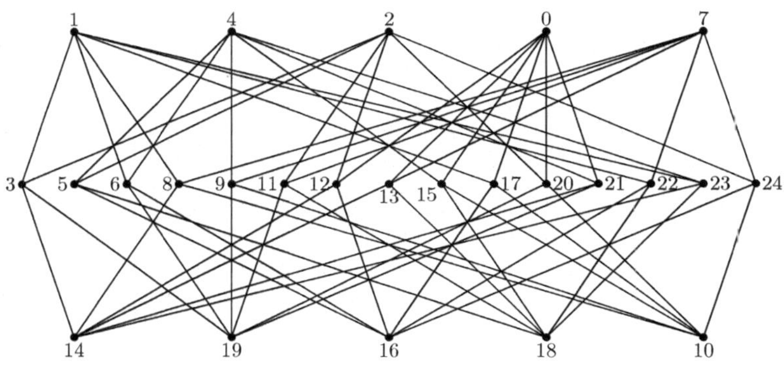


**Fig. 36** Sixty edges between the neighbours and non-neighbours of 25

**Proposition 34** *Four models $T'$, $T''$, $T'''$, $T^{iv}$ of a regular graph of valency 10 on 26 vertices are isomorphic to the SRG T with the automorphism group of order 120.*

***Proof*** Using GAP, the claim was confirmed with the aid of a computer. Second way is (relying on the catalogue of Paulus graphs in [13]) to detect a suitable subgroup of clear symmetries of considered model and to conclude that it can be embedded only into the largest group of order 120. A more direct computer-free way is also pretty clear. Indeed, three first models are using graph $\Delta$, presented in different ways. To compare model $T^{iv}$ with, say, master model $T'$, we may consider subgraphs of neighbours and non-neighbours of vertex $\infty$ in $T^{iv}$ and to detect (even visually) isomorphic subgraphs in $T'$. □

Below, we justify and clarify some structural data about $T$, which is available from [13]. As a rule, the master model $T'$ is crucial for our considerations, though other models might be also of help.

**Proposition 35** *Graph T has 13 maximal cocliques of size 6, which are split into two orbits of length 1 and 12.*

***Proof*** We rely on $T'$. Subset $\mathcal{P} = \{a, b, c, d, e, f\}$ of $V(T)$ induces a 6-coclique. Clearly, according to construction of $T'$, it is invariant with respect to $\mathrm{Aut}(T')$, thus forms an orbit of length 1. To construct one more representative of a 6-coclique in $T$, let us start from $f \in \mathcal{P}$, $f = C(0, 1, 2, 3, 4)[1, 4]$. This is a cycle $C_5$. Construct the complement $\overline{C_5}$. Consider all its arcs in, say, clockwise direction. Then we result in five vertices from $\Omega$, namely, sequence $(17, 5, 13, 3, 9)$ (we use the labelling from Table 1). Check that we get a 6-coclique. Working in counterclockwise order, we obtain another coclique which is containing $f$. Clearly, both cocliques have the same stabilizer in $\mathrm{Aut}(T)$ of order 10, isomorphic to $D_5$ and are interchanged by (normal) involution from $\mathrm{Aut}(T)$. Thus we reach second orbit of length 12. Easy, handy (though a bit routine) inspection shows that all 6-cocliques are expired. □

We will need information about the conjugacy classes of subgroups of $\mathrm{Aut}(T)$. Though it was obtained with the aid of GAP, finally it can be easily justified, relying on information from Sect. 6 (a few times a concept of a subdirect product of two groups is also exploited).

**Proposition 36** *The group* $\mathrm{Aut}(T)$ *has 22 conjugacy classes of subgroups. Main data related to them are presented in Supplement, Table 17.* □

Below, as above, maximal means with respect to inclusion $\subseteq$.

**Proposition 37** *Graph T has 10 maximal cliques of size 4, which form one orbit with respect to* $\mathrm{Aut}(T)$ *and 90 maximal cliques of size 3, which split into two orbits of length 60 and 30.*

One representative of a 4-clique is $X_{4,1} = \{0, 12, 16, 17\}$. In "natural" language of [0,4], it consists of four pairs, having common first element. Clearly, this is a clique; there is no other vertex in $V$, which is adjacent to all vertices in $X_{4,1}$. There are 10 combinatorially equivalent ways to construct such a 4-coclique. In algebraic terms, its stabilizer is group #12, which has index 10 in $\mathrm{Aut}(T)$.

Next one is $X_{3,1} = \{0, 12, 23\}$, which consists of a pentagon and two its arcs, stemming from the same vertex. There are $2 \cdot 6 \cdot 5 = 60$ such maximal cliques, their stabilizer is isomorphic to subgroup #4.

$X_{3,2} = \{0, 19, 20\}$ consists from a pentagon and two opposite arcs, forming its edge. This is again a maximal clique (inspect remaining eight arcs in the relevant pentagon). Combinatorial counting leads to $6 \cdot 5 = 30$ such cliques, the stabilizer is subgroup #7.

Altogether, we have 90 3-cliques. According to [13], we are already done. The same was confirmed by GAP. An enough interested reader may confirm that the list is complete by reasonably easy hand computations. □

**Proposition 38** *Besides 13 maximal cocliques of size 6, the graph T has also 12 maximal cocliques of size 5 and 210 maximal cocliques of size 4.*

***Proof*** The proof is similar to the proofs of the previous propositions however, is more routine, because the amount of orbits of $\mathrm{Aut}(T)$ on the non-edges is larger than the amount on edges. We again present representative of each orbit (this time on cocliques), briefly discuss the combinatorial interpretation relevant to it, and structure of the stabilizer. As before, provided results agree with the data in [13], our own computations, and, definitely, may be confirmed without the use of a computer.

There is just one orbit of 5-cocliques with representative $Y_{5,1} = \{0, 1, 2, 4, 7\}$. It corresponds to five arcs of a pentagon $p \in \mathcal{P}$, which form a directed cycle of length 5, namely, $C(0, 1, 2, 3, 4)[1]$. Clearly, we have $2 \cdot 6 = 12$ such 5-cocliques, their stabilizer is group #9.

Now we start discussion of 7 orbits of $\mathrm{Aut}(T)$ on its 4-cocliques.

$Y_{4,1} = \{0, 1, 4, 11\}$ consists of three arcs of a directed triangle and one more arc, disjoint from the triangle. There are $2 \cdot \binom{5}{3} = 20$ such cocliques, stabilizer #11.

$Y_{4,2} = \{0, 1, 4, 24\}$ consists of one pentagon and suitable three arcs of its complement. There are $6 \cdot 5 \cdot 2 = 60$ possibilities to get such a coclique with the set stabilizer #3.

$Y_{4,3} = \{0, 1, 5, 7\}$ consists of four arcs of a directed quadrangle. There are $5 \cdot 3 \cdot 2 = 30$ ways of its selection with stabilizer #6.

$Y_{4,4} = \{0, 1, 11, 14\}$ consists of 3 consecutive arcs of a directed triangle and one more arc, disjoint from the triangle. There are $2 \cdot \binom{5}{3} = 20$ such cocliques, stabilizer #11.

$X_{4,5} = \{0, 2, 22, 25\}$. Here, first, select arbitrarily two pentagons, they have two common edges $\{0, 3\}$, $\{1, 2\}$. Select two arcs between the ends of these edges. We get $2 \cdot \binom{6}{2} = 30$ such cocliques, stabilizer #8.

$X_{4,6} = \{0, 5, 24, 25\}$: the same description like for $X_{4,5}$.

$X_{4,7} = \{0, 22, 24, 25\}$. Here, first, select arbitrary directed edge, say $(0, 1)$ and find all three pentagons, which do not contain $\{0, 1\}$. There are $5 \cdot 4 = 20$ options, stabilizer #11.

Thus, altogether we have $20 + 60 + 30 + 20 + 30 + 30 + 20 = 210$ cocliques of size 4. According to the explained pattern, the proof is finished. □

*Remark 4* Note that cocliques $Y_{4,1}$ and $Y_{4,4}$ as well as $X_{4,5}$ and $X_{4,6}$, cannot be distinguished by combinatorial or algebraic arguments. Thus, we face an occurrence of combinatorial chirality, the concept borrowed from mathematical chemistry. It is used to reflect similarity between right and left, see, e.g. [71].

Using GAP we also enumerated all induced cycles in graph $T$. General results are presented in Table 12. Here for each possible length $\ell$ of cycles, we provide their

**Table 12** Induced cycles in graph $T$

| $\ell$ | Total amount | Amount of orbits | Multiset of orbits |
|---|---|---|---|
| 3 | 90 | 2 | 30, 60 |
| 4 | 420 | 10 | $30^6, 60^4$ |
| 5 | 1080 | 20 | $30^4, 60^{16}$ |
| 6 | 980 | 22 | $20, 30^{10}, 60^{11}$ |
| 7 | 570 | 10 | $30, 60^9$ |
| 8 | 840 | 15 | $30^2, 60^{13}$ |
| 9 | 0 | 0 | ∅ |
| 10 | 36 | 3 | $12^3$ |

total number, amount of orbits with respect to subgroup $A_5$ of $\mathrm{Aut}(T)$ and multiset of lengths of orbits.

Clearly, cycles of length 3 coincide with subgraphs $K_3$, also, absence of cocliques of size larger than 6 implies the absence of induced cycles of length $\ell = 13$. In fact, there are no cycles of length $\ell = 11$ or $\ell = 12$.

One of the motivations of this activity was to check the existence of a combinatorial map on graph $T$ with high symmetry. Recall that a *combinatorial map* $\mathcal{M}$ on graph $T$ is a collection of induced cycles in $T$, such that each edge of $T$ appears exactly in two cycles. Note that with this definition we disregard possibilities of the embedding of such a map into a suitable surface. The group $\mathrm{Aut}(\mathcal{M})$ consists of all permutations in $\mathrm{Aut}(T)$, which preserve $\mathcal{M}$ as the entire set.

**Proposition 39** *There is no combinatorial map $\mathcal{M}$ on graph $T$, which is invariant with respect to subgroup $A_5$ of* $\mathrm{Aut}(T)$.

***Proof*** Graph $T$ contains $\frac{1}{2} \cdot 26 \cdot 10 = 130$ edges. Thus $\mathcal{M}$ should collectively contain 260 edges. If $\mathcal{M}$ exists, then it is formed by some orbits of $A_5$ on the induced cycles in $T$. Easy inspection of Table 12 shows that the number of edges in cycles of arbitrary orbit is a multiple of 3. Thus the same will be true for $\mathcal{M}$, contradicting to the fact that 260 is not divisible by 3. □

We were not trying to find a map with smaller symmetry, because such an object would be less "nice", while its search requires more efforts. In fact, for each orbit of induced cycles its representative was determined. Below we consider such an information just for the cycles of length 10, because it will be used in the next section. The representatives of orbits are

$$C_{10,1} = C(0, 13, 7, 11, 5, 4, 10, 1, 3, 12)[1, 9],$$

$$C_{10,2} = C(0, 15, 10, 1, 6, 5, 11, 7, 14, 12)[1, 9],$$

$$C_{10,3} = C(0, 21, 8, 22, 9, 25, 6, 24, 2, 20)[1, 9].$$

Cycle $C_{10,1}$ can be described with the aid of an auxiliary directed pentagon $C(0, 1, 3, 2, 4)[1]$. Take its 5-arcs, put on distance 2 in desired 10-gon and fill remaining 5 places by suitably defined elements from $\Omega$.

Cycle $C_{10,2}$ can be obtained similarly from $C(0, 1, 3, 2, 4)[1]$. Each time we have $6 \cdot 2 = 12$ options to get all 10-gons in the orbit. These two orbits are chiral to each other.

Cycle $C_{10,3}$ can be obtained from the pentagon #23 in $\mathcal{P}$. Order cyclically remaining 5 cycles in $\mathcal{P}$ with respect to the rotation group of the selected pentagon. Fill remaining 5 positions in 10-gon by suitable elements from $\Omega$. Again, there are 12 possibilities to construct 10-gon in this orbit.

## 25 Hoffman Cocliques and Association Schemes on 260 Copies of Graph $T$

We restart with group $PSL(2, 25)$ in its 2-transitive action of degree 26 on $V = PG(1, 25)$. The group $PSL(2, 25)$ has two conjugacy classes of subgroups $A_5$ acting intransitively with two orbits of length 20 and 6. The size of each class is equal to 65. As we are aware, from each copy of group $(A_5, V)$, we get two copies of graphs, isomorphic to $T$. These copies are switching equivalent. Thus, altogether we obtain 130 copies of graph $T$, which belong to the same two-graph $\mathcal{D}$. Let us denote the set of these graphs by $\Psi_1$. Similarly, we get the set $\Psi_2$ of other 130 copies of $T$, which belong to two-graph $\bar{\mathcal{D}}$. Denoting by $\Psi = \Psi_1 \cup \Psi_2$ the union of disjoint sets $\Psi_1$ and $\Psi_2$, we reach possibility to consider intransitive action $(PSL(2, 25), \Psi)$ of degree 260 with 2 orbits of length 130. Let $Y = (\Psi, 2 - orb(PSL(2, 25), \Psi))$ be the Schurian coherent configuration with two fibres of size 130. We constructed this coherent configuration and investigated it, using COCO and GAP.

**Proposition 40** *The following holds:*

*(i) The coherent configuration $Y$ has rank 28.*

*(ii) The 2-orbits $(T_0, T_i)$ with first reference copy $T_0$ of $T$ are described with the aid of the following data, presented in Table 13 in traditional* COCO *notation. Here $T_0$ is our master copy of graph $T$, the graphs $T_i$, corresponding to label $i$, are described in file, available from the authors.*

*(iii) The next 14 2-orbits with $(T_{130}, T_i)$ as a representative, are introduced in the similar manner. (Note that two-graph $D$ and $\bar{D}$ are isomorphic and invariant with respect to the same group $PSL(2, 25)$. The information about reference graph $T_{130}$ is provided in Table 14.*

*(iv) The followings are pairs of antisymmetric 2-orbits:* $(2, 5)$, $(9, 14)$, $(10, 15)$, $(11, 16)$, $(12, 18)$, $(13, 17)$, $(24, 26)$. *All other 2-orbits are symmetric.*

*(v) The group $(PSL(2, 25), \Psi)$ is 2-closed.*

***Proof*** As was explained, this proposition essentially depends on the use of computer algebra packages. Nevertheless, it is possible to explain a part of obtained results on a human-friendly level (in the sense of [77]). Part of such explanations are briefly outlined below. First nine 2-orbits $R_0 - R_8$ are inducing pairs of isomorphic graphs, which belong to the same two-graph $D$. This means that for each representative $(T_0, T_i) \in R_i, 0 \leq i \leq 8$, the graph $T_i$ is switching equivalent to our canonical copy $T_0$ of graph $T$. Switching between two graphs requires a suitable equitable partition with two cells. For consideration of equitable partitions see, e.g. [74].

To continue proof, we need a couple of helpful facts from spectral graph theory.

**Proposition 41** *Let $\Gamma$ be a connected SRG with eigenvalues $k$, $r$, $s$ and multiplicities 1, $f$, $g$, respectively. Assume that $\Gamma$ is primitive, thus $r \neq 0$, and $C$ is a coclique in $\Gamma$. Then*

$$|C| \leq \frac{v}{1 - \frac{k}{s}}$$

*with equality if and only if $\{C, \bar{C}\}$ forms an equitable partition of $V(\Gamma)$.*

Usually, this inequality is called *Hoffman ratio bound.* In case of equality, the corresponding coclique is called *Hoffman coclique.*

***Proof*** We refer to [16] for two proofs. The second one uses statistical arguments of "quadratic counting". □

Now we need to exploit the full set of parameters of $T$:

$$(v, k, \lambda, \mu, r, s, f, g) = (26, 10, 3, 4, 2, -3, 13, 12).$$

**Table 13** Half of 2-orbits of $(PSL(2, 25), \Psi)$

| $R_i$ | Subdegree | Representative $i$ | Equit. part. |
|---|---|---|---|
| $R_0$ | 1 | 0 | – |
| $R_1$ | 12 | 1 | $A_1$ |
| $R_2$ | 12 | 2 | $A_3$ |
| $R_3$ | 30 | 3 | $A_4$ |
| $R_4$ | 30 | 4 | $A_4$ |
| $R_5$ | 12 | 6 | $A_3$ |
| $R_6$ | 20 | 7 | $A_2$ |
| $R_7$ | 12 | 25 | $A_3$ |
| $R_8$ | 1 | 71 | $A_1$ |
| $R_9$ | 30 | 130 | |
| $R_{10}$ | 30 | 131 | |
| $R_{11}$ | 30 | 132 | |
| $R_{12}$ | 20 | 135 | |
| $R_{13}$ | 20 | 137 | |

**Table 14** Second half of 2-orbits of $(PSL(2, 25), \Psi)$

| $R_i$ | Subdegree | Representative $i$ |
|---|---|---|
| $R_{14}$ | 30 | 0 |
| $R_{15}$ | 30 | 1 |
| $R_{16}$ | 30 | 2 |
| $R_{17}$ | 20 | 6 |
| $R_{18}$ | 20 | 13 |
| $R_{19}$ | 1 | 130 |
| $R_{20}$ | 30 | 131 |
| $R_{21}$ | 30 | 132 |
| $R_{22}$ | 12 | 133 |
| $R_{23}$ | 12 | 137 |
| $R_{24}$ | 12 | 142 |
| $R_{25}$ | 20 | 146 |
| $R_{26}$ | 12 | 152 |
| $R_{27}$ | 1 | 166 |

According to the ratio bound, we get that

$$|C| \leq \frac{26}{1 - \left(\frac{10}{-3}\right)} = \frac{26}{\frac{13}{3}} = 6.$$

Thus the 6-cocliques in $T$ are exactly Hoffman cocliques. As we know, there are exactly 13 such cocliques in $T$, which form two orbits of length 1 and 12.

On next stage let us consider criterion for the case when switching leads to an SRG with the same parameters.

**Proposition 42** *Let $\Gamma$ be an SRG with parameters $(v, k, \lambda, \mu)$ associated with a regular two-graph. Assume that the switching is done with respect to the partition $\{S, \bar{S}\}$ of $V(\Gamma)$. Then $\Gamma$ is switched to an SRG $\Gamma'$ with the same parameters if and only if $\{S, \bar{S}\}$ is an equitable partition such that each vertex in $\bar{S}$ is adjacent to exactly half of vertices in $S$.*

***Proof*** This is Proposition 10.3.3. in [15], which follows from some earlier arguments in that text. □

Now, we come back to Proposition 40. It is easy to enumerate all square matrices of order 2, which correspond to equitable partitions of $T$, implying via switching the same SRG, up to isomorphism. These are matrices

$$A_1 = \begin{pmatrix} 0 & 10 \\ 3 & 7 \end{pmatrix} \text{ with } |S| = 6, \qquad A_2 = \begin{pmatrix} 1 & 9 \\ 4 & 6 \end{pmatrix} \text{ with } |S| = 8,$$

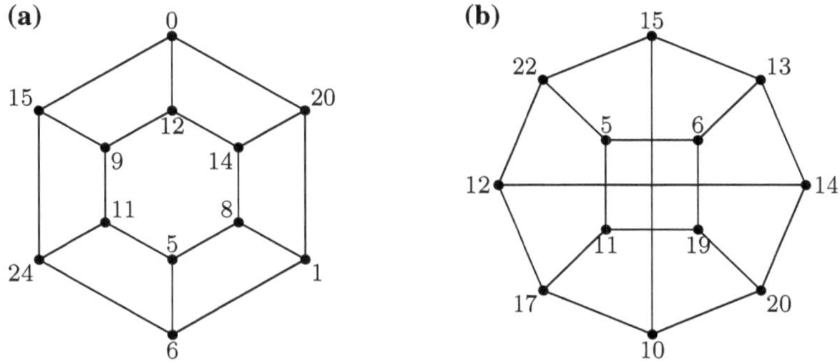


**Fig. 37** Graphs induced on $S$ by $T_3$ and $T_4$

$$A_3 = \begin{pmatrix} 2 & 8 \\ 5 & 5 \end{pmatrix} \text{ with } |S| = 10 \quad \text{and} \quad A_4 = \begin{pmatrix} 3 & 7 \\ 6 & 4 \end{pmatrix} \text{ with } |S| = 12.$$

The case of $A_1$ clearly corresponds to the Hoffman cocliques of size 6. In terms of our coherent configuration $Y$, it is represented by relations $R_8$ and $R_1$ of valency 1 and 12, respectively. The case of $A_2$ corresponds to the induced subgraph $4 \circ K_2$ on the set $S$. As it is shown in Table 13, case $A_3$ appears most frequently, that is 3 times. For all three appearances, the induced subgraph on $S$ is a 10-cycle $C_{10}$. For each of corresponding basic relations of $Y$, the stabilizer of reference copy $(T_0, T_i)$ acts regularly on the set $S$, thus each time the valency of this relation is equal to 12. It is interesting to understand how these 2-orbits $R_2$, $R_5$ and $R_7$ might be discriminated from each other. The case of relation $R_7$ is quite easy. Here the stabilizer of an induced cycle on $S$ has orbits of lengths $1, 5, 10, 10$ on the vertex set $V(T_0)$. More sophisticated is proceeding of the pair of antisymmetric 2-orbits $R_2$ and $R_5$ with representatives $T_2$ and $T_6$. In both cases, lengths of the orbits of the corresponding stabilizers are $1, 5, 5, 5, 10$ and we have to compare the induced cycles $C_{10}$ on $S$ versus the induced subgraphs on orbits of length 5.

Matrix $A_4$ corresponds to relations $R_3$ and $R_4$ both of valency 30, with representatives $T_3$ and $T_4$, respectively. For relation $R_3$ with the representative $T_3$, the induced cubic graph of size 12 is depicted in Fig. 37a. The induced subgraph for $R_4$, obtained from $T_4$ and depicted in Fig. 37b is less symmetric. Note that in [40] all small regular (and, in particular, cubic) graphs were constructively enumerated and their properties were described. Thus, according to this catalogue, our two cubic graphs on 12 vertices have numbers 59 and 78 with groups of order 24 and 8, respectively.

The basic relations $R_{19} - R_{27}$ on the fibre $\Psi_2$ of $Y$ may be explained in the same manner. We omit this information in order to save space and also because of the sketchy style of presentation in this section.

A real challenge appears from the attempts to explain 5 pairs of antisymmetric relations from fibre $\Psi_1$ to $\Psi_2$ and vice versa. Here copies of graph $T$ belong to diverse two-graphs $D$ and $\bar{D}$ and thus can not be obtained by switching. Thus we investigated for each pair $(T_0, T_i)$ the intersection graph $T_0 \cap T_i$, where by intersection we mean intersection of edge sets of $T_0$ and $T_i$. (In fact, similar way is exploited when we speak about the switching with respect to $\{S, \bar{S}\}$. Just here the obtained picture is explained in more clear terms: the intersection consists of the induced subgraphs on $S$ and $\bar{S}$, while knowledge of induced graphs on $S$ was enough in almost all cases.)

It turns out that for all five 2-orbits from $\Psi_1$ to $\Psi_2$ the intersection graph appears to be a suitable regular graph of valency 4 on 26 vertices. The automorphism groups of the intersection graphs and their orbits on the vertex and edge set of $T_0 \cap T_i$ were determined. Unfortunately, the analysis of these graphs in a human-friendly style would be less successful. This is why we are not attempting to discuss such results below. Clearly, the intersection graphs $T_{130} \cap T_i$ for five paired 2-orbits from $\Psi_2$ to $\Psi_1$ are isomorphic to corresponding intersection graphs above. We also found the three traditional groups Aut$(Y)$, CAut$(Y)$ and AAut$(Y)$. As was mentioned, Aut$(Y) \cong PSL(2, 25)$. The group CAut$(Y)$ acts transitively on the set $\Psi$ and has structure $E_4 \times (PSL(2, 25) \rtimes \mathbb{Z}_2) \rtimes \mathbb{Z}_2$. As group of degree 260, it has rank 10. The quotient group CAut$(Y)$/Aut$(Y)$ has order 16 and is isomorphic to the group $\mathbb{Z}_2 \times D_4$.

The group AAut$(Y)$ has order 32 and is isomorphic to $E_4 \times D_4$. Therefore, the group CAut$(Y)$/Aut$(Y)$ is subgroup of index 2 in the group AAut$(Y)$. Thus, in principle, there may exist non-algebraic mergings of $Y$ and they may be non-Schurian. Note that some of them may be not association schemes.

According to COCO, there exist 52 non-trivial mergings of $Y$, which are association schemes, 18 of them are symmetric. Ranks of these mergings vary from 3 to 14. Striking information is that the major part of the mergings (36 from 52) are non-Schurian. Some of them are isomorphic. The smallest rank of non-Schurian mergings is equal to 7. Due to the space and time limitations in the preparation of the current text, we do not provide more details. It is expected to submit results of a more accurate analysis in a forthcoming publication.

## 26 A Model for the Pair of Classical Generalized Quadrangles of Order 5

In this section, we briefly consider the pair of classical generalized quadrangles of order 5, which are usually denoted by $GQ(5)$. Recall that there exists two such structures, one is usually denoted by $W(5)$, while second one by $H(4, q)$. Both structures are *dual* to each other, that is if one is regarded as generalized quadrangle coming from SRG $X$, then other is coming from SRG $Y$, which is the line graph of initial generalized quadrangle, while $X$ is its point graph.

We refer to [95] as a classical source about the generalized quadrangles. Recall that classical generalized quadrangle of order $q$, $q$ is a prime-power, are defined in terms of finite geometries: $W(q)$ is naturally embedded into $PG(3, q)$, while $H(4, q)$ into $PG(4, q)$. If $q$ is even, then $W(q) \cong H(4, q)$, otherwise the structures are not isomorphic. The essential feature of $W(q)$ is that it allows *spreads*, that is partition of the point set into $q^2+1$ disjoint lines, each of size $q+1$. Because the structure $H(4,q)$ is dual to $W(q)$, the automorphism groups of both structures (which are isomorphic to the groups of point and line graphs) are isomorphic as abstract groups. Traditionally, one writes that $\mathrm{Aut}(W(p)) \cong PSp(4,p) : \mathbb{Z}_2$, while $\mathrm{Aut}(H(4,p)) \cong P\Omega(5,p) : \mathbb{Z}_2$, here $p$ is an odd prime. This is related to the classical isomorphism between the finite simple groups $PSp(4,p)$ and $P\Omega(5,p)$.

In what follows, we mainly will be concentrated on the pair of two $GQ(5)$, though many facts can be extended for arbitrary odd prime $p > 5$. It is significant to stress, that all $GQ(q)$ are classified for $q = 2, 3, 4$. Namely, there exists exactly two $GQ(3)$, while $GQ(2)$ and $GQ(4)$ are unique. Thus, case $q = 5$ is the smallest one, when the classification of $GQ(q)$ is not finished: this slightly motivates our presentation below.

Considering Atlas [3], one can observe that the group $PSp(4,5)$ contains two isomorphism classes of maximal subgroups, isomorphic to $PSL(2,25)$. In what follows, we will shed a new light on these classical embeddings of groups. Our presentation in current paper has slight methodological intersection with [78, 83], though we do not rely on any results, considered in these sources.

First, we start from two general models. Let $p$ be an odd prime, $M(p)$ the Miquelian plane, which is $S(3, p+1, p^2+1)$. Consider incidence structure $\Sigma'$, which is defined as follows:

- points of $\Sigma'$ are points and lines of $M(p)$;
- lines of $\Sigma'$ are SRGs $\Delta$, which belong to the pair $\{D, \bar{D}\}$ of two-graphs, from which $M(p)$ was constructed, cf. Sect. 18 for the case $p = 5$, (here $\Delta$ is an SRG with parameters of the Paley graph $Pal(p^2)$);
- each point $X$ of $M(p)$ is incident to an SRG $\Delta \in \{D, \bar{D}\}$, provided $X$ belongs in $M(p)$ to the line, which corresponds to coclique in $\Delta$;
- similarly, each line $\ell$ of $M(p)$ is incident to an SRG $\Delta \in \{D, \bar{D}\}$, provided that $\ell$ is a coclique in $\Delta$.

Let us denote by $\Sigma''$ the incidence structure, which is dual to $\Sigma'$.

**Proposition 43** *The followings hold:*

*(a) The group $PSL(2,p^2)$ acts on $\Sigma'$ (dually on $\Sigma''$) as subgroup of* $\mathrm{Aut}(\Sigma')$ *(dually* $\mathrm{Aut}(\Sigma'')$*).*
*(b) The group $PSL(2,p^2)$ has two orbits on point set and one orbit of line set of $\Sigma'$.*
*(c) Pair of incidence structures $\{\Sigma', \Sigma''\}$ is pair of dual generalized quadrangles of order $p$.*

***Proof*** Parts $(a)$ and $(b)$ follow from properties and our way of construction of $M(p)$. We present a very brief outline of arguments for $(c)$.

The structure $\Sigma'$ has $v' = v + b = p^2 + 1 + p(p^2+1) = (p+1)(p^2+1)$ points. (Here $v$ and $b$ are numbers of points and lines of the Miquelian plane $M(p)$, respectively.) The incidence graph of the pair $(\Sigma', \Sigma'')$ has valency $p+1$. Its coherent closure is non-Schurian rank 5 association scheme. The fact that we get an incidence graph of a generalized quadrangle was confirmed with the aid of a computer for $p = 3, 5$. After that for these cases, the computer-free proof was elaborated. The outline of a general proof was then obtained. □

The results in the proposition above were obtained jointly with Dennis Epple. They are subject of a paper in progress. Below, we provide brief additional information for the case $p = 5$.

First, we start from coherent configuration with two fibres of size 130 and 26, which is invariant with respect to action of $PSL(2, 25)$. It has rank 14 and its 2-closure is twice larger. There exists exactly one association scheme merging of it, which has rank 3 and is Schurian with the group of order 9360000.

After that, we start with intransitive action of the group $PSL(2, 25)$ on the three orbits of length 130, 26 and 156. This action is 2-closed, that is its automorphism group coincides with $PSL(2, 25)$. This is rank 47 action of degree 312. COCO returns 13 association schemes as mergings, all of them are symmetric. Six mergings are of rank 5, other six have rank 4.

As it was already mentioned, a more detailed analysis of these results is postponed to a future paper.

## 27 Miscellanea

Aiming to follow the announced style of the tutorial paper, which mainly concentrates on the numerous facts related, explicitly or implicitly, to the PRT-graph, we were trying to avoid in the main body of the text various deviations, not related evidently to the selected mainstream line of the text. Nevertheless, some of such deviations are briefly touched below, subject of possible interest for a curious reader.

*27.1* The DSRG $\Gamma_1$ and $\Gamma_2$, considered in Sect. 13, are mentioned on the home page of A. Brouwer [13], where reasonably full information about this class of structures is collected. It turns out that the graph $\Gamma_1$ is uniquely determined by its parameter set. The similar question for $\Gamma_2$ seems to be open, but we notice that $\Gamma_2$ is not isomorphic to its reverse, so up to isomorphisms there are known two different graphs, however, in the theory of DSRGs they are regarded as equivalent graphs.

*27.2* Below, we discuss virtually a possible *fractional diagram* of the graph $\tilde{T}$, where $\tilde{T}$ is the restriction of $T$ on the vertex set of $\Delta$. Here each vertex of $\tilde{T}$ is denoted by fraction $i/j$, where $i, j \in [0, 4]$, $i \neq j$. According to the definition of the graph $\tilde{T}$, two of its vertices $i/j$ and $k/l$ are adjacent in one of three cases: $(i = l \wedge k = j) \vee (i = k \wedge j \notin \{k, l\}) \vee (j = l \wedge i \in \{k, l\})$. This shows that the fractional colour number of graph $\tilde{T}$ (as well as of $\Delta$) is equal to 5/2. We refer to [47], Sect. 7, for helpful discussion of elements of fractional graph theory in the framework of AGT.

*27.3* The so-called *Terwilliger algebra* of the dodecahedron $\Delta$ is investigated in [87]. Recall that a DTG $\Delta$ is antipodal, and it generates a P-polynomial, though not Q-polynomial association schemes (see [7] for corresponding definitions). The authors of [87] determine the structure of the Terwilliger algebra $\mathcal{T}(\Delta)$ of $\Delta$ and claim that their result provides a counterexample to certain conjecture posed by P. Terwilliger. Note that in this case, $\mathcal{T}(\Delta)$ has dimension 76. It turns out that $\mathcal{T}(\Delta)$ is a coherent Schurian algebra of rank 76 with 6 fibres (the latter fact is not mentioned in [87]).

*27.4* There are five different ways to embed regular tetrahedron into graph $\Delta$. This fact in some portion was exploited by us. Among a number of diverse reformulations in literature, we refer to a few ones below. Symmetries of such embeddings, literally treated as transformations of the Euclidean 3D-space, are considered in [108]. Some nice visual demonstrations of the properties of the embeddings of tetrahedron into $\Delta$ are presented in the home pages of John Baez [5] and Robert W. Gray [48].

*27.5* So-called *unfoldings* of a polytope are closely related to spanning trees of its skeleton. This question is discussed on a quite general theoretical level in [19]. Concrete calculations are arranged, as an illustration, for a few regular polytopes. In the case of a dodecahedron, it is shown that number of non-equivalent unfoldings is equal to $u(\Delta) = \frac{1}{120} \cdot (120 \cdot 43200 + 15 \cdot 1440) = 43380$.

*27.6* Besides the (0, 2)-graph strictly related to $\Delta$, which was discussed above, there exists one more such regular graph on 20 vertices with valency 6. It is denoted in [14] by 6.7 (see Table 2) and has automorphism group of order 24 with 3 orbits on vertices of lengths 4, 4 and 12.

*27.7* The fact of discovery of some DTGs at Moscow by members of Moscow group of researchers, in particular, by A.A. Ivanov and late V.A. Zaichenko, still remains, in a sense, hidden for modern audience. We refer for more details to the survey [72] in Russian, and, in particular, to Sect. 7 in it.

*27.8* Relatively short consideration of some nice structures on 260 and 312 points, presented in Sects. 25 and 26, may and should be discussed in a more general context for arbitrary value of odd prime $p$ (here $p = 5$). As was mentioned, this is subject of a project in progress, jointly with D. Epple. See also some extra remarks in the next section.

*27.9* The PRT-graph was, in fact, independently discovered by Zhi Gang Sun in [109]. The description of a graph in this text is given in a bit sophisticated terms. The presented graph is invariant with respect to the dihedral group $D_5$ of order 10. Case by case analysis in paper confirms that this is an SRG with parameters (26, 10, 3, 4). We checked the correctness of the construction and presented arguments.

*27.10* A short note [45], written in a quite elementary and friendly style, describes rotation and full groups of $\Delta$ and confirms that they are isomorphic to groups $A_5$ and $A_5 \times \mathbb{Z}_2$, respectively.

*27.11* It is clear that the full group $\mathrm{Aut}(\Delta)$ of the symmetries of graph $\Delta$ is intimately related to the group of symmetries of the smallest non-trivial BIBD with parameters

(6, 3, 2). Among a number of publications related to this topic, we pay the reader's attention to the note [49].

*27.12* The graph $T$ was selected as the logo of the recent conference on AGT *"Symmetry versus regularity"*, which was held at Pilsen (Czech Republic) in July 2018. A short version of the current text is available on the home page of WL 2018. For some readers probably this seed would be more preferred in comparison with the huge bush, which appears above.

*27.13* The search for SRGs with $2p$ vertices, $p$ a prime, stems from a classical problem in permutation group theory: classification of all primitive (rank 3) groups of degree $2p$, $p$ a prime. A few SRGs are known on 10, 26 and 82 vertices such that, in addition, they are so-called *bicirculants*, that is, they admit a semi-regular automorphism of order $p$. Clearly, the Petersen graph is one of them. We refer to [89] as a significant growing point of investigations in this direction.

*27.14* One more interesting class of SRGs is those on $v = p^2 + 1$ vertices, $p$ a prime. Here are many interesting points of intersections with the current text, especially in the case, when $v$ is twice a prime-power, like it happens for $v = 10, 26, 50, 82, 122$, etc. We expect to discuss this phenomenon in further papers.

*27.15* One can ask for the smallest value of $v$, for which there exists a primitive coherent (non-Schurian) algebra of order $v$ (preferably of rank 3), while all such primitive algebras do not have vertex-transitive automorphism group. Clearly, this is the case for $v = 26$. It is easy to understand that the smallest such case is $v = 15$. Indeed, there exists a doubly regular tournament $C$ of order 15, with $\mathrm{Aut}(C)$ of order 21, which is clearly not vertex transitive, see, e.g. [93] for more details.

## 28 Concluding Discussion

The results of investigations, presented in the current paper, imply a number of reasonably challenging questions, which are discussed below.

*28.1* There are four classes of SRGs on 25 and 26 vertices, those which belong to the two-graphs, denoted by A, B, C, D in Sect. 16. We were able to investigate all SRGs, which appear in the class D, without essential use of computers. Fulfilment of a similar task for the remaining classes A, B, C might be of a definite interest.

*28.2* A computer-free confirmation of the fact that, up to isomorphism, there exist exactly four above-mentioned regular two-graphs on 26 vertices, according to our knowledge, remains open. Attention of the reader's interest to this challenging problem was one of the motivations in the creation of the current paper.

*28.3* An interesting "sporting" question for the experts in constructive enumeration of SRGs would be to determine the next parameter set (with respect to the number of vertices) of SRGs, for which there exist putative graphs, however, there are no vertex-transitive graphs.

*28.4* Let us call an SRG $X$ *quasi-vertex transitive*, if the group $\mathrm{Aut}(X)$ has exactly two orbits on the vertex set of $X$. Clearly, our graph $T$ belongs to this class. A number of such graphs with $p^2 + 1$ vertices, $p \geq 7$, will be subject of a forthcoming project, jointly with D. Epple.

*28.5* As was discussed above, the graph $T$ contains extremal, in a concretely formulated sense, Hoffman cocliques of size 6. Similar SRGs on $p^2 + 1$ vertices will be also subject of planned investigations.

*28.6* Uniqueness of inversive planes of prime order $p$ is a quite challenging question in finite geometry. The first open case is $p = 11$. We wish to hope that the links of these geometric structures with coherent configurations and association schemes, which were observed and demonstrated in the current text, might be of some help for researchers working in this area.

*28.7* Similarly, even a more challenging open question is the classification of $GQ(5)$ of order 5, up to isomorphism. Brief drafts, presented in Sects. 25 and 26, of the properties of some coherent configurations on 260 and 312 points, are intimately related to the structures $GQ(5)$, presenting some of their properties from a reasonably fresh viewpoint. Thus, expectations for further progress in this task, clearly, are quite ambitious, though, probably, not hopeless.

*28.8* Many families of non-Schurian association schemes were touched in our text on different levels of detail and rigour. Further, more detailed attention to these structures is (a quite realistic) task for future job. A particular class of such structures, which is related to the action of the groups $PSL(2, q)$ on so-called quasi-projective points (notion due to S. Reichard, though the concept goes back to R. Mathon and A. Neumaier) definitely awaits careful attention.

*28.9* We think that new bipartite-schematic property of Miquelian inversive planes, observed and demonstrated by us (mainly on the level of examples), deserves special attention, especially in the context of all inversive planes.

*28.10* Following the spirit of J.H. Redfield, a few decades ago the Burnside marks were widely used in the classical enumeration theory. Section 3.1 of [42], as well as a couple of other papers at the same collection reflect some facets of these applications. It seems that such a "romantic" line of enumerative combinatorics still has fresh potential for further developments.

*28.11* To find a clever balance between computer aided and hand computations is becoming quite significant in modern AGT. Hopefully, our paper is adding some fresh patterns to the reader, convincing that frequently machine is more skilful and reliable, having less chances for a routine mistake.

*28.12* The same issue is related to the diagrams: they reflected essential part of our arguments. The reader is welcome to decide each time: to inspect correctness of the entire visual portion of information, or simply to observe the global structure, avoiding entering to small and annoying details.

**Acknowledgements** The authors gratefully acknowledge the contribution of the Scientific Grant Agency of the Slovak Republic (VEGA) under the grant 1/0988/16.

MK gratefully recalls his communication with late Jaap Seidel and Ernie Shult, which was crucial for the shaping of his vision of the theory of two-graphs. He is also much obliged to all late and alive members of the former Moscow group for a fruitful communication. Special thanks go in this direction to Igor Faradžev, the creator of the program part of COCO.

Support of Christian Pech, Sven Reichard, as well as of Dennis Epple was very helpful and efficient. We are much obliged to Andries Brouwer, Peter Cameron and Gareth Jones for communications, which were supporting spirit of this project. We thank also Rosemary Bailey, Katie Brodhead, Willem Haemers, Josef Lauri, Mikhail Muzychuk, Andy Woldar and Yaokun Wu for helpful remarks. Personal attention of Don Thompson was crucial for the start of the project.

The initial impulse in the creation of the paper came from Sasha Ivanov, Ján Karabáš, Roman Nedela, Akihiro Munemasa, all our partners in the planning and running the event WL 2018.

Finally, the authors are pleased to thank Jozef Širáň for his attention to this paper, as well as an anonymous referee for a very enthusiastic positive evaluation.

# Supplements

See Tables 15, 16, 17 and 18

**Table 15** 130 circles in $\mathcal{M}(5)$

| | | | |
|---|---|---|---|
| {0, 1, 2, 3, 4, 5} | {0, 1, 9, 14, 19, 24} | {0, 1, 7, 12, 17, 22} | {1, 4, 16, 19, 23, 25} |
| {1, 14, 15, 17, 21, 23} | {0, 4, 10, 11, 18, 19} | {1, 3, 11, 14, 18, 20} | {0, 4, 6, 8, 12, 15} |
| {0, 8, 10, 14, 16, 17} | {1, 9, 10, 12, 16, 18} | {4, 5, 9, 14, 15, 25} | {0, 3, 6, 13, 14, 25} |
| {0, 6, 7, 18, 20, 24} | {1, 7, 11, 13, 24, 25} | {1, 3, 7, 8, 15, 16} | {4, 11, 12, 14, 16, 24} |
| {0, 2, 16, 18, 22, 25} | {0, 9, 11, 12, 23, 25} | {1, 5, 6, 17, 18, 25} | {6, 10, 12, 19, 21, 25} |
| {1, 2, 10, 11, 22, 23} | {4, 6, 9, 18, 21, 23} | {1, 2, 6, 9, 13, 15} | {4, 5, 7, 11, 21, 22} |
| {2, 3, 11, 12, 17, 21} | {0, 5, 15, 16, 23, 24} | {0, 3, 7, 10, 21, 23} | {2, 4, 13, 14, 17, 18} |
| {8, 12, 17, 18, 19, 24} | {2, 4, 7, 8, 23, 24} | {3, 13, 16, 18, 21, 24} | {1, 5, 8, 10, 21, 24} |
| {7, 14, 16, 20, 21, 25} | {3, 4, 6, 16, 17, 22} | {2, 5, 10, 14, 19, 20} | {0, 2, 8, 9, 20, 21} |
| {5, 12, 20, 22, 24, 25} | {0, 13, 15, 19, 21, 22} | {5, 7, 10, 13, 15, 18} | {2, 6, 14, 21, 22, 24} |
| {1, 6, 8, 19, 20, 22} | {7, 8, 9, 14, 18, 22} | {2, 12, 15, 18, 20, 23} | {8, 11, 15, 18, 21, 25} |
| {2, 3, 15, 19, 24, 25} | {3, 6, 7, 9, 11, 19} | {2, 5, 6, 7, 12, 16} | {3, 5, 18, 19, 22, 23} |
| {4, 7, 15, 17, 19, 20} | {6, 10, 13, 16, 20, 23} | {9, 13, 17, 22, 23, 24} | {5, 9, 16, 17, 19, 21} |
| {0, 1, 6, 11, 16, 21} | {0, 1, 10, 15, 20, 25} | {1, 4, 8, 9, 11, 17} | {0, 1, 8, 13, 18, 23} |
| {1, 4, 6, 7, 10, 14} | {0, 4, 7, 9, 13, 16} | {1, 5, 13, 14, 16, 22} | {0, 4, 14, 20, 22, 23} |
| {0, 4, 17, 21, 24, 25} | {0, 2, 7, 11, 14, 15} | {1, 5, 11, 12, 15, 19} | {2, 4, 6, 11, 20, 25} |
| {1, 3, 6, 12, 23, 24} | {3, 4, 7, 12, 18, 25} | {0, 2, 10, 12, 13, 24} | {1, 3, 9, 21, 22, 25} |
| {2, 4, 9, 12, 19, 22} | {10, 14, 18, 23, 24, 25} | {0, 3, 9, 15, 17, 18} | {0, 3, 12, 16, 19, 20} |
| {2, 3, 9, 14, 16, 23} | {0, 3, 8, 11, 22, 24} | {1, 5, 7, 9, 20, 23} | {4, 5, 8, 16, 18, 20} |
| {3, 5, 7, 14, 17, 24} | {8, 12, 16, 21, 22, 23} | {3, 4, 8, 14, 19, 21} | {7, 9, 12, 15, 21, 24} |
| {0, 5, 6, 9, 10, 22} | {1, 2, 7, 18, 19, 21} | {2, 4, 10, 15, 16, 21} | {6, 14, 15, 16, 18, 19} |
| {0, 2, 6, 17, 19, 23} | {2, 5, 13, 21, 23, 25} | {3, 5, 6, 15, 20, 21} | {7, 8, 10, 11, 12, 20} |
| {0, 5, 7, 8, 19, 25} | {2, 3, 7, 13, 20, 22} | {8, 9, 10, 15, 19, 23} | {2, 5, 9, 11, 18, 24} |
| {8, 13, 14, 15, 20, 24} | {12, 13, 15, 16, 17, 25} | {9, 13, 18, 19, 20, 25} | {11, 19, 20, 21, 23, 24} |
| {2, 5, 8, 15, 17, 22} | {6, 7, 8, 13, 17, 21} | {7, 10, 16, 19, 22, 24} | {3, 4, 11, 13, 15, 23} |
| {6, 7, 15, 22, 23, 25} | {7, 11, 16, 17, 18, 23} | {10, 17, 18, 20, 21, 22} | {6, 8, 9, 16, 24, 25} |
| {1, 4, 12, 13, 20, 21} | {4, 8, 10, 13, 22, 25} | {5, 6, 8, 11, 14, 23} | {7, 12, 13, 14, 19, 23} |
| {2, 7, 9, 10, 17, 25} | {3, 10, 12, 14, 15, 22} | {3, 8, 17, 20, 23, 25} | {3, 4, 9, 10, 20, 24} |
| {3, 5, 8, 9, 12, 13} | {9, 11, 15, 16, 20, 22} | {0, 5, 11, 13, 17, 20} | {2, 8, 11, 13, 16, 19} |
| {6, 10, 11, 15, 17, 24} | {1, 4, 15, 18, 22, 24} | {1, 2, 8, 12, 14, 25} | {0, 5, 12, 14, 18, 21} |
| {1, 3, 10, 13, 17, 19} | {4, 5, 10, 12, 17, 23} | {4, 5, 6, 13, 19, 24} | {3, 5, 10, 11, 16, 25} |
| {11, 14, 17, 19, 22, 25} | {6, 9, 12, 14, 17, 20} | {1, 2, 16, 17, 20, 24} | {2, 3, 6, 8, 10, 18} |
| {6, 11, 12, 13, 18, 22} | {9, 10, 11, 13, 14, 21} | | |

**Table 16** Structure constants of $\mathcal{M}$: the element in the $j$th row and $k$th column in the matrix $P_i$ corresponds to constant $p_{i,j}^k$

| | $j\backslash k$ | 0 | 1 | 2 | 3 | 4 | 5 | 6 | 7 | | $j\backslash k$ | 0 | 1 | 2 | 3 | 4 | 5 | 6 | 7 |
|---|---|---|---|---|---|---|---|---|---|---|---|---|---|---|---|---|---|---|---|
| $i=0$ | 0 | 1 | 0 | 0 | 0 | 0 | 0 | 0 | 0 | $i=1$ | 0 | 0 | 1 | 0 | 0 | 0 | 0 | 0 | 0 |
| | 1 | 0 | 1 | 0 | 0 | 0 | 0 | 0 | 0 | | 1 | 0 | 0 | 1 | 1 | 1 | 0 | 0 | 0 |
| | 2 | 0 | 0 | 1 | 0 | 0 | 0 | 0 | 0 | | 2 | 0 | 0 | 1 | 0 | 1 | 1 | 0 | 0 |
| | 3 | 0 | 0 | 0 | 1 | 0 | 0 | 0 | 0 | | 3 | 0 | 0 | 0 | 1 | 1 | 1 | 0 | 0 |
| | 4 | 0 | 0 | 0 | 0 | 1 | 0 | 0 | 0 | | 4 | 3 | 0 | 0 | 0 | 0 | 0 | 2 | 0 |
| | 5 | 0 | 0 | 0 | 0 | 0 | 1 | 0 | 0 | | 5 | 0 | 2 | 0 | 0 | 0 | 0 | 0 | 3 |
| | 6 | 0 | 0 | 0 | 0 | 0 | 0 | 1 | 0 | | 6 | 0 | 0 | 1 | 1 | 0 | 1 | 0 | 0 |
| | 7 | 0 | 0 | 0 | 0 | 0 | 0 | 0 | 1 | | 7 | 0 | 0 | 0 | 0 | 0 | 0 | 1 | 0 |
| | $j\backslash k$ | 0 | 1 | 2 | 3 | 4 | 5 | 6 | 7 | | $j\backslash k$ | 0 | 1 | 2 | 3 | 4 | 5 | 6 | 7 |
| $i=2$ | 0 | 0 | 0 | 1 | 0 | 0 | 0 | 0 | 0 | $i=3$ | 0 | 0 | 0 | 0 | 1 | 0 | 0 | 0 | 0 |
| | 1 | 0 | 0 | 1 | 0 | 1 | 0 | 1 | 0 | | 1 | 0 | 0 | 0 | 1 | 1 | 0 | 1 | 0 |
| | 2 | 3 | 1 | 0 | 0 | 1 | 0 | 0 | 0 | | 2 | 0 | 0 | 0 | 0 | 0 | 1 | 1 | 3 |
| | 3 | 0 | 0 | 0 | 0 | 0 | 1 | 1 | 3 | | 3 | 3 | 1 | 0 | 0 | 1 | 0 | 0 | 0 |
| | 4 | 0 | 1 | 1 | 0 | 0 | 1 | 0 | 0 | | 4 | 0 | 1 | 0 | 1 | 0 | 1 | 0 | 0 |
| | 5 | 0 | 0 | 0 | 1 | 1 | 0 | 1 | 0 | | 5 | 0 | 0 | 1 | 0 | 1 | 0 | 1 | 0 |
| | 6 | 0 | 1 | 0 | 1 | 0 | 1 | 0 | 0 | | 6 | 0 | 1 | 1 | 0 | 0 | 1 | 0 | 0 |
| | 7 | 0 | 0 | 0 | 1 | 0 | 0 | 0 | 0 | | 7 | 0 | 0 | 1 | 0 | 0 | 0 | 0 | 0 |
| | $j\backslash k$ | 0 | 1 | 2 | 3 | 4 | 5 | 6 | 7 | | $j\backslash k$ | 0 | 1 | 2 | 3 | 4 | 5 | 6 | 7 |
| $i=4$ | 0 | 0 | 0 | 0 | 0 | 1 | 0 | 0 | 0 | $i=5$ | 0 | 0 | 0 | 0 | 0 | 0 | 1 | 0 | 0 |
| | 1 | 3 | 0 | 0 | 0 | 0 | 2 | 0 | 0 | | 1 | 0 | 0 | 1 | 1 | 0 | 0 | 1 | 0 |
| | 2 | 0 | 1 | 1 | 0 | 0 | 0 | 1 | 0 | | 2 | 0 | 1 | 0 | 1 | 0 | 0 | 1 | 0 |
| | 3 | 0 | 1 | 0 | 1 | 0 | 0 | 1 | 0 | | 3 | 0 | 1 | 1 | 0 | 0 | 0 | 1 | 0 |
| | 4 | 0 | 1 | 1 | 1 | 0 | 0 | 0 | 0 | | 4 | 0 | 0 | 0 | 0 | 2 | 0 | 0 | 3 |
| | 5 | 0 | 0 | 1 | 1 | 0 | 0 | 1 | 0 | | 5 | 3 | 0 | 0 | 0 | 0 | 2 | 0 | 0 |
| | 6 | 0 | 0 | 0 | 0 | 2 | 0 | 0 | 3 | | 6 | 0 | 1 | 1 | 1 | 0 | 0 | 0 | 0 |
| | 7 | 0 | 0 | 0 | 0 | 0 | 1 | 0 | 0 | | 7 | 0 | 0 | 0 | 0 | 1 | 0 | 0 | 0 |
| | $j\backslash k$ | 0 | 1 | 2 | 3 | 4 | 5 | 6 | 7 | | $j\backslash k$ | 0 | 1 | 2 | 3 | 4 | 5 | 6 | 7 |
| $i=6$ | 0 | 0 | 0 | 0 | 0 | 0 | 0 | 1 | 0 | $i=7$ | 0 | 0 | 0 | 0 | 0 | 0 | 0 | 0 | 1 |
| | 1 | 0 | 2 | 0 | 0 | 0 | 0 | 0 | 3 | | 1 | 0 | 0 | 0 | 0 | 0 | 1 | 0 | 0 |
| | 2 | 0 | 0 | 0 | 1 | 1 | 1 | 0 | 0 | | 2 | 0 | 0 | 0 | 1 | 0 | 0 | 0 | 0 |
| | 3 | 0 | 0 | 1 | 0 | 1 | 1 | 0 | 0 | | 3 | 0 | 0 | 1 | 0 | 0 | 0 | 0 | 0 |
| | 4 | 0 | 0 | 1 | 1 | 0 | 1 | 0 | 0 | | 4 | 0 | 0 | 0 | 0 | 0 | 0 | 1 | 0 |
| | 5 | 0 | 0 | 1 | 1 | 1 | 0 | 0 | 0 | | 5 | 0 | 1 | 0 | 0 | 0 | 0 | 0 | 0 |
| | 6 | 3 | 0 | 0 | 0 | 0 | 0 | 2 | 0 | | 6 | 0 | 0 | 0 | 0 | 1 | 0 | 0 | 0 |
| | 7 | 0 | 1 | 0 | 0 | 0 | 0 | 0 | 0 | | 7 | 1 | 0 | 0 | 0 | 0 | 0 | 0 | 0 |

**Table 17** Main data about subgroups of Aut$(T)$

| # | Order | Structure | Size of class | Generators | Maximal subgroups |
|---|---|---|---|---|---|
| 1 | 1 | Id | 1 | id | – |
| 2 | 2 | $\mathbb{Z}_2$ | 1 | $\langle g_1\rangle$ | $1^1$ |
| 3 | 2 | $\mathbb{Z}_2$ | 15 | $\langle g_2\rangle$ | $1^1$ |
| 4 | 2 | $\mathbb{Z}_2$ | 15 | $\langle g_3\rangle$ | $1^1$ |
| 5 | 3 | $\mathbb{Z}_3$ | 10 | $\langle g_5\rangle$ | $1^1$ |
| 6 | 4 | $E_4$ | 5 | $\langle g_3, g_8\rangle$ | $4^3$ |
| 7 | 4 | $E_4$ | 15 | $\langle g_1, g_2\rangle$ | $2^1, 3^1, 4^1$ |
| 8 | 4 | $E_4$ | 15 | $\langle g_2, g_4\rangle$ | $3^2, 4^1$ |
| 9 | 5 | $\mathbb{Z}_5$ | 6 | $\langle g_6\rangle$ | $1^1$ |
| 10 | 6 | $\mathbb{Z}_6$ | 10 | $\langle g_1, g_5\rangle$ | $2^1, 5^1$ |
| 11 | 6 | $S_3$ | 10 | $\langle g_5, g_7\rangle$ | $3^3, 5^1$ |
| 12 | 6 | $S_3$ | 10 | $\langle g_5, g_{10}\rangle$ | $4^3, 5^1$ |
| 13 | 8 | $E_8$ | 5 | $\langle g_1, g_2, g_4\rangle$ | $6^1, 7^3, 8^3$ |
| 14 | 10 | $\mathbb{Z}_{10}$ | 6 | $\langle g_9\rangle$ | $2^1, 9^1$ |
| 15 | 10 | $D_5$ | 6 | $\langle g_6, g_{11}\rangle$ | $3^5, 9^1$ |
| 16 | 10 | $D_5$ | 6 | $\langle g_6, g_{12}\rangle$ | $4^5, 9^1$ |
| 17 | 12 | $A_4$ | 5 | $\langle g_3, g_5, g_{13}\rangle$ | $5^4, 6^1$ |
| 18 | 12 | $D_6$ | 10 | $\langle g_1, g_5, g_7\rangle$ | $7^3, 10^1, 11^1, 12^1$ |
| 19 | 20 | $D_{10}$ | 6 | $\langle g_1, g_6, g_{14}\rangle$ | $7^5, 14^1, 15^1, 16^1$ |
| 20 | 24 | $A_4 \times \mathbb{Z}_2$ | 5 | $\langle g_4, g_5\rangle$ | $10^4, 13^1, 17^1$ |
| 21 | 60 | $A_5$ | 1 | $\langle g_6, g_{15}\rangle$ | $12^{10}, 16^6, 17^5$ |
| 22 | 120 | $A_5 \times \mathbb{Z}_2$ | 1 | $\langle g_4, g_{16}\rangle$ | $18^{10}, 19^6, 20^5, 21^1$ |

**Table 18** List of generators appearing in Table 17

| |
|---|
| $g_1 = (0, 19)(1, 18)(2, 10)(3, 15)(4, 14)(5, 8)(6, 13)(7, 16)(9, 12)(11, 17)$ |
| $g_2 = (0, 9)(1, 14)(2, 6)(4, 18)(7, 17)(10, 13)(11, 16)(12, 19)(20, 25)(21, 22)$ |
| $g_3 = (0, 12)(1, 4)(2, 13)(3, 15)(5, 8)(6, 10)(7, 11)(9, 19)(14, 18)(16, 17)(20, 25)(21, 22)$ |
| $g_4 = (0, 11)(2, 13)(3, 8)(5, 15)(6, 10)(7, 12)(9, 16)(17, 19)(20, 25)(23, 24)$ |
| $g_5 = (1, 3, 6)(2, 4, 8)(5, 10, 14)(7, 11, 9)(12, 16, 17)(13, 18, 15)(20, 21, 23)(22, 24, 25)$ |
| $g_6 = (0, 3, 10, 5, 7)(1, 4, 11, 13, 12)(2, 8, 16, 19, 15)(6, 9, 18, 14, 17)(20, 21, 24, 23, 22)$ |
| $g_7 = (1, 7)(2, 4)(3, 9)(6, 11)(10, 14)(12, 15)(13, 17)(16, 18)(21, 23)(24, 25)$ |
| $g_8 = (0, 17)(1, 18)(2, 6)(3, 5)(4, 14)(7, 9)(8, 15)(10, 13)(11, 19)(12, 16)(20, 25)(23, 24)$ |
| $g_9 = (0, 15, 10, 8, 7, 19, 3, 2, 5, 16)(1, 14, 11, 6, 12, 18, 4, 17, 13, 9)(20, 21, 24, 23, 22)$ |
| $g_{10} = (0, 19)(1, 16)(2, 14)(3, 12)(4, 10)(5, 8)(6, 17)(7, 18)(9, 15)(11, 13)(21, 23)(24, 25)$ |
| $g_{11} = (1, 11)(2, 8)(3, 7)(5, 10)(6, 9)(12, 13)(15, 16)(17, 18)(20, 21)(22, 24)$ |
| $g_{12} = (0, 19)(1, 17)(2, 5)(3, 16)(4, 14)(6, 12)(7, 15)(8, 10)(9, 13)(11, 18)(20, 21)(22, 24)$ |
| $g_{13} = (0, 16)(1, 14)(2, 10)(3, 8)(4, 18)(5, 15)(6, 13)(7, 19)(9, 11)(12, 17)(21, 22)(23, 24)$ |
| $g_{14} = (0, 2)(1, 9)(3, 15)(4, 6)(5, 16)(7, 8)(10, 19)(11, 17)(12, 18)(13, 14)(20, 23)(21, 24)$ |
| $g_{15} = (0, 1, 5, 14, 9)(2, 7, 15, 17, 6)(3, 11, 13, 10, 16)(4, 12, 19, 18, 8)(20, 22, 21, 25, 23)$ |
| $g_{16} = (0, 6, 7)(1, 8, 17)(3, 14, 12)(4, 9, 15)(5, 11, 18)(13, 16, 19)(20, 23, 24)(21, 25, 22)$ |

## References


1. G.M. Adel'son-Vel'skii, B.Ju. Weisfeiler, A.A. Leman, I.A. Faradžev, An example of a graph which has no transitive group of automorphisms. (Russian) Dokl. Akad. Nauk SSSR **185**, 975–976 (1969)
2. V.L. Arlazarov, A.A. Leman, M.Z. Rosenfeld, Computer aided construction and analysis of graphs with 25, 26 and 29 vertices. (Russian). Moscow, IPU (1975)
3. ATLAS of Finite Group Representations - Version 3.2., http://brauer.maths.qmul.ac.uk/Atlas/
4. L. Babel, I.V. Chuvaeva, M. Klin, D.V. Pasechnik, Algebraic combinatorics in mathematical chemistry. Methods and algorithms. II. Program implementation of the Weisfeiler-Leman algorithm, http://www.arxiv.org/abs/1002.1921
5. J.C. Baez, Home page, http://math.ucr.edu/home/baez/. Accessed June 2018
6. R.A. Bailey, *Association Schemes. Designed Experiments, Algebra and Combinatorics*. Cambridge Studies in Advanced Mathematics, vol. 84. (Cambridge University Press, Cambridge, 2004), pp. xviii+387
7. E. Bannai, T. Ito, *Algebraic Combinatorics: I. Association Schemes* (The Benjamin/Cummings Publishing Co., Inc., Menlo Park, CA, 1984)
8. A. Betten, M. Klin, R. Laue, A. Wassermann, Graphical $t$-designs via polynomial Kramer-Mesner matrices. Discrete Math. **197**(198), 83–109 (1999)
9. N.L. Biggs, *Distance-regular Graphs with Diameter Three*. Algebraic and geometric combinatorics, in *North-Holland Mathematics Studies*, vol. 65 (1982), pp. 69–80
10. N.L. Biggs, The Icosian calculus of today. Proc. R. Irish Acad. **95A**(Supplement), 23–34 (1995)
11. R.C. Bose, Strongly regular graphs, partial geometries and partially balanced designs. Pac. J. Math. **13**, 389–419 (1963)
12. R.C. Bose, T. Shimamoto, Classification and analysis of partially balanced incomplete block designs with two associate classes. J. Am. Stat. Assoc. **47**, 151–184 (1952)
13. A.E. Brouwer, home page: www.win.tue.nl/~aeb
14. A.E. Brouwer, Classification of small (0,2)-graphs. J. Combin. Theory A **113**, 1636–1645 (2006)
15. A.E. Brouwer, W. Haemers, *Spectra of Graphs* (Springer, Berlin, 2012)
16. A.E. Brouwer, A.M. Cohen, A. Neumaier, *Distance Regular Graphs* (Springer, Berlin, 1989)
17. R.H. Bruck, Finite nets II. Uniqueness and imbedding. Pac. J. Math. **13**, 421–457 (1963)
18. R.H. Bruck, Finite nets. I. Numerical invariants. Canad. J. Math. **3**, 94–107 (1951)
19. F. Buekenhout, M. Parker, The number of nets of the regular convex polytopes in dimension $\leq 4$. Discrete Math. **186**(1–3), 69–94 (1998)
20. P.J. Cameron, blog, https://cameroncounts.wordpress.com/2018/07/03/
21. P.J. Cameron, *Permutation Groups*. London Mathematical Society Student Texts, vol. 45 (Cambridge University Press, Cambridge, 1999), p. x+220
22. P.J. Cameron, J.H. van Lint, *Designs, Graphs, Codes and Their Links* (Cambridge University Press, Cambridge, 1991), p. x+240
23. P.J. Cameron, Two remarks on Steiner systems. Geom. Dedicata **4**, 403–418 (1975)
24. L.C. Chang, Association schemes of partially balanced block designs with parameters $v = 28$, $n_1 = 12$, $n_2 = 15$ and $p_{11}^2 = 4$. Sci. Record **4**, 12–18 (1960)
25. Y. Chen, The Steiner systems $S(3, 6, 26)$. J. Geom. **2**, 7–28 (1972)
26. W.H. Clatworthy, *Tables of Two-associate Partially Balanced Designs* (National Bureau of Standards, Washington D.C., 1973)
27. C.J. Colbourn, A. Rosa, *Triple Systems* (Clarendon Press, Oxford, 1999)
28. Conference in Algebraic Graph Theory: *Symmetry vs. Regularity, (the first 50 years since Weisfeiler-Leman stabilization)*, WL2018, July 1–7, 2018, Pilsen, Czech Republic, https://www.iti.zcu.cz/wl2018/index.html
29. J.H. Conway, A. Hulpke, J. McKay, On transitive permutation groups. LMS J. Comput. Math. **1**, 1–8 (1998)

30. D.G. Corneil, R.A. Mathon, Algorithmic techniques for the generation and analysis of strongly regular graphs and other combinatorial configurations. Algorithmic aspects of combinatorics (Conf., Vancouver Island, B.C., Ann. Discrete Math. **2**(1978), pp. 1–32 (1976)
31. H.S.M. Coxeter, A symmetrical arrangement of eleven hemi-Icosahedra, in *North-Holland Mathematics Studies*, vol. 87(C) (1984), pp. 103–114
32. H.S.M. Coxeter, *Regular Polytopes*, 3rd edn. (Dover Publications, USA, 1973)
33. H.S.M. Coxeter, W.O.J. Moser, *Generators and Relations for Discrete Groups*, 4th edn. (Springer, Berlin, 1980)
34. P. Dembowski, *Finite Geometries* (Springer, Heidelberg, 1968)
35. R.H.F. Denniston, Uniqueness of the inversive plane of order 5. Manuscr. Math. **8**, 11–19 (1973)
36. J.D. Dixon, B. Mortimer, *Permutation Groups*. Graduate Texts in Mathematics, vol. 163 (Springer, New York, 1996), p. xii+346
37. A. Duval, A directed version of strongly regular graphs. J. Combin. Theory Ser. A **47**, 71–100 (1988)
38. D. Epple, M. Klin, *Some Strongly Regular Graphs on* $p^2 + 1$ *Vertices* $p \in \{3, 5, 7, 11\}$ *and Related Structures*. Paper in preparation
39. M. Erickson, S. Fernando, W.H. Haemers, D. Hardy, J. Hemmeter, Deza graphs: a generalization of strongly regular graphs. J. Combin. Designs **7**, 395–405 (1999)
40. I.A. Faradžev, *Constructive Enumeration of Combinatorial Objects*. Algorithmic studies in combinatoric (Russian), vol. 185 (Nauka, Moscow, 1978), pp. 3–11
41. I.A. Faradžev, M. Klin, Computer package for computation with coherent configurations, in *Proceedings ISSAC-91*, ed. by S.M. Watt (ACM Press, Bonn, 1991), pp. 219–223
42. I.A. Faradžev, M.H. Klin, M. Muzichuk, Cellular rings and groups of automorphisms of graphs, in *Investigations in algebraic theory of combinatorial objects*, ed. by I.A. Faradžev, A.A. Ivanov, M.H. Klin, A.J. Woldar (Kluwer Academic Publishers, Dordrecht, 1994), pp. 1–152
43. F. Fiedler, M. Klin, M. Muzychuk, Small vertex-transitive directed strongly regular graphs. Discrete Math. **255**, 87–115 (2002)
44. P.H. Fisher, T. Pentilla, Ch. Praeger, G. Royle, Inversive planes of odd order. Eur. J. Combin. **10**, 331–336 (1989)
45. L.L. Foster, On the symmetry group of the dodecahedron. Math. Mag. **63**(2), 106–107 (1990)
46. *GAP - Groups, Algorithms, Programming—A System for Computational Discrete Algebra*, www.gap-system.org
47. C.D. Godsil, G. Royle, *Algebraic Graph Theory* (Springer, New York, 2001)
48. R.W. Gray, home page, http://www.rwgrayprojects.com. Accessed December 2018
49. K. Gray, A.P. Street, Two regular solids holding the twelve (6, 3, 2)-designs. Bull. Inst. Combin. Appl. **57**, 53–59 (2009)
50. R. Grimaldi, *Discrete and Combinatorial Mathematics*, 5th edn. (2007)
51. Ya.Yu. Gol'fand, A.V. Ivanov, M. Klin, Amorphic Cellular Rings, in *Investigations in Algebraic Theory of Combinatorial Objects*, ed. by I.A. Faradžev et al. (Kluwer Academic Publishers, Dordrecht, 1994), pp. 167–186
52. J.L. Gross, T.W. Tucker, *Topological Graph Theory* (Wiley, New York, 1987)
53. Š. Gyürki, M. Klin, M. Ziv-Av, *WL-Logo*, https://www.iti.zcu.cz/wl2018/
54. M. Hall Jr., *The Theory of Groups*. Reprinting of the 1968 edition (Chelsea Publishing Co., New York, 1976), p. xiii+434
55. A. Hanaki, I. Miyamoto, *Tables of Homogeneous Coherent Configurations*, http://kissme.shinshu-u.ac.jp/as/
56. F. Harary, *Graph Theory* (Addison-Wesley Publishing Co., Reading, Mass.-Menlo Park, Calif.-London, 1969), p. ix+274
57. A. Heinze, M. Klin, Loops, Latin squares and strongly regular graphs: An algorithmic approach via algebraic combinatorics, in *Algorithmic Algebraic Combinatorics and Gröbner Bases*, ed. by M. Klin et al. (Springer, Berlin, 2009), pp. 3–65
58. D.G. Higman, Coherent configurations. I. Rend. Sem. Mat. Univ. Padova **44**, 1–25 (1970)

59. J.W.P. Hirschfeld, *Projective Geometries over Finite Fields*, 2nd edn., Oxford Mathematical Monographs (Clarendon Press, Oxford, 1998)
60. D.A. Holton, J. Sheehan, *The Petersen Graph* (Cambridge University Press, Cambridge, 1993)
61. T.C. Holyoke, On the structure of multiply transitive permutation groups. Am. J. Math. **4**, 787–796 (1952)
62. B. Hopkins, Hamiltonian paths on platonic graphs. Int. J. Math. Math. Sci. **29–32**, 1613–1616 (2004)
63. D.R. Hughes, F.C. Piper, *Design Theory* (Cambridge University Press, Cambridge, 1985)
64. Q.M. Husain, On the totality of the solutions for the symmetrical incomplete block designs: $\lambda = 2, k = 5$ or 6. Sankhya **7**, 204–208 (1945)
65. A.A. Ivanov, Construction with the aid of a computer of some new automorphic graphs, in *"Air Physics and Applied Mathematics"* (Moscow, MPhTI, 1981), pp. 144–146 (in Russian)
66. A.A. Ivanov, M.H. Klin, S.V. Tsaranov, S.V. Shpectorov, On the problem of computing subdegrees of a transitive permutation group. UMN **6**, 115–116 (1983). (in Russian)
67. G.A. Jones, *Paley and the Paley Graphs*, arXiv:1702.00285
68. G.A. Jones, *Permutation Groups (Theory and Applications)* (Belianum, Vydavatelstvo Univerzity Mateja Bela, Banská Bystrica, 2015), p. viii+84
69. G.A. Jones, M. Ziv-Av, Petrie duality and the Anstee-Robertson graph. Symmetry **7**, 2206–2223 (2015)
70. A. Kerber, *Applied Finite Group Actions, Algorithms and Combinatorics*, vol. 19, 2nd edn. (Springer, New York, 1999)
71. A. Kerber, R. Laue, M. Meringer, Ch. Rücker, E. Schymanski, *Mathematical Chemistry and Chemoinformatics: Structure Generation, Elucidation and Quantitative Structure-Property Relationships* (De Gruyter, 2014)
72. M.Kh. Klin, I.A. Faradzhev, The method of V-rings in the theory of permutation groups and its combinatoric applications. *Investigations in Applied Graph Theory (Russian)* (Nauka Sibirsk. Otdel., Novosibirsk, 1986), pp. 59–97
73. B. Kim, Mappings which preserve regular dodecahedrons. Int. J. Math. Math. Sci. **31**(4), 515–522 (2006)
74. M. Klin, Š. Gyürki, *Selected Topics from Algebraic Graph Theory* (Belianum, Vydavatelstvo Univerzity Mateja Bela, Banská Bystrica, 2015), p. xi+209
75. M. Klin, I. Kovács, Automorphism groups of rational circulant graphs. Electron. J. Combin. **19**(1), Paper 35, 52 pp (2012)
76. M. Klin, A. Munemasa, M. Muzychuk, P.H. Zieschang, *Directed Strongly Regular Graphs Via Coherent (cellular) Algebras*. Preprint Kyushu-MPS-1997-12 (Kyushu University, 1997)
77. M. Klin, C. Pech, S. Reichard, A. Woldar, M. Ziv-Av., Examples of computer experimentation in algebraic combinatorics. Ars Math. Contemp. **3** (2), 237–258 (2010)
78. M.H. Klin, S. Reichard, A.J. Woldar, Siamese Combinatorial Objects via Computer Algebra Experimentation, in *Algorithmic algebraic combinatorics and Gröbner bases* (Springer, Berlin, 2009), pp. 67–112
79. M. Klin, C. Rücker, G. Rücker, G. Tinhofer, Algebraic combinatorics in mathematical chemistry. Methods and algorithms. I. Permutation groups and coherent (cellular) algebras. MATCH **40**, 7–138 (1999)
80. M. Klin, M. Ziv-Av, Enumeration of Schur rings over the group $A_5$, in *Computer algebra in scientific computing*. Lecture Notes in Computer Science, vol. 8136 (Springer, Berlin, 2013), pp. 219–230
81. MCh. Klin, R. Pöschel, K. Rosenbaum, *Angewandte Algebra für Mathematiker und Informatiker* (Vieweg and Sohn, Braunschweig, 1988)
82. M. Klin, A. Munemasa, M. Muzychuk, P.H. Zieschang, Directed strongly regular graphs obtained from coherent algebras. Linear Algebra Appl. **377**, 83–109 (2004)
83. M. Klin, S. Reichard, A. Woldar, Siamese objects, and their relation to color graphs, association schemes and Steiner designs. Bull. Belg. Math. Soc. Simon Stevin **12**(5), 845–857 (2005)

84. M.H. Klin, D.M. Mesner, A.J. Woldar, A combinatorial approach to transitive extensions of generously unitransitive permutation groups. J. Combin. Des. **18**(5), 369–391 (2010)
85. M. Klin, N. Kriger, A. Woldar, On the existence of self-complementary and non-self-complementary strongly regular graphs with Paley parameters. J. Geom. **107**(2), 329–356 (2016)
86. J. Lauri, R. Scapellato, *Topics in Graph Automorphisms and Reconstruction*, 2nd edn. (Cambridge University Press, Cambridge, 2016)
87. F. Levstein, D. Penazzi, The Terwilliger algebra of the dodecahedron, (English summary). Rev. Un. Mat. Argentina **45**(2), 11–20 (electronic) (2004)
88. E. Lord, *Symmetry and Pattern in Projective Geometry* (Springer, London, 2013)
89. A. Malnič, D. Marušič, P. Šparl, On strongly regular bicirculants. Eur. J. Combin. **28**(3), 891–900 (2007)
90. R. Mathon, 3-class association schemes, in *Proceedings of the Conference on Algebraic Aspects of Combinatorics* (University of Toronto, Toronto, Ont., 1975), pp. 123–155. Congressus Numerantium, No. XIII, Utilitas Math., Winnipeg, Man., 1975
91. B.D. McKay, Nauty user's guide, ver. 1.5, Technical Report TR-CS-90-02 (Computer Science Department, Australian National University, 1990)
92. M. Mulder, $(0, \lambda)$-graphs and $n$-cubes. Disc. Math. **28**, 179–188 (1979)
93. D.V. Pasechnik, Skew-symmetric association schemes with two classes and strongly regular graphs of type $L_{2n-1}(4n-1)$. Interactions between algebra and combinatorics. Acta Appl. Math. **29**(1–2), 129–138 (1992)
94. A.J.L. Paulus, Conference matrices and graphs of order 26, Technische Hogeschool Eindhoven. Report WSK 73/06, (Eindhoven, 1973), pp. I – IV, 0–89
95. S.E. Payne, J.A. Thas, *Finite Generalized Quadrangles*, Research Notes in Mathematics, vol. 110. Pitman (Advanced Publishing Program, Boston, MA, 1984), p. vi+312
96. M.Z. Rozenfeld, The construction and properties of certain classes of strongly regular graphs. Uspehi Mat. Nauk **28**, 197–198 (1973)
97. A. Rudvalis, $(v, k, \lambda)$-graphs and polarities of $(v, k, \lambda)$-designs. Math Z. **120**, 224–230 (1971)
98. J.J. Seidel, *Geometry and Combinatorics. Selected Works of J.J. Seidel Edited and with a preface by D. G. Corneil and R. Mathon* (Academic Press, Inc., Boston, MA, 1991), p. xix+410
99. J.J. Seidel, On two-graphs and Shult's characterization of symplectic and orthogonal geometries over $GF(2)$. T.H.-Report, No. 73-WSK-02 (Department of Mathematics, Technological University Eindhoven, Eindhoven, 1973), p. i+25
100. S.S. Shrikhande, The uniqueness of the $L_2$ association scheme. Ann. Math. Statist. **30**, 781–798 (1959)
101. S.S. Shrikhande, V.N. Bhat, Nonisomorphic solutions of pseudo-(3, 5, 2) and pseudo-(3, 6, 3) graphs. Ann. N. Y. Acad. Sci. **175**, 331–350 (1970)
102. S.S. Shrikhande, V.N. Bhat, Graphs derivable from $L_3(5)$ graphs. Sankhya Ser. A **33**, 315–350 (1971)
103. S.S. Shrikhande, V.N. Bhat, Seidel-equivalence in $LB_3(6)$ graphs. Aequationes Math. **8**, 271–280 (1972)
104. E. Shult, *Supplement to the "Graph Extension Theorem"* (University of Florida, Gainesville, Fla. (mimeographic notes), 1971), pp. 1–100
105. E. Shult, The graph extension theorem. Proc. Am. Math. Soc. **33**, 278–284 (1972)
106. L.H. Soicher, GRAPE: A system for computing with graphs and groups, Groups and computation (New Brunswick, 1991), DIMACS Ser. Discrete Math. Theoret. Comput. Sci. **11**, 287–291; Am. Math. Soc. (Providence, RI, 1993)
107. E. Spence, Two-graphs, in *Handbook of design and analysis of experiments*, ed. by A. Dean, M. Morris, J. Stufken, D. Bingham (Chapman & Hall/CRC Handbooks of Modern Statistical Methods. CRC Press, Boca Raton, FL, 2015), p. xix+940; pp. 875–882
108. H. Stachel, On the tetrahedra in the dodecahedron. KoG **4**, 5–10 (1999)
109. Z.G. Sun, A strong (10, 3, 4) regular graph. Kexue Tongbao (English Ed.) **28**(3), 298–300 (1983)

110. D.B. Surowski, Stability of arc-transitive graphs. J. Graph Theory **38**(2), 95–110 (2001)
111. D.B. Surowski, Automorphism groups of certain unstable graphs. Math. Slovaca **53**(3), 215–232 (2003)
112. D.M. Thompson, Design constructibility: strongly regular graphs and block designs, Ph.D. dissertation (University of Arizona, 1979)
113. D.M. Thompson, Eigengraphs: constructing strongly regular graphs with block designs. Utilitas Math. **20**, 83–115 (1981)
114. Boris Weisfeiler (ed.), *On Construction and Identification of Graphs*, vol. 558. Lecture Notes in Mathematics (Springer, Berlin, 1976)
115. B. Weisfeiler, A.A. Leman, A reduction of a graph to a canonical form and an algebra arising during this reduction. Nauchno-Techn. Inform. Ser. **2**, 9 (1968)
116. H. Wielandt, Group Theory, in *Mathematische Werke/Mathematical works*, vol. 1, ed. by B. Huppert, H. Schneider (Walter de Gruyter & Co., Berlin, 1994)

# Paley and the Paley Graphs

**Gareth A. Jones**

**Abstract** This paper discusses some aspects of the history of the Paley graphs and their automorphism groups.

**Mathematics Subject Classifications** 01A60 · 05-03 · 05B05 · 05B20 · 05E30 · 11E25 · 12E20 · 20B25 · 51M20

## 1 Introduction

Anyone who seriously studies algebraic graph theory or finite permutation groups will, sooner or later, come across the Paley graphs and their automorphism groups. The most frequently cited sources for these are, respectively Paley's 1933 paper [66], and Carlitz's 1960 paper [14]. It is remarkable that neither of those papers uses the concepts of graphs, groups or automorphisms. Indeed, one cannot find these three terms, or any synonyms for them, in those papers: Paley's paper is entirely about the construction of what are now called Hadamard matrices, while Carlitz's is entirely about permutations of finite fields.

The aim of the present paper is to explore how this strange situation came about, by explaining the background to these two papers and how they became associated with the Paley graphs. This involves describing various links with other branches of mathematics, such as matrix theory, number theory, design theory, coding theory, finite geometry, polytope theory and group theory. The paper is organised in two main parts, the first covering the graphs and the second their automorphism groups, each largely in historical order. However, in order to establish basic concepts, we start with the definition and elementary properties of the Paley graphs.

---

G. A. Jones (✉)
School of Mathematics, University of Southamton, Southampton SO17 1BJ, UK
e-mail: G.A.Jones@maths.soton.ac.uk



G. A. Jones et al. (eds.), *Isomorphisms, Symmetry and Computations in Algebraic Graph Theory*, Springer Proceedings in Mathematics & Statistics 305,
https://doi.org/10.1007/978-3-030-32808-5_5

## 2 Definition and Properties of the Paley Graphs

The Paley graph $P(q)$ has vertex set $V = \mathbb{F} := \mathbb{F}_q$, a field of prime power order $q = p^e \equiv 1 \bmod (4)$, with two vertices $u$ and $v$ adjacent if and only if $u - v$ is an element of the set

$$S = \{x^2 \mid x \in \mathbb{F},\ x \neq 0\}$$

of quadratic residues (non-zero squares) in $\mathbb{F}$. It is thus the Cayley graph [15] for the additive group of $\mathbb{F}$, with $S$ as the connection set. The choice of $q$ ensures that $-1 \in S$, so $S = -S$ and $P(q)$ is an undirected graph; the fact that $S$ generates the additive group ensures that $P(q)$ is connected. The neighbours of a vertex $v$ are the elements of $S + v$, so its valency is $|S| = (q-1)/2$. (If $q \equiv 3 \bmod (4)$ this construction gives a directed graph, in fact a tournament, since each pair of vertices $u \neq v$ are joined by a unique arc $u \to v$, where $v - u \in S$.)

For example, Fig. 1 shows $P(9)$ drawn on a torus, formed by identifying opposite sides of the outer square. Here $\mathbb{F}_9 = \mathbb{F}_3[i]$ where $i^2 = -1$; in other words, this map is the quotient of a Cayley map for the additive group $\mathbb{Z}[i]$ of Gaussian integers, modulo the ideal (3). In the notation of Coxeter and Moser [18, Chap. 8], this is the map $\{4,4\}_{3,0}$. There is an analogous chiral pair of torus embeddings of $P(13)$ as the triangular maps $\{3,6\}_{3,1}$ and $\{3,6\}_{1,3}$; Fig. 2 shows the former, with opposite sides of the outer hexagon identified.

Figure 3 shows $P(13)$, exhibiting its dihedral symmetry under the automorphisms $v \mapsto \pm v + b$, $b \in \mathbb{F}_{13}$. Vertices are identified with $0, 1, \ldots, 12$ in cyclic order, and edges $uv$ are coloured black, blue or red as $u - v = \pm 1, \pm 3$ or $\pm 4$, respectively.

It is clear from the definition that further combinatorial properties of the graphs $P(q)$ will depend on the properties of quadratic residues in finite fields. For any odd prime power $q$, let $Q$ be the *Jacobsthal matrix* for $\mathbb{F} = \mathbb{F}_q$: this has rows and columns indexed by the elements of $\mathbb{F}$, with $(u, v)$ entry $\chi(v - u)$ for all $u, v \in \mathbb{F}$,

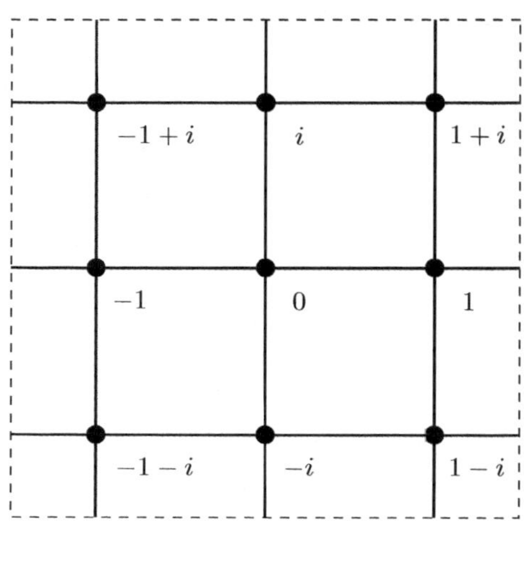


**Fig. 1** $P(9)$ drawn on a torus

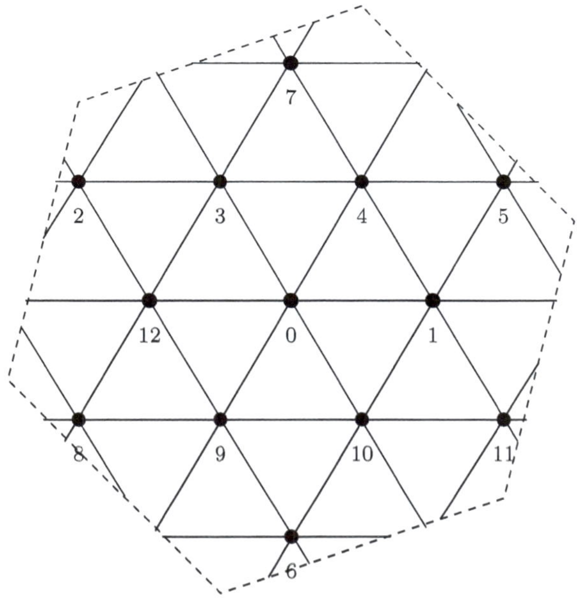


**Fig. 2** $P(13)$ drawn on a torus

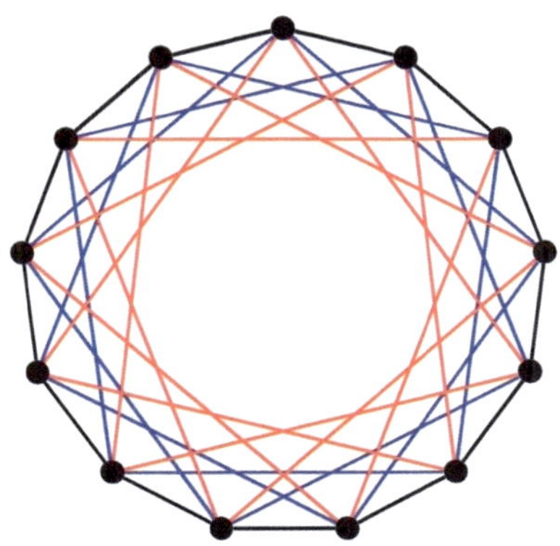

**Fig. 3** $P(13)$, showing dihedral symmetry

where $\chi : \mathbb{F} \to \mathbb{C}$ is the *quadratic residue character* of $\mathbb{F}$, defined by

$$\chi(x) = \begin{cases} 0 & \text{if } x = 0, \\ 1 & \text{if } x \in S, \\ -1 & \text{otherwise.} \end{cases}$$

Thus the restriction of $\chi$ to the multiplicative group $\mathbb{F}^* := \mathbb{F} \setminus \{0\}$ is a group epimorphism $\mathbb{F}^* \to \{\pm 1\}$ with kernel $S$. When $q$ is a prime $p$, $\chi(x)$ is the Legendre symbol $(\frac{x}{p})$.

The matrix $Q$ is symmetric or skew-symmetric as $q \equiv 1$ or 3 mod (4), with

$$QJ = JQ = 0 \quad \text{and} \quad QQ^T = qI - J,$$

where $J$ is the matrix of order $q$ with all entries equal to 1. The first two equations here are obvious. For the last equation, note that $QQ^T$ has $(u, v)$ entry $\sum_{w \in \mathbb{F}} \chi(w - u)\chi(w - v)$ given by the following lemma:

**Lemma 1** *If $u, v \in \mathbb{F}$ then*

$$\sum_{w \in \mathbb{F}} \chi(w - u)\chi(w - v) = \begin{cases} -1 & \text{if } u \neq v. \\ q - 1 & \text{if } u = v. \end{cases} \tag{1}$$

***Proof*** If $u \neq v$ we have

$$\begin{aligned} \sum_{w \in \mathbb{F}} \chi(w - u)\chi(w - v) &= \sum_{w \neq u,v} \chi(w - u)^2 \chi(x) \quad \left(x := \frac{w - v}{w - u}\right) \\ &= \sum_{w \neq u,v} \chi(x) \\ &= \sum_{x \neq 0,1} \chi(x) \\ &= \sum_{x \neq 0} \chi(x) - \chi(1) \\ &= -1, \end{aligned}$$

since clearly $\sum_{x \neq 0} \chi(x) = 0$. The case $u = v$ is obvious. □

If $q \equiv 1$ mod (4) then $P(q)$ has adjacency matrix

$$A = \frac{1}{2}(Q - I + J),$$

obtained from $Q$ by replacing every entry $-1$ with 0. By squaring $A$, we obtain the following lemma:

**Lemma 2** *If $q \equiv 1$ mod (4), and $u$ and $v$ are distinct vertices of $P(q)$, then the number $|(S + u) \cap (S + v)|$ of common neighbours of $u$ and $v$ is*

$$\frac{1}{4}(q - 3 - 2\chi(u - v)) = \begin{cases} (q - 5)/4 & \text{if } u - v \in S, \\ (q - 1)/4 & \text{if } u - v \notin S. \end{cases} \tag{2}$$

(When $q \equiv 3$ mod (4) we find that $|(S + u) \cap (S + v)| = (q - 3)/4$ for all pairs $u \neq v$; see also [56, Exercise 1.22].)

This shows that $P(q)$ is a strongly regular graph, with parameters $v = q$ (the number of vertices), $k = (q-1)/2$ (their common valency), $\lambda = (q-5)/4$ and $\mu = (q-1)/4$ (the number of common neighbours of an adjacent or non-adjacent pair of vertices).

There are several other ways to derive Lemma 2, for instance from a result of Jacobsthal [44, 45] that if $p$ is an odd prime, then the Legendre symbol satisfies

$$\sum_{x=1}^{p} \left(\frac{x^2+c}{p}\right) = \begin{cases} -1 & \text{if } c \not\equiv 0 \bmod (p), \\ p-1 & \text{if } c \equiv 0 \bmod (p); \end{cases} \tag{3}$$

the same proof gives the corresponding result for all finite fields of odd order. One can also deduce Lemma 2 from results of Perron [70] and of Kelly [53] on the distribution of quadratic residues (with analogous results for $q \equiv 3 \bmod (4)$); they both prove their results only in the case where $q$ is prime, though Kelly notes that his arguments are also valid for all odd prime powers, as indeed are those of Perron. Basile and Brutto give a geometric proof in [2].

In fact, by the following lemma one can deduce the strong regularity of $P(q)$, and the values of $k$, $\lambda$ and $\mu$, merely from the facts that $P(q)$ is self-complementary (under the isomorphism $P(q) \to \overline{P(q)},\ v \mapsto av$ for a non-residue $a$), and is arc-transitive (under the automorphisms $v \mapsto av + b$ where $a \in S$ and $b \in \mathbb{F}$).

**Lemma 3** *Any self-complementary arc-transitive graph is strongly regular, with parameters* $v = 4t+1$, $k = 2t$, $\lambda = t-1$ *and* $\mu = t$ *for some integer* $t$.

***Proof*** Since the graph is arc-transitive, its automorphism group acts transitively on the vertices, so they all have the same valency $k$. The stabiliser of each vertex is transitive on its neighbours, and hence, since the graph is self-complementary, also on its non-neighbours, so the graph is strongly regular. The complement of a strongly regular graph with parameters $(v, k, \lambda, \mu)$ is also strongly regular, with parameters $(v, v-k-1, v-2-2k+\mu, v-2k+\lambda)$. Since the graph is self-complementary we can equate these parameters, giving

$$v = 2k+1 \quad \text{and} \quad \mu = \lambda + 1.$$

In any strongly regular graph, double counting of the edges between the neighbours and non-neighbours of a particular vertex gives

$$(v-k-1)\mu = k(k-\lambda-1),$$

so substituting for $v$ and cancelling $k$ gives

$$\mu = k - \lambda - 1.$$

Solving the two simultaneous equations for $\lambda$ and $\mu$ give the result, with $t = \mu$. □

In particular, we see that $P(q)$ has parameters $v = q$, $k = (q-1)/2$, $\lambda = (q-5)/4$ and $\mu = (q-1)/4$.

## 3 Jacobsthal and Sums of Squares

The property of quadratic residues expressed by Eq. (1) can be traced back to the work of Jacobsthal on the representation of primes as sums of squares.

In 1625 Girard stated that each prime $p \equiv 1 \bmod (4)$ can be written as a sum

$$p = a^2 + b^2 \quad (a, b \in \mathbb{Z}) \tag{4}$$

of two squares. A few years later Fermat claimed to have a proof, which has never been found, and Euler eventually provided one in 1749. (See [24, Chap. VI] for a detailed account of the history of this theorem, and [79, Sect. 6.8] for a concise summary.) There are several fairly elementary proofs of this result: see [47, Sects. 10.1, 10.6] or [79, Sect. 6.7]. For instance, one can use simple area calculations to show that if $u^2 \equiv -1 \bmod (p)$ then the lattice $\{(x, y) \in \mathbb{Z}^2 \mid y \equiv ux \bmod (p)\}$ in $\mathbb{R}^2$ has a non-zero element $(a, b)$ within the disc $x^2 + y^2 < 2p$. Zagier has given an elegant one-sentence proof in [89]. However, these proofs are not constructive. There are interesting discussions of constructive proofs in [19, Sect. V.3] and [83]; these include Gauss's simple but hardly practical solution

$$a = \left\langle \frac{1}{2}\binom{2k}{k} \right\rangle, \quad b = \langle (2k)!\, a \rangle,$$

where $p = 4k + 1$ and $\langle n \rangle$ is the residue of $n$ mod $(p)$ closest to 0, and also in [83] some efficient modern algorithms for solving (4).

In 1907 Jacobsthal [45] published explicit formulae for integers $a$ and $b$ satisfying (4), based on work in his thesis [44]. Specifically, he took

$$a = \frac{\varphi(r)}{2}, \quad b = \frac{\varphi(n)}{2}$$

for any residue $r$ and non-residue $n$ mod $(p)$, where

$$\varphi(e) := \sum_{m=1}^{p} \chi(m)\chi(m^2 + e).$$

(For typographic convenience we write $\chi$ here, rather than the Legendre symbol used by Jacobsthal.) It is easy to see that $\varphi(e)$ is even. Now $\varphi(e) = 0$ if $e \equiv 0 \bmod (p)$, and otherwise, since $\varphi(e) = \chi(x)\varphi(ex^2)$ for all $x \not\equiv 0 \bmod (p)$, $\varphi^2(e)$ depends only on whether $e$ is congruent to a residue or a non-residue mod $(p)$; thus

$$\sum_{e=1}^{p} \varphi^2(e) = \frac{p-1}{2}\left(\varphi^2(r) + \varphi^2(n)\right)$$

where $r$ and $n$ are any residue and non-residue. On the other hand, calculating the left hand side directly, using the definition of $\varphi$ and summing first over $e$, leads via

Eq. (1) to

$$\sum_{e=1}^{p} \varphi^2(e) = p(p-1)\,(1+\chi(-1))\,.$$

It follows that if $p \equiv 1 \bmod (4)$, so that $\chi(-1) = 1$, then

$$\left(\frac{\varphi(r)}{2}\right)^2 + \left(\frac{\varphi(n)}{2}\right)^2 = p,$$

as required.

For example, if $p = 13$ we can take $r = 1$ and $n = 2$, with $\varphi(1) = 6$ and $\varphi(2) = -4$, so that $13 = 3^2 + (-2)^2$.

It is tempting to speculate whether Gauss was aware of (1) in some form when $q = p$. The author has no direct evidence for this (it does not appear in the *Disquisitiones*), but the Gauss expert Franz Lemmermeyer has commented [57] that he 'would have seen the proof in a second', either by counting $\mathbb{F}_p$-rational points on the curve $x^2 - y^2 = 1$, or along the following lines.

Use Euler's criterion $\chi(x) \equiv x^{(p-1)/2} \bmod (p)$ and the Binomial Theorem to express each summand in the left hand side $L$ of (1) as a polynomial in $\mathbb{Z}_p[w]$, and then use the fact [30, Sect. 19] that if $k \in \mathbb{N}$ then

$$\sum_{w\in\mathbb{Z}_p} w^k = \begin{cases} p-1 & \text{if } k \equiv 0 \bmod (p-1), \\ 0 & \text{otherwise,} \end{cases}$$

to show that $L$, regarded as an element of $\mathbb{Z}$, is congruent to $-1 \bmod (p)$. If $u \neq v$ then two of the summands in $L$ are equal to 0, and the remaining $p-2$ are each $\pm 1$, so $|L| \leq p-2$ and hence $L = -1$. The case $u = v$ is trivial.

## 4 Perron, Brauer, Hopf and Schur

In 1952 Perron [70] studied the distribution of quadratic residues modulo a prime $p$. His theorems that are relevant here can be stated concisely as follows:

**Theorem 4.1** *Let $\mathbb{F} = \mathbb{F}_p$ for a prime $p = 4n \pm 1$, define $S_0 = S \cup \{0\}$, and let $a \in \mathbb{F}^*$. Then*

$$|(S_0 + a) \cap S_0| = \begin{cases} n & \text{if } p = 4n-1 \text{ or } a \notin S, \\ n+1 & \text{if } p = 4n+1 \text{ and } a \in S. \end{cases}$$

His similar results for the set of non-residues follow immediately on taking complements, and the corresponding result for $S$ rather than $S_0$ can be deduced from the equation

$$|(S+a)\cap S| = |S_0+a)\cap S_0| - |\{\pm a\}\cap S|.$$

In the same year Brauer [8], writing on the distribution of quadratic residues with applications to Hadamard matrices (or Hadamard determinants as he called them — see Sect. 5), noted that some of Perron's results were corollaries to a theorem in Jacobsthal's thesis [44], that if $p$ is an odd prime, and $c \not\equiv 0 \bmod (p)$, then

$$\sum_{x=1}^{p} \chi(x^2+c) = -1. \tag{5}$$

Referring to the case $p = 4n - 1$ of Theorem 4.1, Brauer wrote: 'As early as 1920, H. Hopf showed me his proof of this theorem using (5), and its application to the construction of Hadamard determinants of order $p+1 = 4n$. However, he never published it since I. Schur already knew this result at that time. Independently this theorem and its application to Hadamard determinants were published by R. E. A. Paley [66] in 1933.'

This story was repeated by Dembowski in [23, p. 97], where he defined Paley designs as examples of Hadamard designs (see his footnote (4)). Dembowski asserted that, according to Brauer, Schur already knew of these designs; however, this is not clear, since Brauer did not mention designs in [8]. Certainly, the designs are implicit in the matrices, but the first to make an explicit connection seems to have been Todd [81], in 1933 (see Sect. 7.4).

Since Schur and Frobenius were Jacobsthal's advisors for his 1906 doctoral thesis, one can assume that Schur actually knew the result Brauer refers to much earlier than 1920. Theorem 4.1 extends in the obvious way to finite fields $\mathbb{F}_q$ of any odd order $q$. In particular, if $q \equiv 1 \bmod (4)$ we see that

$$|(S+a)\cap S| = \begin{cases} (q-5)/4 & \text{if } a \in S, \\ (q-1)/4 & \text{if } a \notin S, \end{cases}$$

giving the parameters $\lambda$ and $\mu$ for the strongly regular graph $P(q)$.

## 5 Hadamard Matrices and Designs

Here we briefly discuss Hadamard matrices and designs, mentioned in the preceding section. A *Hadamard matrix* of order $m$ is an $m \times m$ matrix $H$, with all entries equal to $\pm 1$, and with mutually orthogonal rows, so that $HH^T = mI$. These matrices are named after Jacques Hadamard (1865–1963), who proved in 1893 [34] that an $m \times m$ complex matrix $H = (h_{ij})$, with $|h_{ij}| \le 1$ for all $i$ and $j$, satisfies $|\det H| \le m^{m/2}$; in the case where each $h_{ij}$ is real, $H$ attains this bound if and only if it is a Hadamard matrix. Hadamard matrices have many modern applications, in areas such as engineering, coding theory, cryptography, physics and statistics (see [39, 41, 75], for example).

As early as 1867 Sylvester [80], as part of a study of orthogonal matrices, gave a recursive construction of what later became known as Hadamard matrices, of each order $m = 2^e$: he started with $H = I_1 = (1)$ and used a Hadamard matrix $H$ of order $m$ to construct the Kronecker product

$$\begin{pmatrix} + & + \\ + & - \end{pmatrix} \otimes H = \begin{pmatrix} H & H \\ H & -H \end{pmatrix},$$

a Hadamard matrix of order $2m$. (It is typographically convenient to write entries 1 and $-1$ as $+$ and $-$.) Connoisseurs of Victorian English literary style and social attitudes will, no doubt, appreciate the way in which Sylvester commended his ideas to his readers: he described the many possible applications of his theory, listed at some length in the title of his paper, as '... furnishing interesting food for thought, or a substitute for the want of it, alike to the analyst at his desk and the fine lady in her boudoir.'

In 1898 Scarpis [73] gave a construction of Hadamard matrices of certain orders, but there seems to have been little further progress in their construction until Paley's paper [66] in 1933. This was motivated by problems in combinatorics and geometry, raised by his Cambridge contemporaries Todd [81] and Coxeter [16] in papers that appeared in the same volume.

It is easy to show [48, Lemma 6.28] that, apart from trivial examples of order 1 or 2, if a Hadamard matrix of order $m$ exists then $m \equiv 0 \bmod (4)$. The Hadamard Conjecture is that the converse is also true. For example, in 1933 Paley wrote [66, p. 312] 'It seems probable that, whenever $m$ is divisible by 4, it is possible to construct an orthogonal matrix of order $m$ composed of $\pm 1$, but the general theorem has every appearance of difficulty.' (At that time, a square matrix was called orthogonal if it had mutually orthogonal rows; nowadays we impose the extra requirement that all rows have unit length, though it might be more consistent to call such a matrix orthonormal.)

The following process shows that Hadamard matrices lead naturally to certain block designs. Multiplying various rows and columns of a Hadamard matrix by $-1$ yields a *normalised* Hadamard matrix, one in which all entries in the first row and column are equal to 1. Deleting this row and column leaves a square matrix of order $m - 1$; its columns and rows can be identified with the points and blocks of a block design, with entries $\pm 1$ indicating incidence or non-incidence of points and blocks. If $m \geq 4$ there are $m - 1$ points and blocks, each block has size $(m - 2)/2$, and any two blocks have $(m - 4)/4$ points in common. A block design with these properties is called a *Hadamard design*. As noted by Todd [81] (see Sect. 7.4), this process is reversible, so that every Hadamard design corresponds to a normalised Hadamard matrix.

## 6 Paley

Raymond Edward Alan Christopher Paley was born in Bournemouth, UK, on 7 January 1907, the son of an army officer who died before Paley was born. He attended Eton College, where he was a King's Scholar, entitling him to reduced fees. This school, founded in 1440 by King Henry VI, is noted for having educated 20 British prime ministers, together with one Fields medallist and a number of fictional characters ranging from Captain Hook, via Bertie Wooster, to James Bond. He then studied mathematics at Trinity College, Cambridge, taking his Ph.D. under the supervision of J. E. Littlewood.

In his short life, Paley's main mathematical contributions were in analysis, and many were of considerable significance: Littlewood-Paley theory, the Paley-Wiener Theorem, and the Paley-Zygmund inequality, for example. A footnote in [66, p. 318], citing papers by Walsh, Kaczmarz and himself, suggests that Paley's expertise in constructing Hadamard matrices arose partly from his work on orthogonal functions. In addition to Littlewood, he collaborated with Zygmund, who spent the year 1930–31 in Cambridge; Zygmund's 1935 book *Trigonometric Series* drew heavily on their joint work. In 1932 Paley obtained a research fellowship to allow him to work with Wiener at MIT. While in the USA, he also collaborated with Pólya, who was visiting Princeton. Some of his collaboration with Coxeter and Todd, which gave rise to his constructions of Hadamard matrices in [66], may have taken place in the USA, as they visited Princeton in 1932–1933 and 1933–1934, respectively.

In the foreword to his edition of *Littlewood's Miscellany*, Béla Bollobás has written that 'Paley ... was one of the greatest stars in pure mathematics in Britain, whose young genius frightened even Hardy.' However, after a highly promising start to his career, Paley died on 7 April 1933 at the age of 26, caught in an avalanche while skiing near Banff. Wiener wrote in [87] that '... he was already recognised as the ablest of the group of young English mathematicians who have been inspired by the genius of G. H. Hardy and J. E. Littlewood. In a group notable for its brilliant technique, no one had developed this technique to a higher degree than Paley. Nevertheless he should not be thought of primarily as a technician, for with his ability he combined creative power of the first order.' MathSciNet lists 23 publications by Paley, including a reprint and a Russian translation of his work with Wiener on Fourier transforms.

## 7 Paley's Hadamard Matrix Constructions

In [66] Paley gave several constructions of Hadamard matrices, based on the combinatorial properties of quadratic residues such as equation (1). Perhaps surprisingly, the name Hadamard does not appear in his paper, nor in the accompanying papers by Todd [81] and Coxeter [16], apart from once in the title of a bibliographic reference (to an abstract [31] of a talk by Gilman) added by Paley after submitting his paper:

Paley and Coxeter referred to what we now call Hadamard matrices as 'U-matrices', while Todd had no special name for them.

Paley described two constructions, based on finite fields, which give Hadamard matrices of order $m = q + 1$ or $2(q + 1)$ for each prime power $q \equiv 3$ or $1 \bmod (4)$, respectively. He gave proofs only in the cases where $q$ is prime, crediting Todd and Coxeter for the proof when $q \equiv 3 \bmod (4)$, and Davenport (another Cambridge contemporary) for pointing out the crucial property (1) of the Legendre symbol; however, later in his paper he noted [66, p. 316] that his proofs generalise easily to odd prime powers. He then showed in [66, Table 1] that combinations of these constructions and that of Sylvester yield Hadamard matrices of all orders $m \equiv 0 \bmod (4)$ up to and including 200, with the exceptions of 92, 116, 156, 184 and 188. Subsequently, these and many other orders $m$ have been dealt with, but the conjecture is still open. (Around the same time as Paley, Gilman showed how to construct Hadamard matrices of order $2^{\nu} n_1^{\nu_1} \dots n_k^{\nu_k}$ where each $n_i \equiv 0 \bmod (4)$, each $n_i - 1$ is prime, and each $\nu_i \geq 1$; reference [31], cited by Paley, is an abstract of a lecture on this subject given by Gilman, but his work does not seem to have been published.)

Paley stated the above results as lemmas. In another construction, stated as the only theorem in his paper, Paley proved that if $m$ is a power of 2 one can partition the $2^m$ possible rows of $m$ entries $\pm 1$ into $2^m/m$ sets, each set forming the rows of a Hadamard matrix of order $m$ (see Sect. 7.3 for an outline proof of this result).

## 7.1 *Paley's First Construction*

Let $Q$ be the Jacobsthal matrix for the field $\mathbb{F} = \mathbb{F}_q$, where $q$ is an odd prime power, and let $R$ be the row vector $(1, 1, \dots, 1)$ of length $q$. In [66, Lemma 2 and p. 316], Paley showed that if $q \equiv 3 \bmod (4)$ then

$$H = \begin{pmatrix} 1 & R \\ R^T & Q - I \end{pmatrix}$$

is a Hadamard matrix of order $m = q + 1$, where $I$ denotes the identity matrix of order $q$. The fact that distinct rows are orthogonal follows immediately from Eq. (1). These matrices are now known as *Paley-Hadamard matrices of type I.*

## 7.2 *Paley's Second Construction*

In [66, Lemma 3] Paley started with the matrix

$$(B_{ij}) = \begin{pmatrix} 0 & R \\ R^T & Q \end{pmatrix},$$

where $q \equiv 1 \bmod (4)$, and then replaced each entry $B_{ij} = \pm 1$ or 0 with the $2 \times 2$ matrix

$$\pm\begin{pmatrix} + & + \\ + & - \end{pmatrix} \quad \text{or} \quad \begin{pmatrix} + & - \\ - & - \end{pmatrix}$$

respectively. This gives a symmetric Hadamard matrix of order $m = 2(q+1)$. The proof is similar to that for his first construction. These matrices are now known as *Paley-Hadamard matrices of type II.*

## 7.3 Paley's Third Construction

In another construction, stated as the only theorem in his paper, Paley proved that if $m$ is a power of 2 one can partition the $2^m$ possible rows of $m$ entries $\pm 1$ into $2^m/m$ sets, each set forming the rows of a Hadamard matrix of order $m$. Since this result seems to be rather less well-known, and since it was subsequently used by Todd [81] and Coxeter [16], we give an outline proof, following Paley's notation (though with more modern terminology).

Let $m = 2^k$, let $i, j \in \{0, 1, \ldots, 2^k - 1\}$ have binary representations

$$i = \sum_{\lambda=0}^{k-1} \eta_\lambda 2^\lambda, \quad j = \sum_{\lambda=0}^{k-1} \zeta_\lambda 2^\lambda$$

where $\eta_\lambda, \zeta_\lambda \in \{0, 1\}$, and write

$$A_{ij} = \prod_{\lambda=0}^{k-1} (-1)^{\eta_\lambda \zeta_{\lambda-1-\lambda}}.$$

The matrix $M = (A_{ij})$ is the $k$th Kronecker power of the matrix $\begin{pmatrix} + & + \\ + & - \end{pmatrix}$.

Then $A_{i_1 j} A_{i_2 j} = A_{ij}$, where the binary representation of $i$ is the term by term mod (2) sum of those for $i_1$ and $i_2$. Thus, the $m$ rows $(A_{ij})$, $i$ fixed, $j = 0, \ldots, m-1$, form an elementary abelian group under term by term multiplication. This multiplication rule also shows that distinct rows of $M$ are orthogonal, so $M$ is a Hadamard matrix.

For each of the $2^m$ sequences $B = (B_j) \in \{\pm 1\}^m$ we have a Hadamard matrix $(A_{ij} B_j)$. As shown by Paley, if two of these matrices have a row in common (possibly in different positions), they have all their rows in common, so these matrices $(A_{ij} B_j)$ are partitioned into $2^m/m$ sets of size $m$, those in the same or different sets having all or none of their rows in common. Choosing one matrix from each set proves the theorem.

## 7.4 Todd's Paper

Todd used Paley's theorem in the related paper [81]. Motivated by work of Coxeter [16] on polytopes, Todd was interested in the problem of finding $4n-1$ subsets of size $2n-1$ in a $(4n-1)$-element set, each pair having an intersection of size $n-1$. (Such an arrangement is now called a Hadamard design with parameters $(4n-1, 2n-1, n-1)$, and the chosen subsets are called blocks.) As explained in Sect. 5, this is equivalent to finding a Hadamard matrix $H$ of order $m=4n$. Normalising $H$, then removing the first row and column, and finally replacing all entries $-1$ with 0 gives the incidence matrix of the design.

One obvious solution to Todd's problem, corresponding to the matrix $M$ of order $m=2^k$ in the proof of Paley's main theorem [66] (see Sect. 7.3), is to take the blocks to be the hyperplanes in the $(k-1)$-dimensional projective geometry $PG(k-1,2)$ over the field $\mathbb{F}_2$, with $n=2^{k-2}$; the points and lines of the Fano plane correspond to the simplest case, $k=3$. Another solution, corresponding to Paley's Lemmas 2 and 4, is to take the blocks to be the translates of the set $S$ of quadratic residues in the field $\mathbb{F}_q$ of order $q\equiv 3 \bmod (4)$, with $n=(q+1)/4$. (Dembowski called this a *Paley design* [23, p. 97].) Todd also constructed other examples, for instance classifying them all in the case where there are $4n-1=15$ points.

In addition, Todd considered the automorphism group of such a design, that is, the largest subgroup of $S_{4n-1}$, acting on the points, which permutes the blocks. In the finite geometry example, this is the collineation group $PGL_k(2)$ of the geometry (see [23, p. 31] for notation for groups of projective transformations). In the quadratic residue example, the automorphism group contains the subgroup

$$A\Delta L_1(q) := \{v \mapsto av^{\gamma}+b \mid a\in S,\ b\in \mathbb{F}_q,\ \gamma\in \mathrm{Gal}\,\mathbb{F}_q\}$$

of index 2 in $A\Gamma L_1(q)$ (see also Sect. 9.1 for this group); in some cases, such as when $q=19$, 23 or 27 (in fact for all $q\geq 19$, by a later result of Kantor [50]) this is the whole automorphism group, but in other cases the automorphism group is larger, for instance isomorphic to $PSL_2(q)$ (acting with degree $q$) when $q=7$ or 11. (Although Todd did not mention this, when $4n-1=7$ the isomorphism between the projective geometry and quadratic residue solutions illustrates the isomorphism $PGL_3(2)\cong PSL_2(7)$.)

## 7.5 Coxeter's Paper

In the other related paper [16], Coxeter was interested in generalising the well-known partitions of the vertices of the cube or of the dodecahedron into those of two or five tetrahedra. Schoute [74] had already given similar examples of such compound polytopes in dimension 4, and Coxeter wanted to construct examples in dimensions $m\geq 5$. In this case, the possibilities are more restricted since the only regular polytopes are the simplex $\alpha_m$, the cross-polytope $\beta_m$, and its dual, the 'measure polytope', or $m$-cube $\gamma_m$. (When $m\geq 5$ there are no analogues of the dodecahedron {5, 3} and the icosahedron {3, 5} for $m = 3$, or of the 24-cell {3, 4, 3}, the 120-cell {5, 3, 3} and the 600-cell {3, 3, 5} for $m = 4$; here the brackets {...} denote Schläfli symbols for polytopes, see [17].)

Coxeter defined a *compound polytope* in dimension $m$ to be a set of $D$ concentric, finite, convex, $m$-dimensional polytopes $\Pi$, which are transitively permuted by the symmetry group of the set. Let $\Pi$ have $V$ vertices and $F$ faces. The compound polytope is *vertex-regular* if the $DV$ vertices of its components are the $v$ vertices of a regular polytope $\pi_1$, each taken $d_1$ times for some $d_1 \geq 1$, so that $DV = d_1 v$. For example, when $m = 3$ one could take $\pi_1$ to be a cube $\gamma_3 = \{4, 3\}$ (see Fig. 4) or dodecahedron {5, 3}, with its vertices partitioned into those of $D = 2$ or 5 tetrahedra $\Pi = \alpha_3 = \{3, 3\}$; here $V = F = 4$, $v = 8$ or 20, and $d_1 = 1$. For an example where $d_1 > 1$ take the two possible partitions for the dodecahedron, mirror images of each other, so that $D = 10$ and $d_1 = 2$.

Dually, Coxeter defined a compound to be *face-regular* if the $DF$ bounding spaces (hyperplanes spanned by faces) of its components are the $f$ bounding spaces of a regular polytope $\pi$, each taken $d$ times, so that $DF = df$. For example, $\pi$ could be the octahedron $\pi = \beta_3 = \{3, 4\}$ or icosahedron {3, 5}, regarded as the intersection of $D = 2$ or 5 tetrahedra, so that $d = 1$. In the latter case one could also take the two possible sets of five tetrahedra, giving $D = 10$ and $d = 2$.

A simple argument, counting vertices, shows that if $m \geq 5$ then in any vertex-transitive compound, $\pi_1$ must be $\gamma_m$, with a dual result $\pi = \beta_m$ for face-transitive compounds. It follows that the only possibilities are (using a slightly more consistent version of Coxeter's notation):

1. $d_1\gamma_m[D\beta_m]$, a vertex-transitive compound of $D$ cross-polytopes $\Pi = \beta_m$, their vertices $d_1$ at a time forming an $m$-cube $\pi_1 = \gamma_m$;
2. its dual $[D\gamma_m]d\beta_m$, a face-transitive compound of $D$ hypercubes $\Pi = \gamma_m$, their bounding spaces $d$ at a time forming a cross-polytope $\pi = \gamma_m$;
3. $d_1\gamma_m[D\alpha_m]d\beta_m$, a self-dual vertex- and face-transitive compound of $D$ simplices $\Pi = \alpha_m$, their vertices and faces $d_1$ and $d$ at a time forming $\pi_1 = \gamma_m$ and $\pi = \beta_m$.

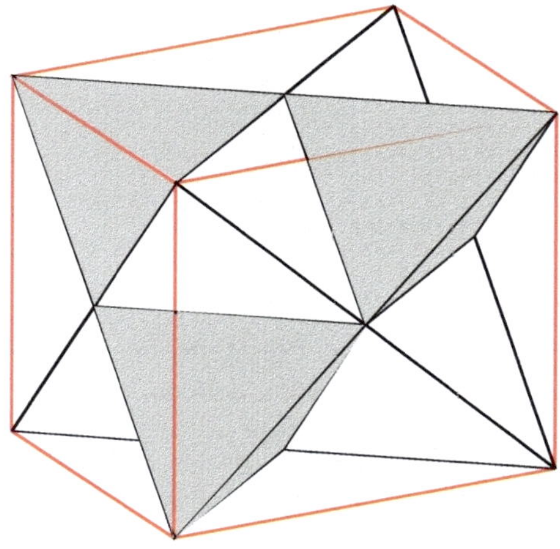

**Fig. 4** A stella octangula, inscribed in a cube

For example, Kepler's stella octangula, shown in Fig. 4, is the compound

$$\gamma_3[2\alpha_3]\beta_3 = \{4, 3\}[2\{3, 3\}]\{3, 4\}.$$

This is formed as above from two tetrahedra $\alpha_3 = \{3, 3\}$, inscribed in a cube $\gamma_3 = \{4, 3\}$ and intersecting in an octahedron $\beta_3 = \{3, 4\}$, with $d_1 = d = 1$.

Coxeter showed that compounds of type (1) and (2) exist, in dimension $m$, if and only if a compound of type (3) exists, in dimension $m - 1$, with the same values of $d_1$, $d$ and $D$. Moreover, the existence of such compound polytopes is equivalent to that of certain Hadamard matrices, as follows:

The points in $\mathbb{R}^m$ with all coordinates $\pm 1$ form the vertices of an $m$-cube $\gamma_m$. If $H$ is a Hadamard matrix of order $m$, then the rows of $H$ and $-H$ are the vertices of a cross-polytope $\beta_m$ inscribed in $\gamma_m$, and this is just one component of a compound polytope of type (1); the dual compound has type (2). Similarly, given a normalised Hadamard matrix of order $m$, deleting the first column gives $m$ rows, the vertices of a simplex $\alpha_{m-1}$ inscribed in $\gamma_{m-1}$, and this leads to a compound of type (3), but in dimension $m - 1$. The converse is also true, that for each of these three types, any such compound arises in this way from a Hadamard matrix.

Thus, apart from trivial cases where $m \leq 2$, $m$ must be divisible by 4. Conversely, Coxeter deduced from Paley's work [66] on Hadamard matrices that compounds of types (1) and (2) exist in dimensions $4, 8, 12, \ldots, 88$, and those of type (3) exist in dimensions $3, 7, 11, \ldots, 87$, although the next cases 92 and 91 were at that time still undecided. (A Hadamard matrix of order 92 was found in 1962 by Baumert, Golomb and Hall [3], using a construction due to Williamson [88] and a great deal of computing.) Similarly, when $m = 2^k$ the $2^m/m$ Hadamard matrices given by Paley's main theorem yield a compound

$$\gamma_{2^k-1}[2^{2^k-k-1}\alpha_{2^k-1}]\beta_{2^k-1}$$

for each $k \geq 2$, generalising the stella octangula for $k = 2$.

As shown by Todd [81], Hadamard matrices of order $m = 4n$ are equivalent to certain block designs, or arrangements of subsets, on $4n - 1$ points, so Todd's results on the orders of their automorphism groups can be used to consider the possibilities for the parameters $d_1$ and $D$. If the automorphism group has order $N$ there are $(4n - 1)!/N$ different designs. If we take the $4n - 1$ elements as coordinate places, each block corresponds to a point in $\mathbb{R}^{4n-1}$ with coordinates respectively $\pm 1$ at its elements and non-elements. Each design, therefore, gives $4n - 1$ points which, together with $(1, 1, \ldots, 1)$, are the vertices of a simplex $\alpha_{4n-1}$ inscribed in $\gamma_{4n-1}$. The $(4n - 1)!/N$ different designs thus yield a compound of each type (1), (2) and (3), with $d_1 = (4n - 1)!/N$ and hence

$$D = \frac{d_1 v}{V} = \frac{2^m d_1}{2m} = \frac{2^{4n-3} d_1}{n} = \frac{2^{4n-3}(4n-1)!}{nN}.$$

For example, for $n = 2$ the design based on quadratic residues mod (7) has automorphism group $PSL_2(7)$ of order $N = 168$, giving such compounds with $d_1 = 30$

and $D = 480$. Similarly, for $n = 3$, using quadratic residues mod (11) gives $N = |PSL_2(11)| = 660$, $d_1 = 60480$ and $D = 10321920$.

In his text [17] on regular polytopes, Coxeter considered compound polytopes of dimensions 3 and 4 in considerable detail, but unfortunately when it came to higher dimensions [17, pp. 287–288] he wrote simply 'To save space, we have disregarded the possibility of compounds in more than four dimensions', and after giving a few examples, 'The theory of these compounds is connected with orthogonal matrices of $\pm 1$'s', with no references to Hadamard, Sylvester, Paley or Todd.

## 8 The Origin of the Paley Graphs

The concept of a graph does not appear in Paley's paper [66], nor in the accompanying papers by Todd [81] and Coxeter [16]. This is not surprising since they had no need of graphs for their work, and in any case, graphs were little known and rarely studied in the 1930s; indeed, the first textbook on graph theory, by König [55], was not published until 1936, three years after Paley's death. The graphs which eventually carried his name first appeared in the literature nearly 30 years later, in two highly influential and almost simultaneous papers, one by Sachs [72], and the other by Erdős and Rényi [28].

### *8.1 Sachs*

In 1962 Sachs [72] introduced the concept of a self-complementary graph, one which is isomorphic to its complement. In this paper, he was particularly interested in such graphs which are also regular (meaning that all vertices have the same valency) and cyclic (invariant under a cyclic permutation of the vertices). As examples, he constructed the Paley graphs $P(q)$ for primes $q = p \equiv 1 \bmod (4)$, using elementary properties of quadratic residues and Legendre symbols to verify that they satisfy these conditions. For instance, multiplying all vertices by a fixed non-residue induces an isomorphism $P(q) \to \overline{P(q)}$. Sachs did not consider the automorphism group, apart from noting that translation by 1 confirms the cyclic property. Nor did he consider the general case of prime powers $q$ (in this case, $P(q)$ is again self-complementary and regular, but it is not cyclic unless $q$ is prime). He did not name these graphs in his paper, nor did he cite Paley, or any other source, for them.

### *8.2 Erdős and Rényi*

In 1963 Erdős and Rényi [28] considered the following problem (among many others): given an asymmetric graph $G$ (one with no non-identity automorphisms), what is the minimum number $A(G)$ of edge-changes (deletions or insertions of edges) required to allow a non-identity automorphism? If $\Delta_{uv}$ denotes the number of vertices $w \neq u, v$ adjacent to just one of the distinct vertices $u$ and $v$ in $G$, then making $\Delta_{uv}$ edge-changes allows an automorphism transposing $u$ and $v$, and fixing all other vertices, so

$$A(G) \leq \min_{u \neq v} \Delta_{uv}. \tag{6}$$

Simple counting arguments show that if $G$ has order $n$ then

$$\min_{u \neq v} \Delta_{uv} \leq \lfloor \frac{n-1}{2} \rfloor, \tag{7}$$

so that

$$A(G) \leq \lfloor \frac{n-1}{2} \rfloor. \tag{8}$$

(See [1] for a much stronger bound $A(G) \leq 5$ for planar graphs.) In their paper, Erdős and Rényi conjectured that no asymmetric graph attains the upper bound in (8). However, they constructed the Paley graphs $P(q)$ (without referring to Paley) as examples (far from asymmetric) of what they called $\Delta$-graphs, those which attain equality in (7): indeed, it follows immediately from (2) that $\Delta_{uv} = (q-1)/2$ for all pairs $u \neq v$ in $P(q)$.

They constructed $P(q)$ first [28, p. 301] for primes $q \equiv 1 \bmod (4)$, referring to Lagrange, Perron [70] and Kelly [53] for the fact, equivalent to (2), that if $a \neq 0$ in $\mathbb{F}$ then $S + a$ contains $(q-1)/4$ quadratic non-residues. (It is frustrating that they gave no citation for Lagrange.) Later [28, p. 302] they extended the construction to all prime powers $q \equiv 1 \bmod (4)$, again quoting Kelly [53] for the required form of (2) in this more general context. They did not consider the full automorphism group of $P(q)$, merely noting that translation by 1 is always a non-identity automorphism.

In remarks added after submission, Erdős and Rényi referred to a forthcoming paper by Bose (presumably [7]) on strongly regular graphs, and pointed out that any $\Delta$-graph is strongly regular, of order $n \equiv 1 \bmod (4)$ and valency $k = (n-1)/2$, though their assertion about the numbers $\lambda$ and $\mu$ (in modern notation) of common neighbours of two adjacent and non-adjacent vertices is clearly incorrect (see (2) for the correct values). They also remarked that in the case where $q$ is prime, their $\Delta$-graphs $P(q)$ coincide with the self-complementary graphs constructed by Sachs in [72] (see Sect. 8.1).

## 8.3 Naming the Paley Graphs

Although the graphs $P(q)$ were known, and in the literature, by the early 1960s, Paley's name does not seem to have been associated with them until the early 1970s. The first appearance in the literature the author has found for the term 'Paley graph' is in the book by Cameron and van Lint [12, p. 14], published in 1975, where it is introduced as if it was already accepted terminology. It appears in the 2nd edition (1993) of the book by Biggs [5] on Algebraic Graph Theory, but not the first, published in 1974. The earliest title in MathSciNet containing the term is a paper [6] by Blass, Exoo and Harary, published in 1981. On the other hand, the term ‘Paley design’ was used by Dembowski in 1968, in his book on Finite Geometries [23], and the following year by Kantor in [50], so perhaps the term ‘Paley graph’ evolved naturally from this. Several colleagues have suggested Jaap Seidel as the originator of the term, but he is unfortunately no longer with us to confirm or deny this. His highly influential 1965 paper with van Lint [82] cited that of Paley [66], and used Paley’s matrices to construct equilateral point sets in elliptic geometry, while Andries Brouwer has confirmed that the term was standard and understood by all in Eindhoven in the 1970s. Whoever originated the term, whether or not it is justified is discussed in Sect. 10.

### *8.4 Pseudo-Paley Graphs*

The Paley graphs are strongly regular graphs with parameters $v = q, k = (q-1)/2$, $\lambda = (q-5)/4$, $\mu = (q-1)/4$. However, these properties do not characterise them. A *pseudo-Paley graph* is a strongly regular graph with the same parameters $v, k, \lambda, \mu$ as a Paley graph. In 2001 Peisert [69] constructed a new infinite class of such graphs, now called *Peisert graphs*, as follows.

Let $\mathbb{F} = \mathbb{F}_q$ where $q$ is an even power of a prime $p \equiv 3 \bmod (4)$, and let $P^*(q)$ be the Cayley graph for the additive group of $\mathbb{F}$ with respect to the generating set $\{\omega^j \mid j \equiv 0 \text{ or } 1 \bmod (4)\}$, where $\omega$ is a primitive root in $\mathbb{F}$ (that is, a generator of the group $\mathbb{F}^*$). This is an undirected graph, which is (up to isomorphism) independent of the choice of $\omega$. It is a pseudo-Paley graph, and like $P(q)$ it is self-complementary and arc-transitive, but it is not isomorphic to a Paley graph. Indeed, Peisert showed that, apart from one other graph of order $23^2$, the graphs $P(q)$ and $P^*(q)$ are the only pseudo-Paley graphs which are self-complementary and arc-transitive.

More recently Klin, Kriger and Woldar [54] have used association schemes based on affine planes to construct pseudo-Paley graphs of order $q = p^2$ for odd primes $p$. Most of these are neither self-complementary (for $p \geq 17$) nor arc-transitive (for $p \geq 11$). The numbers of both self-complementary and non-self-complementary examples grow rapidly as $p \to \infty$.

## 9 The Automorphism Group of a Paley Graph

### *9.1 Characterising the Automorphisms*

When they introduced the graph $P(q)$, both Sachs [72] and Erdős and Rényi [28] noted that $v \mapsto v + 1$ is an automorphism of the graph. In Sachs’s case, $q$ is prime, so this implies that the graph is vertex-transitive, but neither of these papers contains any further discussion of the automorphism group.

In fact, it is clear from its construction that $P(q)$ is invariant under translation by any element of $\mathbb{F}$, under multiplication by any element of $S$, and under any field automorphism of $\mathbb{F}$. For any odd prime power $q$, these transformations generate the subgroup

$$A\Delta L_1(q) := \{v \mapsto av^\gamma + b \mid a \in S,\ b \in \mathbb{F},\ \gamma \in \mathrm{Gal}\,\mathbb{F}\}$$

of order $q(q-1)e/2$ and of index 2 in $A\Gamma L_1(q)$, so for each $q \equiv 1 \bmod (4)$ we have

$$A\Delta L_1(q) \leq \mathrm{Aut}\ P(q).$$

For example, the automorphisms of $P(9)$ induced by the additive, multiplicative and Galois groups of the field $\mathbb{F} = \mathbb{F}_9$ can be seen in Fig. 1 as translations, rotations about 0 and reflection in the horizontal axis. Similarly, automorphisms $v \mapsto 4v$ and $v \mapsto v + 1$ of order 6 and 13 of $P(13)$, generating $A\Delta L_1(13)$, can be seen in Figs. 2 and 3.

In fact, the elements of $A\Delta L_1(q)$ are the only automorphisms:

**Theorem 9.1** *If* $q \equiv 1 \bmod (4)$ *then*

$$\mathrm{Aut}\ P(q) = A\Delta L_1(q).$$

***Proof*** We have already established one inclusion. To prove the reverse inclusion, let $\alpha$ be any automorphism of $P(q)$. Since the subgroup $A\Delta L_1(q)$ of Aut $P(q)$ acts transitively on the arcs of $P(q)$, by composing $\alpha$ with a suitable element of this subgroup we may assume that $\alpha$ fixes 0 and 1. As an automorphism of $P(q)$, $\alpha$ satisfies

$$\chi(\alpha(u) - \alpha(v)) = \chi(u - v) \tag{9}$$

for all $u, v \in \mathbb{F}$. In 1960 Carlitz [14] proved that if $q$ is a power of an odd prime $p$, then any permutation of $\mathbb{F}_q$ fixing 0 and 1 and satisfying (9) has the form $v \mapsto v^{p^i}$ for some $i$. This implies that $\alpha$ is a field automorphism, so $\alpha \in A\Delta L_1(q)$. □

In the terminology introduced by Wielandt in [86], Theorem 9.1 asserts that the permutation group $A\Delta L_1(q)$ is 2-closed, that is, it is the full automorphism group of the set of binary relations on $\mathbb{F}_q$ which it preserves.

To be more precise about [14], the theorem Carlitz proved was as follows:

**Theorem 9.2** *If $q$ is a power of an odd prime $p$, then any permutation polynomial $\mathbb{F}_q \to \mathbb{F}_q$, which fixes* 0 *and* 1 *and satisfies (9), has the form $v \mapsto v^{p^i}$ for some $i$.*

However, simple counting shows that any function $\mathbb{F}_q \to \mathbb{F}_q$ can be represented by a polynomial (of degree less than $q$), and as Carlitz later wrote (see his errata [14, p. 999] and Hall's review [35]), any function satisfying (9) must be a permutation, so his theorem actually applies to any function fixing 0 and 1 and satisfying (9).

Carlitz's theorem was motivated by a problem in finite geometry. In [14], he simply wrote that it 'answers a question raised by W. A. Pierce in a letter to the writer,' without giving any further details. However Hall, in his review [35] of the paper, stated that 'The result is somewhat negative in its applications to the theory of projective planes since it shows that a construction by Pierce, analogous to the Moulton construction of non-Desarguesian planes, can yield only Desarguesian planes in the prime case.' Clearly, this was a reference to Pierce's paper [71], which was published a year later and which used Carlitz's result to extend Moulton's construction [63]. (For further background, see comments in the introduction of the paper [62] by McConnel, who was a student of Carlitz.)

## 9.2 *Carlitz*

Leonard Carlitz (1907–1999) completed his doctorate at the University of Pennsylvania in 1930. After a year working with E. T. Bell at Caltech, he spent the academic year 1931–1932 as an International Research Fellow in Cambridge, where Hardy had just returned after eleven years in Oxford. According to Hayes's obituary of Carlitz [38], 'This was the era when Hardy and Littlewood led one of the great centres of research in number theory, and Carlitz found the mathematical atmosphere there exhilarating. His work in additive number theory derives from that period.' In [60] Bollobás has written'In December 1931, Hardy and Littlewood announced weekly meetings of a conversation class to start in January 1932 in Littlewood's rooms. According to E. C. Titchmarsh, "this was a model of what such a thing should be. Mathematicians of all nationalities and ages were encouraged to hold forth on their own work, and the whole exercise was conducted with a delightful informality that gave ample scope for free discussion after each paper." Nevertheless, as Dame Mary Cartwright wrote, a little later there was a metamorphosis of Littlewood's conversation class into a larger gathering run by Hardy.'

Paley was still in Cambridge that year, so it seems inevitable that, as common members of the group around Hardy and Littlewood, he and Carlitz would have met and come to know each other. To what extent, if any, they influenced each other, is unknown. However, it seems likely that when, nearly 30 years later, Carlitz proved Theorem 9.2, he was completely unaware of any possible connection with Paley and his work.

After his year in Cambridge, Carlitz took up a position at the recently founded Duke University, in North Carolina. Indeed, it seems likely that it was a strong reference from Hardy which got him this position: who else could have been the 'Oxford don' who, according to Durden [27] (see also [61]), wrote that Carlitz was 'fully master of the technique of his trade' and 'better equipped in the analytic theory of numbers than anyone else in America'? Carlitz spent the rest of his career at Duke University, editing the Duke Mathematical Journal and becoming one of the most prolific mathematicians of the 20th century.

## 9.3 Subsequent Proofs

Carlitz's proof of Theorem 9.2 is technically quite difficult, and it involves no graph theory or group theory, just calculations with polynomials over finite fields. However, in the case $q = p$, Theorem 9.1 is a straightforward consequence of Burnside's theorem [11] that a simply transitive group of prime degree $p$ must be solvable, and hence (by a result of Galois, see [43, Satz II.3.6]) a subgroup of $AGL_1(p)$, as pointed out by Bruen [9] in 1972. More generally, since Aut $P(q)$ contains $A\Delta L_1(q)$ and cannot be doubly transitive, it is a rank 3 permutation group with suborbit-lengths 1, $(q-1)/2$, $(q-1)/2$. It is, therefore primitive, since $1 + (q-1)/2$ does not divide $q$. In the case $q = p^2$ Theorem 9.1, therefore, follows easily from Wielandt's classification [86] of simply primitive groups of degree $p^2$ (see also [49, Theorem B′]): these are either rank 3 subgroups of the wreath product $S_p \wr S_2$, with suborbit-lengths $1, 2(p-1), (p-1)^2$, or subgroups of $AGL_2(p)$. Comparing suborbit-lengths rules out the first possibility, and in the second case linear algebra gives the result. See also results of Dobson and Witte [25] on automorphism groups of graphs with $p^2$ vertices. Later we will show how group theory can also deal with higher powers of $p$.

It took some time before the significance of Theorem 9.2 for Paley graphs was realised. In 1969 Kantor used it, and cited [14], in proving [50, Corollary 8.2]; like Carlitz's theorem, his result was purely about permutations of finite fields, and even though it was in a paper on automorphisms of designs there was no application to automorphism groups. As late as 1972, Shult, having defined $P(q)$ (but not named it) in [77, Example 2], wrote that it was an open question whether Aut $P(q)$ was equal to the group we have called $A\Delta L_1(q)$ or larger, though he noted that certain cases could be handled, citing Higman's paper [40]. Shult gave credit to Kantor for this example, so clearly, the link between Carlitz's theorem and the Paley graphs was not widely understood among group theorists in the early 1970s. Dembowski [23], writing in 1968, included Carlitz's paper [14] in his bibliography, possibly to support a citation of Pierce's paper [71] on p. 233, but after two careful searches of the whole book, the present author has not found any citation of [14].

## 9.4 Generalisations of the Main Theorem

In 1963 McConnel [62] generalised Theorem 9.2, for any prime power $q$, as follows:

**Theorem 9.3** *Let $d$ be a proper divisor of $q-1$, and for $x \in \mathbb{F}$ let $\phi(x) := x^m$ where $m = (q-1)/d$. Then a function $f : \mathbb{F} \to \mathbb{F}$ satisfies*

1. $f(0) = 0$ *and* $f(1) = 1$,
2. $\phi(f(u) - f(v)) = \phi(u - v)$ *for all* $u, v \in \mathbb{F}$,

*if and only if $f(x) = x^{p^j}$ for some $j$ where $d$ divides $p^j - 1$.*

In particular, if $q$ is odd and $d = 2$, we have $\phi = \chi$, giving Theorem 9.2. As in [14], neither graph theory nor group theory was used in [62].

In 1972, using the fact that the functions $f$ satisfying condition (2) of Theorem 9.3, or equivalently satisfying

$$\frac{f(u) - f(v)}{u - v} \in D := \{x \mid x^m = 1\} \text{ for all } u \neq v \in \mathbb{F},$$

form a group under composition, Bruen gave a simple algebraic proof of McConnel's theorem in the case where $q = p$. This was based (as in the case $d = 2$) on Burnside's theorem on permutation groups of prime degree. Bruen did not explicitly name or describe this group (let us call it $G(d)$), but it is clear that its elements are the $mqh$ transformations

$$x \mapsto ax^{p^j} + b \quad (a \in D,\ b \in \mathbb{F},\ d \mid p^j - 1),$$

where $q = p^e$ and $h = \gcd(m, e)$; those elements also satisfying condition (1), or equivalently $a = 1$ and $b = 0$, form the subgroup fixing 0 and 1.

In 1973 Bruen and Levinger [10] extended Bruen's algebraic proof of McConnel's theorem to the case of all prime powers $q$, using ideas taken from Wielandt's proof of Burnside's theorem given in Passman's book [67, Theorem 7.3]. (Dress, Klin and Muzychuk have given an elementary and largely geometric proof of Burnside's theorem in [26], together with a detailed survey of alternative proofs by Burnside, Schur, Wielandt and others.) An essential ingredient in Bruen and Levinger's proof is the vector space of all functions $\mathbb{F} \to \mathbb{F}$ (represented as polynomials of degree less than $q$), and the actions on it of various groups of permutations of $\mathbb{F}$. Some of these ideas overlap with those involving invariant relations and functions, developed by Wielandt in [86]. Taking $d = 2$, the paper [10] seems to be the first to give an explicit description of the elements of Aut $P(q)$, and thus to give an implicit statement of Theorem 9.1.

In 1990 Lenstra [58] gave a rather shorter proof of McConnel's theorem, based on that of Bruen and Levinger. He stated the theorem in the slightly more elegant form that

$$G(d) = \{x \mapsto ax^{\gamma} + b \mid a \in D,\ b \in \mathbb{F},\ \gamma \in \operatorname{Gal}\mathbb{F},\ \phi^{\gamma} = \phi\},$$

where now $\phi$ is an epimorphism $\mathbb{F}^* \to E$ for some group $E$ (necessarily cyclic) of order $d$. In addition, he considered those functions $f : \mathbb{F} \to \mathbb{F}$ for which there is a permutation $\kappa$ of $E$ such that

$$\phi(f(u) - f(v)) = \kappa\phi(u - v) \text{ for all } u \neq v \in \mathbb{F}. \tag{10}$$

He showed that these functions $f$ form the normaliser $N(G(d))$ of $G(d)$ in the symmetric group on $\mathbb{F}$. In order to describe the elements of this group, let $K$ denote the subfield of $\mathbb{F}$ generated by $D$, and define a $K$*-semilinear automorphism* of $\mathbb{F}$ to be an automorphism $\beta$ of the additive group of $\mathbb{F}$ for which there is a field automorphism $\gamma$ of $K$ satisfying $\beta(xy) = (\gamma x)(\beta y)$ for all $x \in K$ and $y \in \mathbb{F}$. Then $N(G(d))$ consists of the transformations $x \mapsto x^{\beta} + b$ of $\mathbb{F}$ such that $b \in \mathbb{F}$ and $\beta$ is a $K$-semilinear automorphism of $\mathbb{F}$.

As we have remarked, Carlitz's paper contains no references to graphs, groups or automorphisms, or to Paley. Some of these later generalisations by McConnel, Bruen, Levinger and Lenstra use group theory, to a varying extent, but none of them mentions graphs or Paley. The first proof to do that is the subject of the next section.

## 9.5 *Muzychuk's Proof of the Main Theorem*

In 1987 Muzychuk [64] independently gave a full proof of Theorem 9.1. At that time he was a Ph.D. student in Kiev, supervised by V.A.Ustimenko. This was towards the end of a long period during which contacts between Soviet mathematicians and those in the West were almost non-existent, so it is not surprising that the results of Carlitz and his successors were not known there, and were not cited in this paper; indeed, the only citations were to the Russian translations of the book by Cameron and van Lint [13] for the definition (and name) of the Paley graphs, and of that by Serre [76] for some basic properties of finite fields. As in the case of some of the earlier proofs, the main argument involves the ingenious use of polynomials over finite fields. The paper was written in Russian, and published in a journal difficult to access outside the former Soviet Union, but as international contacts became much easier in the 1990s it became more widely known, with several recent citations listed in MathSciNet. An English translation, including an extension of the main theorem to cover cyclotomic schemes, is included in this volume [65].

## 9.6 *Generalised Paley Graphs*

In 2009 Lim and Praeger [59] introduced a class of graphs which generalise the Paley graphs, and in certain cases, they found their automorphism groups. For consistency with earlier sections, we have changed their notation slightly.

Let $\mathbb{F} = \mathbb{F}_q$ for any prime power $q$, let $m$ be any divisor of $q - 1$, and let $D$ be the unique subgroup of order $m$ in $\mathbb{F}^*$. Like $P(q)$, a *generalised Paley graph* $P = P^{(m)}(q)$ has vertex-set $\mathbb{F} = \mathbb{F}_q$, but with vertices $u$ and $v$ adjacent if and only if $u - v \in D$; in other words, $P$ is the Cayley graph for the additive group of $\mathbb{F}$, with connection set $D$. In order that $D = -D$, giving an undirected graph, we need to impose the following restriction:

- if $q$ is odd then $m$ is even.

Here, unlike Lim and Praeger, we will also assume that

- $D$ generates the additive group of $\mathbb{F}$,

so that $P$ is connected. For example, if $q \equiv 1 \bmod (4)$ and $m = (q-1)/2$, then $P$ is the Paley graph $P(q)$.

It is clear that the transformations

$$x \mapsto ax^{\gamma} + b \quad (a \in D,\ b \in \mathbb{F},\ \gamma \in \mathrm{Gal}\,\mathbb{F}) \tag{11}$$

of $\mathbb{F}$, which form a subgroup of index $d = (q-1)/m$ in $A\Gamma L_1(q)$, are all automorphisms of $P$. Unfortunately, the various extensions of Carlitz's theorem which we have discussed do not provide a converse, since if $m < (q-1)/2$ then condition (2) of Theorem 9.3 is too restrictive: we need the weaker condition that if $\phi(u-v) = 1$ then $\phi(f(u) - f(v)) = 1$. Similarly, Lenstra's condition in Eq. (10) does not help in this situation.

However, Lim and Praeger [59] proved a partial converse, showing that the transformations in (11) are the only automorphisms of $P$ provided $D$ is 'large' in the following sense:

– the index $d = |\mathbb{F}^* : D|$ divides $p - 1$, where $q$ is a power of the prime $p$.

(This implies that $|D \cup \{0\}| > q/p$, so that $P$ is connected. However, there are examples where $P$ is connected, but $d$ does not divide $p-1$ and $\mathrm{Aut}\,P$ is larger: for instance, if $q = p^2$ and $m = 2(p-1)$ then $P$ is a Hamming graph and $\mathrm{Aut}\,P$ is a wreath product $S_p \wr S_2$, of order $2(p!)^2$.) Their proof of this partial converse (unlike the rest of the results in [59]) depends on the classification of finite simple groups, and it would be interesting to find a more elementary proof. For further properties of generalised Paley graphs, see [46] for their regular surface embeddings, and [68] for their product decompositions; see also [68, Remark 1.2] for applications of these graphs to topics such as Ramsey theory and synchronizing groups.

## 9.7 The Automorphism Group of the Paley Tournament

The analogue of the Paley graph for a prime power $q \equiv 3 \bmod (4)$ is a directed graph, called the *Paley digraph*, or *quadratic residue digraph*. This is a tournament since every distinct pair of vertices are joined by a single arc. The automorphism group of any finite tournament has odd order (since no element can transpose two vertices), so by the Feit-Thompson Theorem [29] it is solvable. This makes the study of automorphism groups of tournaments relatively straightforward.

In 1970, Goldberg [33] used results on permutation groups to prove that the Paley digraph has automorphism group $A\Delta L_1(q)$. (There is no reference to Paley in this paper.) In a late note, the author wrote that his theorem was a special case of an unpublished result of Kantor, stated without proof in Dembowski's book [23, p. 98] (see also Kantor's results on 2-homogeneous groups [52], published in 1972). In 1972 Berggren [4] also proved this theorem and showed that the Paley digraphs are the only finite symmetric (vertex- and arc-transitive) tournaments.

## 9.8 Automorphism Groups of Hadamard Matrices

The automorphisms of a Hadamard matrix $H$ were defined by Hall [36] to be the ordered pairs $(P, Q)$ of monomial matrices $P$ and $Q$, with non-zero entries $\pm 1$, such that $PHQ^T = H$; these form a group Aut $H$. The element $(-I, -I)$ is a central involution $\sigma \in \text{Aut } H$, and the quotient $\overline{\text{Aut}}\, H = \text{Aut } H/\langle\sigma\rangle$ acts faithfully on the union of the sets of rows and columns of $H$.

The Paley-Hadamard matrices $H$ of type I have order $m = q + 1$ for prime powers $q \equiv 3 \bmod (4)$. In [36] Hall showed that for these matrices, $\overline{\text{Aut}}\, H$ contains $PSL_2(q)$ (not $P\Sigma L_2(q)$, as asserted in [51], though it does indeed contain this group), acting on both the rows and the columns as a group of Möbius transformations of the projective line $PG(1, q) = \mathbb{P}^1(\mathbb{F}_q) = \mathbb{F}_q \cup \{\infty\}$. He also showed that if $q = 11$ then $\overline{\text{Aut}}\, H$ is strictly larger, acting as the Mathieu group $M_{12}$ on the rows and columns, whereas in 1969 Kantor [51] showed that $\overline{\text{Aut}}\, H = P\Sigma L_2(q)$ whenever $q > 11$. By contrast, the automorphism group of the corresponding Paley design is $A\Delta L_1(q)$, a subgroup of index $q + 1$ in $P\Sigma L_2(q)$, if $q \geq 19$, whereas it is $PSL_2(q) = P\Sigma L_2(q)$ if $q = 7$ or 11 (see Sect. 7.4); the corresponding Paley graph $P(q)$ has automorphism group $A\Delta L_1(q)$ for all $q$ (see Sect. 9.1).

The Paley-Hadamard matrices $H$ of type II have order $m = 2(q + 1)$ for prime powers $q \equiv 1 \bmod (4)$. If $q = 5$ then $H$ is equivalent to the Paley-Hadamard matrix of type I and order 12, discussed above. In 2008 De Launey and Stafford [21] showed that if $q > 5$ then Aut $H$ has a subgroup of index 2 isomorphic to $\Gamma L_2(q)/S$, where we identify $S$ with the group of scalar matrices $\lambda I$ $(\lambda \in S)$ in $GL_2(q)$; the full group is obtained by adjoining an element of order 4, whose square is the central involution in $\Gamma L_2(q)/S$, corresponding the matrices $\lambda I$ $(\lambda \in \mathbb{F}^* \setminus S)$. Their proof uses the classification of finite simple groups, via the classification of 2-transitive finite permutation groups.

# 10 Attribution and Terminology

We have seen that the papers by Paley [66] and Carlitz [14], frequently cited in connection with the Paley graphs and their automorphism groups, do not in fact mention graphs, groups or automorphisms. Indeed, an inspection of their publication records suggests that neither of these mathematicians had much interest in either graph theory or group theory. This, therefore, raises the question of whether it is appropriate to refer to the graphs $P(q)$ as 'Paley graphs', or to attribute Theorem 9.1, describing their automorphism groups, to Carlitz.

In view of our remarks in the preceding sections, to refer to Theorem 9.1 as 'Carlitz's Theorem' seems slightly overgenerous, even though he did the hard work in providing most of the argument for the difficult half of the proof. (In any case, Carlitz is hardly short of recognition: MathSciNet lists 732 publications by him, together with (at the time of writing) 2308 citations of his work, and 260 publications with his name in the title.) Perhaps it would be more correct to reserve this term for the result he actually proved in [14], namely Theorem 9.2, and to refer to Theorem 9.1 as a straightforward corollary to his theorem.

At first sight, Paley's connection with the graphs $P(q)$ seems to be rather tenuous. As was later shown, Hadamard matrices, including those constructed by him, give rise to graphs, in the sense that normalising the matrix, deleting the first row and column, and replacing each entry $-1$ with 0, produces the adjacency matrix of a graph. Paley's first construction yields directed graphs of prime power order $q \equiv 3 \bmod (4)$ (the Paley tournaments), rather than the undirected graphs of order $q \equiv 1 \bmod (4)$ which bear his name, while his second construction yields undirected graphs of order $2q + 1$, where $q \equiv 1 \bmod (4)$. However, one ingredient of this second construction [66, p. 314] is a (non-Hadamard) matrix $(B_{ij})$ of order $q + 1$ such that deleting its first row and column yields the Jacobsthal matrix $Q = (\chi(j - i))$ for $\mathbb{F}_q$, and hence (after replacing each entry $-1$ with 0) the adjacency matrix for the graph $P(q)$. In this sense, the Paley graphs do arise naturally from Paley's paper, though not directly from the Hadamard matrices he constructed.

Whether this justifies naming these graphs after Paley is debatable. The first person to describe these graphs in the literature seems to have been Sachs [72], in 1962, who restricted attention to prime values of $q$, followed independently in 1963 by Erdős and Rényi [28], who considered the general case. In 1971 Higman [40] constructed the Paley graphs as examples of strongly regular graphs, but did not name them or refer to Paley; he referred to certain rank 3 permutation groups acting on $P(q)$ as 'of Singer type', a terminology which does not seem to have survived.

The main link between Paley and the graphs $P(q)$ is the use of quadratic residues, and in particular their combinatorial property given by Eq. (1). However, this result can be traced back at least to Jacobsthal [44, 45], a generation earlier. Whoever coined the term 'Paley graph' probably did so as shorthand for a more accurate but clumsy phrase such as 'graph based on Paley's construction', rather than as a deliberate attribution. As suggested to the author by Mikhail Muzychuk, the terms 'Paley graph' and 'Carlitz's theorem' appear to be further instances of Stigler's Law of Eponymy (which is, of course, itself also a misattribution [78]).

**Acknowledgements** The author is grateful to Brian Alspach, Norman Biggs, Béla Bollobás, Andries Brouwer, Peter Cameron, Chris Godsil, Willem Haemers, Joshua Insley, Mikhail Muzychuk, Cheryl Praeger and Don Taylor for helpful comments, information and suggestions, and in particular to Mikhail Klin and Franz Lemmermeyer for their valuable advice on the extensive literature related to the Paley graphs and on the mathematical works of Gauss.

## References

1. V.A. Aksionov, O.V. Borodin, L.S. Mel'nikov, G. Sabidussi, M. Stiebitz, B. Toft, Deeply asymmetric planar graphs. J. Combin. Theory Ser. B **95**, 68–78 (2005)
2. A. Basile, P. Brutto, Proprietà delle ovali di un piano desarguesiano finito, Rend. Circ. Mat. Palermo (2) **24**, 233–243 (1975)
3. L.D. Baumert, S.W. Golomb, M. Hall Jr., Discovery of a Hadamard matrix of order 92. Bull. Am. Math. Soc. **68**, 237–238 (1962)

4. J.L. Berggren, An algebraic characterisation of finite symmetric tournaments. Bull. Australian Math. Soc. **6**, 53–59 (1972)
5. N.L. Biggs, *Algebraic Graph Theory*, 2nd edn. 1993 (Cambridge University Press, Cambridge, 1974)
6. A. Blass, G. Exoo, F. Harary, Paley graphs satisfy all first-order adjacency axioms. J. Graph Theory **5**, 435–439 (1981)
7. R.C. Bose, Strongly regular graphs, partial geometries and partially balanced designs. Pac. J. Math. **13**, 389–419 (1963)
8. A. Brauer, On a new class of Hadamard determinants. Math. Z. **58**, 219–225 (1952)
9. A. Bruen, Permutation functions on a finite field. Canad. Math. Bull. **15**, 595–597 (1972)
10. A. Bruen, B. Levinger, A theorem on permutations of a finite field. Canad. J. Math. **25**, 1060–1065 (1973)
11. W. Burnside, On some properties of groups of finite order. Proc. Lond. Math. Soc. **33**, 162–185 (1900)
12. P.J. Cameron, J.H. van Lint, *Graph Theory, Coding Theory and Block Designs*, vol. 19. London Math. Soc. Lecture Notes (Cambridge University Press, Cambridge, 1975)
13. P.J. Cameron, J.H. van Lint, *Graph Theory, Coding Theory and Block Designs* [Russian], Translated from the English by B.S. Stečkin (Nauka, Moscow, 1980)
14. L. Carlitz, A theorem on permutations in a finite field. Proc. Am. Math. Soc. **11**, 456–459 (1960). Errata ibid. 999–1000
15. A. Cayley, Desiderata and suggestions: No. 2. The theory of groups: graphical representation, Am. J. Math. **1**, 174–176 (1878)
16. H.S.M. Coxeter, Regular compound polytopes in more than four dimensions. J. Math. Phys. **12**, 334–345 (1933)
17. H.S.M. Coxeter, *Regular Polytopes*, 3rd edn. (Dover, 1974)
18. H.S.M. Coxeter, W.O.J. Moser, *Generators and Relations for Discrete Groups*, 4th edn. (Springer, Berlin, 1979)
19. H. Davenport, *The Higher Arithmetic*, 5th edn. (Cambridge University Press, Cambridge, 1982)
20. W. de Launey, R.M. Stafford, On cocyclic weighing matrices and the regular group actions of certain Paley matrices. Discrete Appl. Math. **102**, 63–101 (2000)
21. W. de Launey, R.M. Stafford, On the automorphisms of Paley's type II Hadamard matrix. Discrete Math. **308**, 2910–2924 (2008)
22. P. Delsarte, J.M. Goethals, J.J. Seidel, Orthogonal matrices with zero diagonal, II. Canad. J. Math. **23**, 816–832 (1971)
23. P. Dembowski, *Finite Geometries* (Springer, Berlin, 1968)
24. L.E. Dickson, *History of the Theory of Numbers, II: Diophantine Analysis* (Stechert, New York, 1934)
25. E. Dobson, D. Witte, Transitive permutation groups of prime-squared degree. J. Algebraic Combin. **16**, 43–69 (2002)
26. A.W.M. Dress, M.H. Klin, M.E. Muzychuk, On $p$-configurations with few slopes in the affine plane over $\mathbb{F}_p$ and a theorem of W. Burnside's, Bayreuth. Math. Schr. **40**, 7–19 (1992)
27. R.F. Durden, *The Launching of Duke University, 1924–1949* (Duke University Press, 1993)
28. P. Erdős, A. Rényi, Asymmetric graphs. Acta Math. Acad. Sci. Hungary **14**, 295–315 (1963)
29. W. Feit, J.G. Thompson, Solvability of groups of odd order. Pac. J. Math. **13**, 755–1029 (1963)
30. C.F. Gauss, Theoria residuum biquadraticorum, Commentatio prima. Comment. Soc. Regiae Sci. Göttingen **6** (1828). German translation: C.F. Gauss, H. Maser, *Untersuchungen über höhere Arithmetik (Disquisitiones Arithmeticae and Other Papers on Number Theory)*, 2nd edn. (Chelsea, New York, 1965), pp. 511–533
31. R.E. Gilman, Abstract 68: on the Hadamard determinant theorem and orthogonal determinants. Bull. Am. Math. Soc. **37**, 30–31 (1931)
32. J.-M. Goethals, J.J. Seidel, Orthogonal matrices with zero diagonal. Canad. J. Math. **19**, 1001–1010 (1967)
33. M. Goldberg, The group of the quadratic residue tournament. Canad. Math. Bull. **13**, 51–54 (1970)

34. J. Hadamard, Résolution d'une question relative aux déterminants. Bull. Sci. Math. **17**, 240–248 (1893)
35. M. Hall, Jr, Review of [14]. Math. Reviews MR0117223 (22 8005), revised 1961
36. M. Hall Jr., Note on the Mathieu group $M_{12}$. Arch. Math. **13**, 334–340 (1962)
37. M. Hall Jr., Automorphisms of Hadamard matrices. SIAM J. Appl. Math. **7**, 1094–1101 (1969)
38. D.R. Hayes, Leonard Carlitz (1907–1999). Notices Am. Math. Soc. **48**, 1322–1324 (2001)
39. A.S. Hedayat, W.D. Wallis, Hadamard matrices and their applications. Ann. Stat. **6**, 1184–1238 (1978)
40. D.G. Higman, Solvability of a class of rank 3 permutation groups. Nagoya Math. J. **41**, 89–96 (1971)
41. K.J. Horadam, *Hadamard Matrices and Their Applications* (Princeton University Press, Princeton, 2007)
42. D.R. Hughes, F.C. Piper, *Design Theory* (Cambridge University Press, Cambridge, 1985)
43. B. Huppert, *Endliche Gruppen I* (Springer, Berlin, 1979)
44. E. Jacobsthal, Anwendung einer Formel aus der Theorie der quadratischen Reste, Dissertation, Berlin, 1906
45. E. Jacobsthal, Über die Darstellung der Zahlen der Form $4n + 1$ als summe zweier Quadrate. J. reine angew. Math. **132**, 238–250 (1907)
46. G.A. Jones, Characterisations and Galois conjugacy of generalised Paley maps. J. Combin. Theory Ser. B **103**, 209–219 (2013)
47. G.A. Jones, J.M. Jones, *Elementary Number Theory* (Springer, London, 1998)
48. G.A. Jones, J.M. Jones, *Information and Coding Theory* (Springer, London, 2000)
49. G.A. Jones, K.D. Soomro, On a theorem of Wielandt concerning simply primitive groups. Math. Proc. Camb. Phil. Soc. **92**, 419–423 (1982)
50. W.M. Kantor, 2-transitive symmetric designs. Trans. Am. Math. Soc. **146**, 1–28 (1969)
51. W.M. Kantor, Automorphism groups of Hadamard matrices. J. Combin. Theory **6**, 279–281 (1969)
52. W.M. Kantor, $k$-homogeneous groups. Math. Z. **124**, 261–265 (1972)
53. J.B. Kelly, A characteristic property of quadratic residues. Proc. Am. Math. Soc. **5**, 38–46 (1954)
54. M. Klin, N. Kriger, A. Woldar, On the existence of self-complementary and non-self-complementary strongly regular graphs with Paley parameters. J. Geom. **107**, 329–356 (2016)
55. D. König, *Theorie der endlichen und unendlichen Graphen* (Akademische Verlagsgesellschaft, Leipzig, 1936); reprinted Chelsea, New York, 1950
56. F. Lemmermeyer, *Reciprocity Laws: from Euler to Eisenstein* (Springer, Berlin, 2000)
57. F. Lemmermeyer, Private Communication (2016)
58. H.W. Lenstra Jr., Automorphisms of finite fields. J. Number Theory **34**, 33–40 (1990)
59. T.K. Lim, C.E. Praeger, On generalised Paley graphs and their automorphism groups. Michigan Math. J. **58**, 293–308 (2009)
60. J.E. Littlewood, *Littlewood's Miscellany* ed. B. Bollobás (Cambridge University Press, Cambridge, 1986)
61. MacTutor History of Mathematics archive, http://www-history.mcs.st-and.ac.uk
62. R. McConnel, Pseudo-ordered polynomials over a finite field. Acta Arith. **8**, 127–151 (1963)
63. F.R. Moulton, A simple non-Desarguesian plane. Trans. Am. Math. Soc. **3**, 192–195 (1902)
64. M.E. Muzychuk, Group of automorphisms of the Paley graph (Russian). *Problems in group theory and homological algebra (Russian)*, pp. 64–69, Matematika, Yaroslav. Gos. Univ., Yaroslavl', 1987, Amsterdam - London, 1973
65. M.E. Muzychuk, *Automorphism Groups of Paley Graphs and Cyclotomic Schemes*, (this volume)
66. R.E.A.C. Paley, On orthogonal matrices. J. Math. Phys. **12**, 311–320 (1933)
67. D. Passman, *Permutation Groups* (Benjamin, New York, 1968)
68. G. Pearce, C.E. Praeger, *Generalised Paley Graphs with a Product Structure*. Ann. Combin. **23**, 171–182 (2019)
69. W. Peisert, All self-complementary symmetric graphs. J. Algebra **240**, 209–229 (2001)

70. O. Perron, Bemerkungen über die Verteilung der quadratischen Reste. Math. Z. **56**, 122–130 (1952)
71. W.A. Pierce, Moulton planes. Canad. J. Math. **13**, 427–436 (1961)
72. H. Sachs, Über selbstkomplementäre Graphen. Publ. Math. Debrecen **9**, 270–288 (1962)
73. U. Scarpis, Sui determinanti di valore massimo. Rendiconti della R. Instituto Lombardo di Scienze e Lettere (2) **31**, 1441–1446 (1898)
74. P.H. Schoute, *Mehrdimensionale Geometrie, 2: Die Polytope* (G. J. Göschen, Leipzig, 1905)
75. J. Seberry, J. Wysocki, T.A. Wysocki, On some applications of Hadamard matrices. Metrika **62**, 221–239 (2005)
76. J.-P. Serre, *A Course in Arithmetic* [Russian], Translated from the French by A.I. Skopin (Mir, Moscow, 1972)
77. E. Shult, The graph extension theorem. Proc. Am. Math. Soc. **33**, 278–284 (1972)
78. S.M. Stigler, Stigler's law of eponymy. Trans. New York Acad. Sci. **39**, 147–158 (1980)
79. J. Stillwell, *Elements of Number Theory* (Springer, New York, 2003)
80. J.J. Sylvester, Thoughts on inverse orthogonal matrices, simultaneous sign-successions, and tessellated pavements in two or more colours, with applications to Newton's rule, ornamental tile-work, and the theory of numbers. Phil. Mag. (4) **34**, 461–475 (1867)
81. J.A. Todd, A combinatorial problem. J. Math. Phys. **12**, 321–333 (1933)
82. J.H. van Lint, J.J. Seidel, Equilateral point sets in elliptic geometry. Nederl. Akad. Wetensch. Proc. Ser. A **69** (= Indag. Math. **28**), 335–348 (1966)
83. S. Wagon, Editor's corner: the Euclidean algorithm strikes back. Am. Math. Monthly **97**, 125–129 (1990)
84. W.D. Wallis, A. Penfold Street, J. Seberry Wallis, *Combinatorics: Room Squares, Sum-free Sets, Hadamard Matrices*, vol. 292. Lecturer Notes in Mathematics (Springer, Berlin, 1972)
85. H. Wielandt, *Finite Permutation Groups* (Academic Press, New York, 1964)
86. H. Wielandt, *Permutation Groups through Invariant Relations and Invariant Functions*, Lecture notes (Ohio State University, Columbus, 1969)
87. N. Wiener, R E A C Paley - in memoriam. Bull. Am. Math. Soc. **39**, 476 (1933)
88. J. Williamson, Hadamard's determinant theorem and the sum of four squares. Duke Math. J. **11**, 65–81 (1944)
89. D. Zagier, A one-sentence proof that every prime $p \equiv 1 \pmod 4$ is a sum of two squares. Am. Math. Monthly **97**, 144 (1990)

# Automorphism Groups of Paley Graphs and Cyclotomic Schemes

**M. E. Muzychuk**

**Abstract** The paper gives a new proof of the old result of McConnell (Acta Arithm 8:127–151, 1963 [9]) stating that the automorphism group of a cyclotomic scheme over the finite field $\mathbb{F}_q$ is a subgroup of $A\Gamma L_1(q)$.

**Keywords** Automorphism group · Cyclotomic schemes · Paley graphs

## 1 Introduction

More than 30 years ago the author was asked by M. Klin whether is it possible to find an elementary proof of the fact that the automorphism group of the Paley graph is contained in $A\Gamma L_1(\mathbb{F}_q)$. As a result, the paper [11] was published where such a proof was presented. Since the journal where the paper appeared was not translated into English, the result was unavailable for all mathematicians working in algebraic graph theory. Last year during the Pilsen conference M. Klin and G. Jones convinced me to submit the English translation of my old result to the Proceedings of the Pilsen Conference on “Algebraic Graph Theory”.

Preparing the English version of [11] I realized that the technique used there works in more general setting. More precisely, it is possible to give a self-contained proof of the well-known fact that the automorphism group of any cyclotomic association scheme of order $q$ is contained in $A\Gamma L_1(q)$ (see [9]).

It seems that the proof of the McConnel's Theorem presented here is a new one. An excellent survey paper [7] written by G. Jones overviews all known proofs of the McConell's Theorem. It also discusses in detail many related results in this area.

Since the automorphism group of the Paley graph coincides with the automorphism group of the cyclotomic scheme with two classes, our main result immediately

---

The author was supported by the Israeli Ministry of Absorption.

---

M. E. Muzychuk (✉)
Department of Mathematics, Ben Gurion University of the Negev, Beer sheva, Israel
e-mail: muzychuk@bgu.ac.il



G. A. Jones et al. (eds.), *Isomorphisms, Symmetry and Computations in Algebraic Graph Theory*, Springer Proceedings in Mathematics & Statistics 305,
https://doi.org/10.1007/978-3-030-32808-5_6

implies that the full automorphism group of the Paley graph of order $q$ is contained in $A\Gamma L_1(q)$.

If $q$ is prime, then the McConnel's result is equivalent to the well-known Burnside's Theorem [4] which states that a simple transitive permutation group of prime degree $q$ is a subgroup of $AGL_1(q)$. A number of alternative proofs of the Burnside's Theorem were proposed since it was published (one of them is presented in [6]). We refer the reader to a nice note of Müller [10] which provides one of those proofs and presents a short overview of all known proofs of the Burnside's Theorem.

It is worth to mention that cyclotomic schemes over near fields were studied in [1, 5]. The complete classification of their automorphism groups was obtained in [5].

The paper is organized as follows. The main section consists of two subsections. The first part collects the necessary results about finite field functions with small number of slopes. All of the statements there except the last one are just English translations of some statements of [11] relevant to the main Theorem. Although the results presented in Sect. 2.1 are weaker than the ones of [2] we use them because they are elementary and make the paper self-contained.

The second subsection is devoted to the proof of the main result.

## 2 Cyclotomic Schemes and Their Automorphisms

Let $\mathbb{F}_q$ be the finite field with $q = p^n$ elements and let $C < \mathbb{F}_q^*$ be a proper nontrivial subgroup. A *cyclotomic scheme* [3] with $e := [\mathbb{F}_q^* : C]$ classes is an association scheme with the point set $\mathbb{F}_q$ the basic non-identical relations of which $R_1, \dots, R_e$ are in one-to-one correspondence with multiplicative cosets of $C$. More precisely, the relation $R_i$, $1 \le i \le e$ consists of all pairs $(x, y) \in \mathbb{F}_q$ satisfying the property $x - y \in C\omega^{i-1}$, where $\omega$ is a generator of $\mathbb{F}_q^*$. The automorphism group of the cyclotomic scheme consists of all permutations $\varphi \in \mathrm{Sym}(\mathbb{F}_q)$ which leaves invariant every relation $R_i$. In what follows we denote this group as $G(q, e)$.

Notice that all basic (di)graphs $\Gamma_i = (\mathbb{F}_q, R_i)$ are pairwise isomorphic. They are undirected if and only if $-1 \in C$. If $e = 2$ and $q \equiv 1 (\mathrm{mod}\ 4)$, then $\Gamma_1$ is the well-known Paley graph (see [7]). It's automorphism group coincides with $G(q, 2)$.

In the case of $e > 2$ the graph $\Gamma_1$ was called the *generalized Paley graph* in [8]. Its automorphism group was studied by Lim and Praeger in [8]. It was shown there that if $\Gamma_1$ is connected and not isomorphic to a Hamming graph, then $\mathrm{Aut}(\Gamma_1) \le AGL_n(p)$. The authors formulated an open problem asking under which conditions on $e$ and $q$ the group $\mathrm{Aut}(\Gamma_1)$ is contained in $A\Gamma L_1(q)$. The problem remains open since then.

There are three types of evident automorphisms of the cyclotomic scheme: additive, multiplicative and field automorphisms. The additive and multiplicative automorphisms generate a subgroup isomorphic to the semidirect product $\mathbb{Z}_p^n \rtimes C$ which acts on $\mathbb{F}_q$ in a natural way $x \mapsto ax + b$ where $a \in C, b \in \mathbb{F}_q$. Every linear transformation of this form belongs to $G(q, e)$.

Some of the field automorphisms, presented as permutations $x \mapsto x^{p^i}$, $i = 0, \ldots, n-1$, may also be automorphisms of the cylclotomic scheme. More precisely, the field automorphism $x \mapsto x^{p^i}$ is a scheme automorphism if and only if it fixes setwise each cyclotomic class $C_k = C\omega^k$, $k = 0, \ldots, e-1$. This is equivalent to $p^i \equiv 1(\text{mod } e)$. Thus the group $G(q, e)$ contains all permutations of the form

$$x \mapsto ax^{p^i} + b \text{ where } b \in \mathbb{F}_q, a \in C \text{ and } p^i \equiv 1(\text{mod } e). \tag{1}$$

The main result of this note states that those are the only automorphisms of the scheme. It is easy to check that a permutation from $A\Gamma L_1(\mathbb{F}_q)$ is an automorphism of the cyclotomic scheme only if it satisfies the condition (1). So, the main result is an immediate consequence of the following.

**Theorem 1** $G(q, e) \leq A\Gamma L_1(\mathbb{F}_q)$.

We start with some definitions. If $\varphi$ is any function from $\mathbb{F}_q$ to $\mathbb{F}_q$, define

$$A_\varphi = \left\{ \frac{\varphi(x) - \varphi(y)}{x - y} \,\middle|\, x, y \in \mathbb{F}_q, x \neq y \right\}.$$

The following statement is straightforward (see also the remark on p. 389 of [3]).

**Proposition 1** *A permutation* $\varphi \in \mathrm{Sym}(\mathbb{F}_q)$ *belongs to* $G(q, e)$ *if and only if* $A_\varphi \subseteq C$.

## 2.1 Field Functions with Small $|A_\varphi|$

Let $\varphi : \mathbb{F}_q \to \mathbb{F}_q$ be an arbitrary function. For each natural number $k \leq q$ we denote by $\sigma_k(\varphi)$ the $k$-th elementary symmetric function of the elements $\varphi(\alpha)$, $\alpha \in \mathbb{F}_q$, that is,

$$\sigma_k(\varphi) = \sum \varphi(\alpha_1)\varphi(\alpha_2)\ldots\varphi(\alpha_k),$$

where the summation is over all subsets $\{\alpha_1, \ldots, \alpha_k\} \subseteq \mathbb{F}_q$ of cardinality $k$.

We denote by $L(\mathbb{F}_q)$ the vector space of all functions from $\mathbb{F}_q$ to $\mathbb{F}_q$. We define on it a symmetric bilinear form

$$(f, g) = \sum_{\alpha \in \mathbb{F}_q} f(\alpha)g(\alpha).$$

**Proposition 2** (Proposition 1, [11]) *The functions* $e_i(x) = x^i$, *for* $i = 0, 1, \ldots, q-1$, *form a basis for the vector space* $L(\mathbb{F}_q)$, *with*[1]

[1] The function $e_0 = x^0$ is the constant function equal to 1.

$$(e_i, e_j) = \begin{cases} 0, & \textit{if } i = 0 \textit{ and } j = 0; \\ 0, & \textit{if } i + j \not\equiv 0 \,(\mathrm{mod}\, q-1), \\ -1, & \textit{if } i + j \equiv 0 \,(\mathrm{mod}\, q-1) \textit{ and } (i, j) \neq (0, 0). \end{cases}$$

***Proof*** The functions $e_i$ are obviously linear independent, and since $L(\mathbb{F}_q)$ has dimension $q$ they form a basis. The formula for $(e_i, e_j)$ follows from the well-known formula for the sum of powers of the elements of a finite field (cf. [12]). □

The next proposition establishes properties of functions in $L(\mathbb{F}_q)$ which are permutations of the set $\mathbb{F}_q$.

**Proposition 3** (Proposition 2, [11]) *The following are equivalent:*

*(i)* $\varphi$ *is a permutation of* $\mathbb{F}_q$*;*
*(ii)* $\prod_{\alpha\in\mathbb{F}_q}(z - \varphi(\alpha)) = z^q - z$*;*
*(iii)* $\sigma_k(\varphi) = \begin{cases} 0, & \textit{if } k \neq q-1 \textit{ and } 1 \leq k \leq q, \\ -1, & \textit{if } k = q-1; \end{cases}$
*(iv)* $\sum_{\alpha\in\mathbb{F}_q} \varphi^k(\alpha) = \begin{cases} 0, & \textit{if } 0 \leq k \leq q-2, \\ -1, & \textit{if } k = q-1 \end{cases}.$

***Proof*** (ii) $\iff$ (iii). This is a consequence of applying Vieta's Theorem to the polynomial $\prod_{\alpha\in\mathbb{F}_q}(z - \varphi(\alpha))$.

(i) $\iff$ (ii). If $\varphi$ is a permutation then

$$\prod_{\alpha\in\mathbb{F}_q}(z - \varphi(\alpha)) = \prod_{\alpha\in\mathbb{F}_q}(z - \alpha) = z^q - z.$$

Conversely, if $\prod_{\alpha\in\mathbb{F}_q}(z - \varphi(\alpha)) = z^q - z$ then $\varphi$ is a permutation, since this polynomial has $q$ distinct roots in $\mathbb{F}_q$.

(i) ⇒ (iv) This follows from the formula for sums of powers [12].

(iv) ⇒ (i). Let $\{y_1, \ldots, y_m\}$ be the image of the function $\varphi$. Then for each $x \in \mathbb{F}_q$ the equation

$$(\varphi(x) - y_1)(\varphi(x) - y_2)\ldots(\varphi(x) - y_m) = 0$$

holds. This means that the function $\varphi^m$ can be expressed as a linear combination of $\varphi^{m-1}, \ldots, \varphi, 1$. However, $\varphi^l$ can then be expressed as a linear combination of $\varphi^{m-1}, \ldots, \varphi, 1$ for all $l \geq m$. In particular, if $q - 1 \geq m$, then

$$\varphi^{q-1}(\alpha) = \sum_{i=0}^{q-2} \beta_i \varphi^i(\alpha) \quad (\alpha \in \mathbb{F}_q).$$

Summing both sides of this equation over all $\alpha \in \mathbb{F}_q$, we get

$$-1 = \sum_{i=0}^{q-2} \beta_i \sum_{\alpha\in\mathbb{F}_q} \varphi^i(\alpha).$$

By our assumption, $\sum_{\alpha \in \mathbb{F}_q} \varphi^i(\alpha) = 0$ for all $i \leq q-2$. Thus $-1 = 0$ in the case of $q-1 \geq m$. Hence $m-1 = q-1$, as required. □

**Proposition 4** (Proposition 3(i), [11]) *Let* $\varphi \in L(\mathbb{F}_q)$. *Then* $\alpha \in A_\varphi \iff \varphi(x) - \alpha x$ *is not a permutation;*

***Proof*** (i) By definition, $\alpha \in A_\varphi$ if and only if there exist elements $x, y$ $(x \neq y)$ for which

$$\frac{\varphi(x) - \varphi(y)}{x - y} = \alpha.$$

Rewriting this equation as

$$\varphi(x) - \alpha x = \varphi(y) - \alpha y,$$

we see that $\alpha \in A_\varphi$ if and only if the function $x \mapsto \varphi(x) - \alpha x$ is not injective. □

The next statement provides an additional information about functions with small $|A_\varphi|$.

**Lemma 1** ([11], Lemma 1) *Suppose that* $\varphi \in L(\mathbb{F}_q)$, $\varphi$ *is not a permutation, and* $|A_\varphi| \leq (q-1)/2$. *Then* $\sum_{\alpha \in \mathbb{F}_q} \varphi^k(\alpha) = 0$ *for all natural numbers k.*

***Proof*** For any $a \in \mathbb{F}_q$ define a function $\varphi_a(x) = \varphi(x) - ax$. Clearly, $P_k(a) := \sigma_k(\varphi_a)$ is a polynomial in $a$. Its leading coefficient coincides with the $k$th elementary symmetric function of the elements of the field $\mathbb{F}_q$. Therefore $P_k$ has degree at most $k-1$ if $k < q-1$. On the other hand, it follows from Propositions 4 and 3(iii) that $P_k(a) = 0$ if $a \in \mathbb{F}_q \setminus A_\varphi$ and $k < q-1$, that is, $P_k$ has at least $q - |A_\varphi| \geq (q+1)/2$ roots. Therefore, the polynomial $P_k(a)$ is identically equal to 0 for all $k \leq (q+1)/2$. In particular, $\sigma_k(\varphi) = P_k(0) = 0$ for all $k \leq (q+1)/2$.

Now let us consider the polynomial

$$\psi(z) = \prod_{\alpha \in \mathbb{F}_q} (z - \varphi(\alpha)).$$

It follows from the preceding part of the proof that $\psi(z) = z^q + f(z)$ where

$$f(z) = a_{(q-1)/2} z^{(q-1)/2} + \cdots + a_0.$$

Let $\{y_1, y_2, \ldots, y_m\}$ be the image of the function $\varphi$. We can see from the definition of the polynomial $\psi(z)$ that $\psi(y_i) = 0$ for all $i = 1, \ldots, m$. Also, every element of the field $\mathbb{F}_q$ is a root of the equation $z^q = z$. Therefore $y_1, \ldots, y_m$ are roots of the polynomial $f(z) + z$. Since $\varphi$ is not a permutation, $f(z) + z$ is a non-zero polynomial (Proposition 3(ii)). Taking into account that its degree is at most $(q-1)/2$, we see that $m \leq (q-1)/2$.

Let us denote by $n_i$ the cardinality of the pre-image $\varphi^{-1}(y_i)$ of the element $y_i$. It is clear that $n_i$ is the multiplicity of $y_i$ as a root of the polynomial $\psi$. Then the multiplicity of $y_i$ as a root of the derivative $\psi'$ is at least $n_i - 1$. However,

$$\sum_{i=1}^{m}(n_i - 1) = q - m \geq \frac{q+1}{2},$$

while the degree of $\psi' = f'$ is at most $(q-3)/2$. Therefore $\psi' = f'$ is the zero polynomial. This is possible only in the case where $\psi = h^p$ for a suitable polynomial $h$. On the other hand,

$$\psi(z) = \prod_{i=1}^{m}(z - y_i)^{n_i}.$$

Using the uniqueness of factorisation into a product of irreducible factors, we get $n_i \equiv 0 \pmod{p}$. The equations

$$\sum_{\alpha\in\mathbb{F}_q} \varphi^k(\alpha) = \sum_{i=1}^{m} n_i y_i^k = 0.$$

allow us to complete the proof. □

**Corollary 1** ([11], Corollary 1.1) *Let $\varphi \in L(\mathbb{F}_q)$, and suppose that $|A_\varphi| \leq (q-1)/2$. Then for all $\alpha \in \mathbb{F}_q$,*

$$\sum_{\alpha\in\mathbb{F}_q}(\varphi(\alpha) - a\alpha)^k = \begin{cases} 0, & if\ 1 \leq k \leq q-2, \\ -1, & if\ k = q-1\ and\ a \notin A_\varphi, \\ 0, & if\ k = q-1\ and\ a \in A_\varphi. \end{cases}$$

***Proof*** If $a \notin A_\varphi$ then $\varphi_a(x) := \varphi(x) - ax, x \in \mathbb{F}_q$ is a permutation, and the result follows from Proposition 3. If $a \in A_\varphi$ then $|A_{\varphi_a}| = |A_\varphi| \leq (q-1)/2$ and we have to use Lemma 1. □

**Corollary 2** ([11], Corollary 1.2) *Let $\varphi \in L(\mathbb{F}_q)$, and suppose that $|A_\varphi| \leq (q-1)/2$. Then $k \cdot (\varphi, e_{k-1}) = 0$ for all $k$ such that $1 \leq k \leq q-2$.*

***Proof*** It is clear from Corollary 1 that

$$P_k(a) = \sum_{\alpha\in\mathbb{F}_q}(\varphi(\alpha) - a\alpha)^k = 0$$

for all $1 \leq k \leq q-2$ and $a \in \mathbb{F}_q$. Representing this expression in powers of $a$, we find that

$$
\begin{aligned}
P_k(a) &= \sum_{\alpha \in \mathbb{F}_q} \left( \sum_{l=0}^{k} \binom{k}{l} (-1)^l \varphi^{k-l}(\alpha) a^l \alpha^l \right) \\
&= \sum_{l=0}^{k} (-1)^l \binom{k}{l} \left( \sum_{\alpha \in \mathbb{F}_q} \varphi^{k-l}(\alpha) \alpha^l \right) a^l \\
&= \sum_{l=0}^{k} (-1)^l \binom{k}{l} (\varphi^{k-l}, e_l) a^l
\end{aligned}
$$

(here we have used the definition of the form $(*, *)$).

Now $P_k(a)$ is a polynomial in $a$ of degree at most $q-2$. On the other hand, $P_k(a)$ has at least $q$ roots. Therefore, it is identically equal to 0, that is, all its coefficients are equal to zero. In particular, the coefficient $k \cdot (\varphi, e_{k-1})$ of $a^{k-1}$ is 0. □

**Lemma 2** *Let $\varphi \in L(\mathbb{F}_q)$ be a function which fixes 0 and satisfies $|A_\varphi| \le (q-1)/2$. Then there exists a function $\varphi_0 \in \langle e_1, \ldots, e_{q/p} \rangle$ such that $\varphi(x) = \varphi_0(x^p)$.*

***Proof*** Let

$$\varphi(x) = \sum_{i=0}^{q-1} c_i e_i$$

be the representation of the function $\varphi$ with respect to the basis $\{e_i\}_{i=0}^{q-1}$ (see Proposition 2). Notice that $\varphi(0) = 0$ implies $c_0 = 0$.

Pick an arbitrary $k \in \{1, \ldots, q-2\}$ coprime to $p$. By Corollary 2 $k(\varphi, e_{k-1}) = 0$ implying $(\varphi, e_{k-1}) = 0$. Therefore $\sum_{i=1}^{q-1} c_i(e_i, e_{k-1}) = 0$. It follows from Proposition 2 that there is a unique $i \in \{1, \ldots, q-1\}$ with $(e_i, e_{k-1}) \neq 0$, namely: $i = q-k$. Therefore $c_{q-k} = 0$ for every $k \in \{1, \ldots, q-2\}$ coprime to $p$. This implies that $\varphi(x) = a_1 x + \sum_{i=1}^{q/p-1} b_i x^{pi}$ for suitable scalars $a_1, b_1, \ldots, b_{q_p-1} \in \mathbb{F}_q$. Taking now $\varphi_0(x) := a_1 x^{q/p} + \sum_{i=1}^{q/p-1} b_i x^i$ we obtain that $\varphi(x) = \varphi_0(x^p)$ with $\varphi_0 \in \langle e_1, \ldots, e_{q/p} \rangle$. □

## 2.2 *Proof of the Theorem 1*

In this subsection we denote by $G_0$ the stabilizer of 0 in the group $G(q, e)$. We are going to show that every element of $G_0$ has a form $x \mapsto a x^{p^i}$ for suitable $a \in C$ and $i \in \{0, 1, \ldots, n-1\}$.

Denote the permutation $x \mapsto e_{q-2}(x) = x^{q-2}, x \in \mathbb{F}_q$ by $t$. It is easy to see that $t(0) = 0$, and that $t(x) = 1/x$ for all $x \neq 0$, so $t$ is an involution. The next statement plays a significant role in the proof of Theorem 1.

**Proposition 5** (Proposition 4, [11]) *The permutation $t$ normalizes $G_0$.*

***Proof*** Pick an arbitrary $\psi \in G_0$. Then $\psi(0) = 0$ and $\frac{\psi(\alpha)-\psi(\beta)}{\alpha-\beta} \in C$ for each pair $\alpha \neq \beta \in \mathbb{F}_q$.

Let us abbreviate

$$\theta(x) := \psi(x^{q-2})^{q-2} = \begin{cases} \psi(x^{-1})^{-1}, & x \neq 0, \\ 0, & x = 0 \end{cases}$$

Pick an arbitrary pair $\alpha \neq \beta \in \mathbb{F}_q$. Since $A_\psi \subseteq C$, we conclude that $\frac{\psi(\alpha)-\psi(\beta)}{\alpha-\beta} \in C$. We have to show that $\frac{\theta(\alpha)-\theta(\beta)}{\alpha-\beta} \in C$.

Assume first that one of $\alpha, \beta$ is zero. Without loss of generality we may assume that $\beta = 0$. Then $\alpha \neq 0$ and

$$\frac{\theta(\alpha)-\theta(\beta)}{\alpha-\beta} = \psi(\alpha^{-1})^{-1}/\alpha = \alpha^{-1}/\psi(\alpha^{-1}) = \left(\frac{\psi(\alpha^{-1})-\psi(\beta)}{\alpha^{-1}-\beta}\right)^{-1} \in C^{-1} = C.$$

Assume now that both $\alpha$ and $\beta$ are non-zeroes. Then

$$\frac{\theta(\alpha)-\theta(\beta)}{\alpha-\beta} = \frac{\psi(\alpha^{-1})^{-1}-\psi(\beta^{-1})^{-1}}{\alpha-\beta} = \frac{\psi(\beta^{-1})-\psi(\alpha^{-1})}{\beta^{-1}-\alpha^{-1}} \cdot \frac{\alpha^{-1}}{\psi(\alpha^{-1})} \cdot \frac{\beta^{-1}}{\psi(\beta^{-1})}$$

Since each of the three factors in the right side belongs to $C \leq \mathbb{F}_q^*$, their product also belongs to $C$, hereby proving $\frac{\theta(\alpha)-\theta(\beta)}{\alpha-\beta} \in C$. □

*Proof of Theorem* 1. Let $\varphi$ be an arbitrary permutation in $G_0$. By Lemma 2, $\varphi(x) = \varphi_0(x^p)$ with $\varphi_0 \in \langle e_1, \ldots, e_{q/p}\rangle$. It follows from Proposition 5 that $\psi(x) := \left(\varphi(x^{q-2})\right)^{q-2} \in G_0$. By Lemma 2 there exists $\psi_0 \in \langle e_1, \ldots, e_{q/p}\rangle$ such that $\left(\varphi(x^{q-2})\right)^{q-2} = \psi_0(x^p)$.

When $x \neq 0$ we have

$$\left(\varphi(x^{q-2})\right)^{q-2} = \frac{1}{\varphi(1/x)}.$$

Therefore for all $x \in \mathbb{F}_q^*$ the following holds

$$\psi_0(x^p) = \frac{1}{\varphi(1/x)} \iff \psi_0(x^p)\varphi(x^{-1}) = 1 \iff \psi_0(x^p)\varphi_0(x^{-p}) = 1.$$

After substitution $x = y^{p^{n-1}}$, $y \in \mathbb{F}_q^*$ we obtain that $\psi_0(y)\varphi_0(y^{-1}) = 1$ holds for all $y \in \mathbb{F}_q^*$.

It follows from $\varphi_0, \psi_0 \in \langle e_1, \ldots, e_{q/p}\rangle$ that

$$\varphi_0(y) = \sum_{i=1}^{q/p} a_i y^i,\ \psi_0(y) = \sum_{i=1}^{q/p} b_i y^i, \quad y \in \mathbb{F}_q$$

for suitable scalars $a_i, b_i \in \mathbb{F}_q$.

Now the equality $\psi_0(y)\varphi_0(y^{-1}) = 1$ reads as follows

$$\forall y \in \mathbb{F}_q^* : \quad (b_1 y + \cdots + b_{q/p} y^{q/p})(a_1 y^{-1} + \cdots + a_{q/p} y^{-q/p}) = 1$$

Multiplying both sides of this equation by $y^{q/p}$, we obtain

$$\forall y \in \mathbb{F}_q^* : \quad (b_1 y + \cdots + b_{q/p} y^{q/p})(a_1 y^{q/p-1} + \cdots + a_{q/p}) - y^{q/p} = 0.$$

This condition implies that the polynomial

$$(b_1 Y + \cdots + b_{q/p} Y^{q/p})(a_1 Y^{q/p-1} + \cdots + a_{q/p}) - Y^{q/p} \in \mathbb{F}_q[Y]$$

is divisible by $Y^{q-1} - 1$. Since, the degree of the above polynomial is at most $2(q/p) - 1 \leq q - 1$ (the equality holds only if $p = 2$), we conclude that

$$(b_1 Y + \cdots + b_{q/p} Y^{q/p})(a_1 Y^{q/p-1} + \cdots + a_{q/p}) - Y^{q/p} = c(Y^{q-1} - 1)$$

for some $c \in \mathbb{F}_q$. Substituting $Y = 0$ into this equation yields $c = 0$ which, in turn, implies

$$(b_1 Y + \cdots + b_{q/p} Y^{q/p})(a_1 Y^{q/p-1} + \cdots + a_{q/p}) = Y^{q/p}$$

By unique factorization theorem for polynomials each of the factors in the left-hand side is a monomial. Therefore $\varphi_0(y) = a_k y^k$ with $k \in \{1, \ldots, q/p\}$ and $a_k \in \mathbb{F}_q^*$. Hence $\varphi(x) = a_k x^{pk}$. By Proposition 1

$$a_k = \frac{\varphi(1) - \varphi(0)}{1 - 0} \in C.$$

Thus we have shown that any permutation $\varphi \in G(q, e)$ fixing zero has a form $\varphi(x) = a_k x^{pk}$ with $a_k \in C$ and $k \in \{1, \ldots, n/p\}$.

Note that the permutation $\varphi(x + 1) - \varphi(1)$ is also an automorphism of the cyclotomic scheme fixing zero. Therefore $\varphi(x + 1) - \varphi(1) = a_i x^{pi}$ for some $i \in \{1, \ldots, q/p\}$ and $a_i \in C$. This immediately implies

$$\forall_{x \in \mathbb{F}_q} \quad a_k (x + 1)^{pk} - a_k = a_i x^{pi}.$$

Now one can easily show that $pk$ should be a power of $p$. □

**Acknowledgements** The author is very thankful to M. Klin and G. Jones for their enormous help with text preparing and for moral support during working on the paper. The author is also grateful to I. Ponomarenko for valuable comments.

## References


1. J. Bagherian, I. Ponomarenko, A.R. Barghi, On cyclotomic schemes over finite near-fields. J. Algebr. Comb. **27**(2), 173–185
2. S. Ball, A. Blokhuis, A.E. Brouwer, L. Storme, T. Szonyi, On the number of slopes of the graph of a function defined over a finite field. J. Combin. Theory Ser. A **86**, 187–196 (1999)
3. A.E. Brouwer, A.M. Cohen, A. Neumaier, *Distance-Regular Graphs* (Springer-Verlag, Berlin, 1989)
4. W. Burnside, *Theory of Groups of Finite Order*, 2nd edn. (Cambridge Univ. Press, 1911)
5. D.V. Churikov, A.V. Vasil'ev, Automorphism groups of cyclotomic schemes over finite near-fields. Sib. El. Math. Rep. **13**, 1271–1282 (2016)
6. A.W.M. Dress, M.H. Klin, M.E. Muzychuk, On $p$-configurations with few slopes in the affine plane and a theorem of W. Burnside. Bayreuth. Math. Schr **40**, 7–19 (1992)
7. G.A. Jones, *Paley and the Paley Graphs*, https://arxiv.org/pdf/1702.00285.pdf
8. T.K. Lim, C.E. Praeger, On generalized Paley graphs and their automorphism groups. Michigan Math. J. **58**, 293–308 (2009)
9. R. McConnel, Psuedo-ordered polynomials over a finite field. Acta Arithm. **8**, 127–151 (1963)
10. P. Müller, Permutation groups of prime degree, a quick proof of Burnside's theorem. Arch. Math. **95**, 15–17 (2005)
11. M.E. Muzychuk, *The Automorphism Group of the Paley Graphs*, Questions in group theory and homological algebra (Jaroslavl State University, 1987), pp. 64–69
12. J.-P. Serre, *A Course in Arithmetic*, 5th edn. (Springer, 1996)

# Recognizing and Testing Isomorphism of Cayley Graphs over an Abelian Group of Order $4p$ in Polynomial Time

**Roman Nedela and Ilia Ponomarenko**

**Abstract** We construct a polynomial-time algorithm that, for a graph $X$ with $4p$ vertices ($p$ is prime), finds (if any) a Cayley representation of $X$ over the group $C_2 \times C_2 \times C_p$. This result, together with the known similar result for circulant graphs, shows that recognizing and testing isomorphism of Cayley graphs over an abelian group of order $4p$ can be done in polynomial time.

**Keywords** Computational geometry · Graph theory · Hamilton cycles

## 1 Introduction

Under a *Cayley representation* of a graph $X$ over a group $G$, we mean a graph isomorphism from $X$ to a Cayley graph over $G$. Two such representations are *equivalent* if the images are *Cayley isomorphic*, i.e., there is a group automorphism of $G$ which is a graph isomorphism between the images. In the present paper, we consider a special case of the following computational problem (below all the groups and graphs are assumed to be finite).

**Problem CRG** *Given a group $G$ and a graph $X$, find a full set of nonequivalent Cayley representations of $X$ over $G$.*

Here we assume that the group $G$ is given explicitly by the multiplication table, and the graph $X$ is given by a binary relation. The output of an algorithm solving the

R. Nedela (✉)
Department of Mathematics and NTIS, University of West Bohemia,
Pilsen, Czech Republic, Mathematical Instititute, Slovak Academy of Sciences,
Banska Bystrica, Slovakia
e-mail: nedela@savbb.sk

I. Ponomarenko
V. A. Steklov Institue of Mathematics, Russian Academy of Sciences,
St. Petersburg, Moscow, Russia
e-mail: inp@pdmi.ras.ru

School of Mathematics and Statistics, Central China Normal University, Wuhan, China



G. A. Jones et al. (eds.), *Isomorphisms, Symmetry and Computations in Algebraic Graph Theory*, Springer Proceedings in Mathematics & Statistics 305,
https://doi.org/10.1007/978-3-030-32808-5_7

problem is represented by a set of bijections $f$ from the vertex set of $X$ onto $G$ such that $(G_{right})^{f^{-1}}$ is a regular subgroup of the group $\mathrm{Aut}(X)$, where $G_{right} \leq \mathrm{Sym}(G)$ is the group induced by right multiplication of $G$. Using the Babai argument in [1], one can establish a one-to-one correspondence between the regular subgroups of $\mathrm{Aut}(X)$ and Cayley representations of $X$.

In general, the Problem CRG seems to be very hard. Even the question, whether the output is empty or not, leads to the Cayley recognition problem asking whether a given graph is isomorphic to a Cayley graph over the group $G$. Not too much is known about the computational complexity of this problem. There are two other related problems.

**Problem CGCI** *Given a group G, test whether two Cayley graphs over G are Cayley isomorphic.*

**Problem CGI** *Given a group G, a Cayley graph over G, and an arbitrary graph, test whether these two graphs are isomorphic.*

Note that the subproblem of the Problem CGI, in which both input graphs are Cayley graphs over $G$, is equivalent to the Problem CGCI whenever $G$ is a CI-group (this fact can be considered as the definition of the CI-group.)

Suppose we are restricted to a family of Cayley graphs for which the Problem CGCI can be solved efficiently; for instance, one can take $G$ to be a group generated by a set of at most constant size. Then one can see that the Problem CGI is polynomial-time reducible to the Problem CRG. In general case, the reduction can be done in polynomial time in the order of the group $\mathrm{Aut}(G)$.

In [6], a polynomial-time algorithm for the Problem CRG was constructed for the case where $G$ is a cyclic group. Up to now,[1] this is the only published result solving the Problem CGCI for an infinite class of groups. It is quite natural to look for an extension of that result to other classes of abelian groups. The main result of the present paper (Theorem 1) does it for abelian groups of order $4p$, where $p$ is a prime. In view of the above discussion, the Problems CRG and CGI are equivalent in this case.

Now we are ready to present the main results of the paper.

**Theorem 1** *For an abelian group G of order* $n = 4p$ *with prime p, the Problems CRG and CGI can be solved in time* $\mathrm{poly}(n)$.

There are exactly two non-isomorphic abelian groups of order $4p$: the cyclic group $C_{4p}$ and the group $E_4 \times C_p$, where $E_4 = C_2 \times C_2$ is the Klein group. In the former case, Theorem 1 follows from [6]. In the latter case, $G$ is a CI-group [11, Theorem 1.2] and hence every graph has at most one Cayley representation over $G$ (up to equivalence). Thus Theorem 1 is an immediate consequence of the following theorem.

---

[1] After the present paper was completed, the authors got aware about the result in [15], where the Problem CGCI for Cayley graph was independently solved for the abelian groups of order $4p$.

**Theorem 2** *Given a graph $X$ with $n = 4p$ vertices ($p$ is a prime), one can test in time* poly($n$) *whether $X$ is isomorphic to a Cayley graph over the group $G = E_4 \times C_p$ and (if so), find a Cayley representation of $X$ over $G$ within the same time.*

Let us outline the proof of Theorem 2. At the first step, we use the Weisfeiler–Leman algorithm to construct the coherent configuration $\mathcal{X}$ associated with the graph $X$ (for the exact definitions, see Sect. 2). Then

$$K := \mathrm{Aut}(\mathcal{X}) = \mathrm{Aut}(X).$$

Therefore $X$ is a Cayley graph over $G$ if and only if $\mathcal{X}$ is a Cayley scheme over $G$. This reduces our problem to finding the set $\mathrm{Reg}(K, G)$ consisting of all semiregular groups $H \leq K$ isomorphic to $G$, where the group $K$ is not "in hand".

At this point, we use the classification of Schur rings (and hence Cayley schemes) over the group $G$ obtained in [3].[2] This enables us to find in time poly($n$) a larger coherent configuration $\mathcal{X}'$ such that

(a) $\mathrm{Reg}(K, C_p) = \mathrm{Reg}(K', C_p)$, where $K' = \mathrm{Aut}(\mathcal{X}')$ and
(b) $|K'| = \mathrm{poly}(n)$ or $K' \cong \mathrm{Sym}(p')^m$, where $p'm = n$ and $m \leq 4$.

Now, if the group $K'$ has small order, then in view of statement (a), one can easily find the set $\mathrm{Reg}(K, C_p)$ by brute force; this is a part of the Main Subroutine described in Sect. 7. In the remaining case, statement (b) implies that $\mathcal{X}'$ is a coherent configuration of a special type studied in Sect. 3. This fact is used in the Main Subroutine for computing the set $\mathrm{Reg}(K, C_p)$. At the final step, we only need to test whether there exists a group belonging to $\mathrm{Reg}(K, C_p)$, which can be extended to a regular subgroup of $K$. This is done in Sect. 8.

## 2 Coherent Configurations

In this section, we collect some notation and then compile basic definitions and facts concerning coherent configurations. In our presentation, we follow [7].

### 2.1 Notation

Throughout the paper, $\Omega$ denotes a finite set of cardinality $n \geq 1$. The diagonal of the Cartesian product $\Omega \times \Omega$ is denoted by $1_\Omega$. For a set $T \subseteq 2^\Omega$, we denote by $T^\cup$ the set of all unions of the elements of $T$. For a set $S \subseteq 2^{\Omega\times\Omega}$, we set $S^* = \{s^* : s \in S\}$, where $s^* = \{(\beta, \alpha) : (\alpha, \beta) \in s\}$. For a point $\alpha \in \Omega$, we set $\alpha S = \cup_{s\in S}\alpha s$, where $\alpha s = \{\beta \in \Omega : (\alpha, \beta) \in s\}$. For relations $r, s \subseteq \Omega \times \Omega$, we set $r \cdot s = \{(\alpha, \gamma) : (\alpha, \beta) \in r,\ (\beta, \gamma) \in s$ for some $\beta \in \Omega\}$.

[2] Independently, such a classification have recently be obtained in [12].

For an equivalence relation $E$ on $\Omega$, we denote by $\Omega/E$ the set of the classes of $E$. If, in addition, $r \subseteq \Omega \times \Omega$, then we set

$$r_{\Omega/E} = \{(\Lambda, \Delta) \in \Omega/E \times \Omega/E : r_{\Lambda,\Delta} \neq \varnothing\},$$

where $r_{\Lambda,\Delta} = r \cap (\Lambda \times \Delta)$. We also put $r_\Lambda = r_{\Lambda,\Lambda}$.

The group of all permutations of $\Omega$ is denoted by $\mathrm{Sym}(\Omega)$. Given a group $K \leq \mathrm{Sym}(\Omega)$ and a $K$-invariant set $\Delta \subseteq \Omega$, the restrictions $k^\Delta$ of $k \in K$ to $\Delta$ form a subgroup of $\mathrm{Sym}(\Delta)$ denoted by $K^\Delta$.

A bijection $f : \Omega \to \Omega'$, $\alpha \mapsto \alpha^f$, naturally defines a bijection $r \mapsto r^f$ from the relations on $\Omega$ onto the relations on $\Omega'$ and a group isomorphism $g \mapsto g^f$ from $\mathrm{Sym}(\Omega)$ onto $\mathrm{Sym}(\Omega')$. For an equivalence relation $E$ on $\Omega$, the bijection $f$ induces a bijection $f_{\Omega/E} : \Omega/E \to \Omega'/E'$ where $E' = E^f$.

## 2.2 Main Definitions

Let $S$ be a partition of the set $\Omega \times \Omega$. A pair $\mathcal{X} = (\Omega, S)$ is called a *coherent configuration* on $\Omega$ if $1_\Omega \in S^\cup$, $S^* = S$, and given $r, s, t \in S$, the number

$$c_{rs}^t = |\alpha r \cap \beta s^*|$$

does not depend on the choice of $(\alpha, \beta) \in t$. The elements of $\Omega$, $S$, $S^\cup$, and the numbers $c_{rs}^t$ are called the *points*, the *basis relations*, the *relations*, and the *intersection numbers* of $\mathcal{X}$, respectively. The numbers $|\Omega|$ and $|S|$ are called the *degree* and the *rank* of $\mathcal{X}$.

The coherent configuration $\mathcal{X}$ is said to be *trivial* if the set $S$ consists of the reflexive relation $1_\Omega$ and (if $n > 1$) the complement of it in $\Omega \times \Omega$, and is called *discrete* if every element of $S$ is a singleton. Observe that the rank of $\mathcal{X}$ is at most two in the former case, and equals $n^2$ in the latter case.

The set of all equivalence relations $E \in S^\cup$ is denoted by $\mathcal{E} = \mathcal{E}(\mathcal{X})$. The coherent configuration is said to be *primitive* if the only elements of $\mathcal{E}$ are the trivial equivalence relations $1_\Omega$ and $\Omega \times \Omega$.

## 2.3 Fibers

Denote by $\Phi = \Phi(\mathcal{X})$, the set of all $\Delta \subseteq \Omega$ such that $1_\Delta \in S$. Then the set $\Omega$ is the disjoint union of the elements of $\Phi$ called the *fibers* of $\mathcal{X}$. Moreover, for each $r \in S$ there exist uniquely determined fibers $\Delta$ and $\Lambda$ such that $r \subseteq \Delta \times \Lambda$. Thus the set $S$ is the disjoint union of the sets

$$S_{\Delta,\Lambda} = \{s \in S : s \subseteq \Delta \times \Lambda\}.$$

Note that $1_\Delta \in rr^*$ and hence the number

$$n_r := |\alpha r| = c_{rr^*}^{1_\Delta}$$

does not depend on $\alpha \in \Delta$. It is called the *valency* of $r$. For any $T \in S^\cup$, we set $n_T$ to be the sum of all valences $n_t$, where $t$ runs over the basis relations of $\mathcal{X}$ that are contained in $T$.

A coherent configuration $\mathcal{X}$ is said to be *homogeneous* if $1_\Omega \in S$. In this case, $n_r = n_{r^*}$ for all $r \in S$. Observe that a primitive coherent configuration is always homogeneous. A homogeneous coherent configuration which is not primitive is said to be *imprimitive*. One can see that every *commutative* coherent configuration, i.e., one with $c_{rs}^t = c_{sr}^t$ for all $r, s, t$, is homogeneous.

## 2.4 Restrictions and Quotients

Let $E \in \mathcal{E}$ be an equivalence relation. For any class $\Delta$ of $E$, denote by $S_\Delta$ the set of all nonempty relations $s_\Delta = s \cap \Delta^2$ with $s \in S$. Then the pair

$$\mathcal{X}_\Delta = (\Delta, S_\Delta)$$

is a coherent configuration called the *restriction* of $\mathcal{X}$ to the set $\Delta$. In the special case where $E$ is the union of $\Lambda \times \Lambda$, $\Lambda \in \Phi$, and $\Delta \in \Phi$, the restriction $\mathcal{X}_\Delta$ is called the *homogeneous component* of $\mathcal{X}$.

Let $\mathcal{X}$ be a homogeneous coherent configuration. Denote by $S_{\Omega/E}$, the set of all nonempty relations $s_{\Omega/E}$, $s \in S$. Then the pair

$$\mathcal{X}_{\Omega/E} = (\Omega/E, S_{\Omega/E})$$

is a coherent configuration called the *quotient* of $\mathcal{X}$ modulo $E$. Let $F \in \mathcal{E}$ be an equivalence relation contained in $E$. For a class $\Delta \in \Omega/E$, the equivalence relation $F_\Delta = F \cap \Delta^2$ on $\Delta$ obviously belongs to the set $\mathcal{E}(\mathcal{X}_\Delta)$. On the other hand, the set $\Delta/F_\Delta$ is a class of the equivalence relation $E_{\Omega/F}$ being the union of basis relation of the quotient $\mathcal{X}_{\Omega/F}$. We have the following commuting rule:

$$(\mathcal{X}_\Delta)_{\Delta/F_\Delta} = (\mathcal{X}_{\Omega/F})_{\Delta/F_\Delta}. \tag{1}$$

The coherent configuration defined in (1) is denoted by $\mathcal{X}_{\Delta/F}$.

## 2.5 Isomorphisms

Two coherent configurations are called *isomorphic* if there exists a bijection between their point sets that induces a bijection between their sets of basis relations. Each such bijection is called an *isomorphism* between these two configurations. The group of all isomorphisms of a coherent configuration $\mathcal{X}$ to itself contains a normal subgroup

$$\mathrm{Aut}(\mathcal{X}) = \{f \in \mathrm{Sym}(\Omega) :\ s^f = s,\ s \in S\}$$

called the *automorphism group* of $\mathcal{X}$, where $s^f$ is the set of all pairs $(\alpha^f, \beta^f)$ with $(\alpha, \beta) \in s$. Thus by definition, $\mathrm{Aut}(\mathcal{X})$ is the intersection of the automorphism groups of the basis relations of $\mathcal{X}$. It is easily seen that if $\Delta \in \Phi^{\cup}$ and $E \in \mathcal{E}$, then

$$\mathrm{Aut}(\mathcal{X})^{\Delta} \leq \mathrm{Aut}(\mathcal{X}_{\Delta}) \quad \text{and} \quad \mathrm{Aut}(\mathcal{X})^{\Omega/E} \leq \mathrm{Aut}(\mathcal{X}_{\Omega/E}),$$

where $\mathrm{Aut}(\mathcal{X})^{\Delta}$ and $\mathrm{Aut}(\mathcal{X})^{\Omega/E}$ are the permutations groups induced by the actions of the setwise stabilizer of $\Delta$ in $\mathrm{Aut}(\mathcal{X})$ and of the group $\mathrm{Aut}(\mathcal{X})$ on the sets $\Delta$ and $\Omega/E$, respectively.

## 2.6 Direct Sum

Let $\mathcal{X}_1 = (\Omega_1, S_1)$ and $\mathcal{X}_2 = (\Omega_2, S_2)$ be coherent configurations. Denote by $\Omega_1 \sqcup \Omega_2$, the disjoint union of the sets $\Omega_1$ and $\Omega_2$. Further, denote by $S_1 \boxplus S_2$ the disjoint union of the set $S_1 \sqcup S_2$ and the set of all Cartesian products $\Delta \times \Lambda$, where $\Delta \in \Phi(\mathcal{X}_i)$ and $\Lambda \in \Phi(\mathcal{X}_{3-i})$, $i = 1, 2$. Then the pair

$$\mathcal{X}_1 \boxplus \mathcal{X}_2 = (\Omega_1 \sqcup \Omega_2, S_1 \boxplus S_2)$$

is a coherent configuration called the *direct sum* of $\mathcal{X}_1$ and $\mathcal{X}_2$. It is easily seen that

$$\mathrm{Aut}(\mathcal{X}_1 \boxplus \mathcal{X}_2) = \mathrm{Aut}(\mathcal{X}_1) \times \mathrm{Aut}(\mathcal{X}_2), \tag{2}$$

where the direct product on the right-hand side acts on the set $\Omega_1 \sqcup \Omega_2$.

## 2.7 Tensor Product

Let $\mathcal{X}_1 = (\Omega_1, S_1)$ and $\mathcal{X}_2 = (\Omega_2, S_2)$ be coherent configurations. Set $S_1 \otimes S_2 = \{s_1 \otimes s_2 :\ s_1 \in S_1,\ s_2 \in S_2\}$, where $s_1 \otimes s_2$ is the relation on $\Omega_1 \times \Omega_2$ consisting of all pairs $((\alpha_1, \alpha_2), (\beta_1, \beta_2))$ with $(\alpha_1, \beta_1) \in s_1$ and $(\alpha_2, \beta_2) \in s_2$. Then the pair

$$\mathcal{X}_1 \otimes \mathcal{X}_2 = (\Omega_1 \times \Omega_2, S_1 \otimes S_2)$$

is a coherent configuration called the *tensor product* of $\mathcal{X}_1$ and $\mathcal{X}_2$. It is easily seen that

$$\mathrm{Aut}(\mathcal{X}_1 \otimes \mathcal{X}_2) = \mathrm{Aut}(\mathcal{X}_1) \times \mathrm{Aut}(\mathcal{X}_2),$$

where the direct product on the right-hand side acts on the set $\Omega_1 \times \Omega_2$.

Let $\mathcal{X}$ be a commutative coherent configuration. We say that an equivalence relation $E \in \mathcal{E}$ has a *complement* with respect to tensor product if there exists an equivalence relation $F \in \mathcal{E}$ such that

$$E \cap F = 1_\Omega \quad \text{and} \quad E \cdot F = \Omega \times \Omega, \tag{3}$$

and $\mathcal{X}$ is isomorphic to $\mathcal{X}_{\Omega/E} \otimes \mathcal{X}_{\Omega/F}$. It should be noted that for a fixed equivalence relations $E$ and $F$ satisfying conditions (3), the isomorphism $\mathcal{X} \to \mathcal{X}_{\Omega/E} \otimes \mathcal{X}_{\Omega/F}$ exists if and only if $|\Omega/E| \cdot |\Omega/F| = n^2$, see [6, Theorem 2.2].

## 2.8 Generalized Wreath Product

Let $\mathcal{X}$ be a homogeneous coherent configuration, and let $E$ and $F$ be equivalence relations belonging to the set $\mathcal{E}$. We say that $\mathcal{X}$ is the $F/E$-*wreath product* if $E \subseteq F$ and for each $r \in S$,

$$r \cap F = \varnothing \quad \Rightarrow \quad r = \bigcup_{(\Delta,\Lambda)\in r_{\Omega/E}} \Delta \times \Lambda.$$

When the equivalence relations are not relevant, we also say that $\mathcal{X}$ is a *generalized wreath product*. It is said to be *trivial* if $E = 1_\Omega$ or $F = \Omega \times \Omega$. The standard wreath product is obtained as a special case of the generalized wreath product with $F = E$ (see also [17, p. 45]). Finally, we say that an equivalence relation $E \in \mathcal{E}$ has a *complement* with respect to (generalized) wreath product if there exists an equivalence relation $F \in \mathcal{E}$ such that $\mathcal{X}$ is the $F/E$-wreath product.

## 2.9 Algorithms

From the algorithmic point of view, a coherent configuration $\mathcal{X}$ on $n$ points is given by the set $S$ of its basis relations. In this representation, one can check in time poly$(n)$ whether $\mathcal{X}$ is homogeneous, commutative, etc. Moreover, within the same time one can list the fibers of $\mathcal{X}$ and construct the restriction $\mathcal{X}_\Delta$ for any $\Delta \in \Phi^\cup$, and can also find a nontrivial equivalence relation $E \in S^\cup$ (if it exists) and construct the quotient of $\mathcal{X}$ modulo $E$.

## 3 Quasi-trivial Coherent Configurations

Let $\mathcal{X} = (\Omega, S)$ be a coherent configuration. It is said to be *quasi-trivial* if

$$\mathrm{Aut}(\mathcal{X})^{\Delta} = \mathrm{Sym}(\Delta) \quad \text{for all } \Delta \in \Phi,$$

where $\Phi = \Phi(\mathcal{X})$. In particular, the restriction $\mathcal{X}_\Lambda$ with $\Lambda \in \Phi^{\cup}$ of a quasi-trivial configuration is quasi-trivial as well. Thus, every homogeneous component of $\mathcal{X}$ is trivial and homogeneous quasi-trivial coherent configurations are exactly trivial ones.

**Lemma 1** *The coherent configuration $\mathcal{X}$ is quasi-trivial if and only if for any two of its fibers $\Delta$ and $\Lambda$, the set $S_{\Delta,\Lambda}$ is a singleton or contains exactly two elements, one of which is a bijection $f : \Delta \to \Lambda$.*[3]

***Proof*** To prove the "if" part, we make use of [5, Lemma 9.4] implying that if $S_{\Delta,\Lambda}$ contains a bijection $f : \Delta \to \Lambda$, then the restriction map

$$\mathrm{Aut}(\mathcal{X}) \to \mathrm{Aut}(\mathcal{X}_{\Omega\setminus\Lambda})$$

is an isomorphism. Using the induction on $|\Phi|$, one can reduce the general case to the case where $S_{\Delta,\Lambda}$ is a singleton for all $\Lambda \in \Phi$ other than $\Delta$. But then it is easily seen that $\mathcal{X} = \mathcal{X}_\Delta \boxplus \mathcal{X}_{\Omega\setminus\Delta}$. By formula (2), this implies that

$$\mathrm{Aut}(\mathcal{X})^{\Delta} = \mathrm{Aut}(\mathcal{X}_\Delta) = \mathrm{Sym}(\Delta),$$

where the last equality is true, because the coherent configuration $\mathcal{X}_\Delta$ is of rank $|S_{\Delta,\Delta}| \le 2$.

To prove the "only if" part, we assume without loss of generality that $\Delta$ and $\Lambda$ are distinct non-singletons. Then the group $K := \mathrm{Aut}(\mathcal{X})^{\Delta\cup\Lambda}$ is the subdirect product of the groups $\mathrm{Sym}(\Delta)$ and $\mathrm{Sym}(\Lambda)$, i.e.,

$$K = \{(g, h) \in \mathrm{Sym}(\Delta) \times \mathrm{Sym}(\Lambda) : \varphi(g) = \psi(h)\},$$

where $\varphi : \mathrm{Sym}(\Delta) \to M$ and $\psi : \mathrm{Sym}(\Lambda) \to M$ are epimorphisms to a suitable group $M$. Each of the groups $\ker(\varphi)$ and $\ker(\psi)$ is a normal subgroup of a 2-transitive group and hence is transitive or trivial. If one of them is nontrivial, then $S_{\Delta,\Lambda}$ is obviously a singleton and we are done. Now let $\ker(\varphi) = \ker(\psi) = 1$, i.e., $\varphi$ and $\psi$ are group isomorphisms. Then $M \cong K$ and hence

$$\mathrm{Sym}(\Delta) \cong K \cong \mathrm{Sym}(\Lambda).$$

Consequently, $|\Delta| = |\Lambda|$. Denote the latter number by $d$. Note that by the assumption, $d \ge 2$. Therefore,

---

[3]This bijection is treated as the binary relation coinciding with the graph of $f$.

$$K_\alpha \cong \mathrm{Sym}(d-1) \quad \text{for all } \alpha \in \Delta \cup \Lambda$$

and hence every non-singleton orbit of the group $K_\alpha \leq \mathrm{Sym}(\Delta \cup \Lambda)$ is of cardinality at least $d-1$. Therefore, there exists a relation $f \in S_{\Delta,\Lambda}$ such that $n_f = 1$. In other words, $f : \Delta \to \Lambda$ is a bijection taking $\delta \in \Delta$ to a unique point of the set $\delta f$. Finally in view of [17, Corollary 13, p. 86], we have

$$|S_{\Delta,\Lambda}| \leq \frac{1}{2}(|S_{\Delta,\Delta}| + |S_{\Lambda,\Lambda}|) = 2$$

which completes the proof. □

Let the coherent configuration $\mathcal{X}$ be quasi-trivial. We define a binary relation $\sim$ on the set $\Phi$ by setting

$$\Delta \sim \Lambda \quad \Leftrightarrow \quad |S_{\Delta,\Lambda}| = 2. \tag{4}$$

This relation is obviously reflexive and symmetric. Assume that $|S_{\Lambda,\Gamma}| = 2$ for some $\Gamma \in \Phi$. Then by Lemma 1, the set $S_{\Lambda,\Gamma}$ contains a bijection $g : \Lambda \to \Gamma$. It is easily seen that the superposition $fg : \Delta \to \Gamma$ coincides with $f \cdot g$ and hence belongs to $S_{\Delta,\Gamma}$. It follows that $\Delta \sim \Gamma$. Thus, $\sim$ is an equivalence relation.

Denote by $\Phi_1, \ldots, \Phi_m$ the classes of the equivalence relation $\sim$. For each $i = 1, \ldots, m$, set

$$\Delta_i = \Delta_{i1} \cup \cdots \cup \Delta_{im_i},$$

where $\Delta_{i1}, \ldots, \Delta_{im_i}$ are the fibers belonging to the class $\Phi_i$. Then formula (4) immediately implies that

$$\mathcal{X} = \boxplus_{i=1}^{m} \mathcal{X}_{\Delta_i}. \tag{5}$$

By formula (2), this enables us to find the automorphism group of a quasi-trivial coherent configuration. Let $f_{ij} : \Delta_{i1} \to \Delta_{ij}$ be a bijection belonging to the set $S_{\Delta_{i1},\Delta_{ij}}$ (note that $f_{ij}$ is uniquely determined unless $|\Delta_{i1}| = 2$).

**Theorem 3** *Let $\mathcal{X}$ be a quasi-trivial coherent configuration. Then in the above notation,*

$$\mathrm{Aut}(\mathcal{X}) \cong \prod_{i=1}^{m} \mathrm{Sym}(\Delta_{i1}).$$

*Moreover, $g \in \mathrm{Aut}(\mathcal{X})$ if and only if for every $i = 1, \ldots, m$ there exists $g_i \in \mathrm{Sym}(\Delta_{i1})$ such that*

$$g^{\Delta_{ij}} = \begin{cases} g_i, & \text{if } j = 1, \\ f_{ij}^{-1} g_i f_{ij} & \text{if } j > 1. \end{cases} \tag{6}$$

***Proof*** The second statement is an immediate consequence of the first one, which follows from formulas (2) and (5). □

# 4 Cayley Schemes

## 4.1 General Facts

Let $G$ be a group. A coherent configuration $\mathcal{X}$ is called *a Cayley scheme* over $G$ if

$$\Omega = G \quad \text{and} \quad G_{right} \leq \mathrm{Aut}(\mathcal{X}).$$

In this case, the coherent configuration $\mathcal{X}$ is homogeneous, and commutative if the group $G$ is abelian. We note that each basis relation of $\mathcal{X}$ is the arc set of a Cayley graph on $G$. If the group $G$ is cyclic, then $\mathcal{X}$ is said to be a *circulant* scheme.

For every equivalence relation $E \in \mathcal{E}$, denote by $H = H_E$ the class of $E$ containing the identity element of $G$. Then $H$ is a subgroup of $G$ (isomorphic to the setwise stabilizer of $H$ in $G_{right}$) and the classes of $E$ equal the right $H$-cosets of $G$. In particular,

$$n_E = |H| \quad \text{and} \quad |\Omega/E| = |G/H|.$$

In what follows, we set $\mathcal{X}_{G/H} = \mathcal{X}_{\Omega/E}$. In the case where $\mathcal{X}$ is the $F/E$-wreath product for some $E, F \in \mathcal{E}$, we also say that $\mathcal{X}$ is the $U/L$-wreath product, where $U = H_F$ and $L = H_E$.

We set

$$\mathfrak{S} = \mathfrak{S}(\mathcal{X}) = \{H_E \leq G : E \in \mathcal{E}(\mathcal{X})\}.$$

For a group $H \in \mathfrak{S}$, we denote by $E_H$ the equivalence relation $F$ on $G$, for which $H = H_F$. Thus, $E_{H_E} = E$ and $H_{E_H} = H$. The following statement is a consequence of [9, Theorem 5.6].

**Theorem 4** *Let $\mathcal{X}$ be a Cayley scheme over an abelian group $G$. Suppose that $\mathcal{X}$ is the $U/L$-wreath product for some $U, L \in \mathfrak{S}$ such that*

$$\mathrm{Aut}(\mathcal{X}_{G/L})^{U/L} = \mathrm{Aut}(\mathcal{X}_U)^{U/L}. \tag{7}$$

*Then* $\mathrm{Aut}(\mathcal{X})^U = \mathrm{Aut}(\mathcal{X}_U)$ *and* $\mathrm{Aut}(\mathcal{X})^{G/L} = \mathrm{Aut}(\mathcal{X}_{G/L})$.

***Proof*** Set $\Delta_U = \mathrm{Aut}(\mathcal{X}_U)$ and $\Delta_0 = \mathrm{Aut}(\mathcal{X}_{G/L})$. Then according to [9, Sect. 5.2], equality (7) enables us to define the canonical generalized wreath product

$$K = \Delta_U \wr_{U/L} \Delta_0.$$

Since $\mathrm{Aut}(\mathcal{X})^U \leq \Delta_U$ and $\mathrm{Aut}(\mathcal{X})^{G/L} \leq \Delta_0$, we conclude by [9, Corollary 5.5] that $\mathrm{Aut}(\mathcal{X}) \leq K$. The reverse inclusion follows from [9, Corollary 5.4]. Thus, $\mathrm{Aut}(\mathcal{X}) = K$. This implies that

$$\mathrm{Aut}(\mathcal{X})^U = K^U = \Delta_U = \mathrm{Aut}(\mathcal{X}_U)$$

and similarly $\mathrm{Aut}(\mathcal{X})^{G/L} = \mathrm{Aut}(\mathcal{X}_{G/L})$. □

A Cayley scheme $\mathcal{X} = (G, S)$ is said to be *cyclotomic* if there exists a group $K \leq \mathrm{Aut}(G)$ such that

$$S = \mathrm{Orb}(G_{right}K, G \times G).$$

The cyclotomic scheme $\mathcal{X}$ is called *proper* (respectively, *trivial*) if $K$ is a proper subgroup (respectively, the identity subgroup) of the group $\mathrm{Aut}(G)$. Note that for every cyclotomic scheme $\mathcal{X}$, the set $\mathfrak{S}(\mathcal{X})$ contains all characteristic subgroups of $G$.

## 4.2 Cayley Schemes over $E_4 \times C_p$

The following theorem was proved in [3, Sect. 6.2] in the language of S-rings. Though the statement of the theorem was not formulated explicitly there, a careful analyses of the proof reveal cases (1) and (2) below, see [12] as well.

**Theorem 5** *Let $\mathcal{X}$ be a Cayley scheme over the group $G = E_4 \times C_p$. Then one of the following statements holds*

*(1) $\mathcal{X}$ is trivial or cyclotomic and*
*(2) $\mathcal{X}$ is a nontrivial tensor or generalized wreath product.*

**Corollary 1** *Let $\mathcal{X}$ be a Cayley scheme over the group $G$ and $H$ a minimal subgroup in the set $\mathfrak{S} = \mathfrak{S}(\mathcal{X})$. Then the Cayley scheme $\mathcal{X}_H$ is trivial or isomorphic to a proper cyclotomic scheme over $C_p$.*

***Proof*** Clearly, $|H|$ is a divisor of $4p$ other than 1. Next, by the minimality of $H$ the scheme $\mathcal{X}_H$ is primitive. If $|H| = 4p$, then $\mathcal{X}_H = \mathcal{X}$ and $\mathfrak{S} = \{1, H\}$. This implies that the scheme $\mathcal{X}$ is neither cyclotomic nor a nontrivial tensor or generalized wreath product. By Theorem 5, we conclude that $\mathcal{X}$ and hence $\mathcal{X}_H$ is trivial. Now let $|H| = 4$. Then $\mathcal{X}_H$ is a primitive scheme over $E_4$ and hence is trivial. In the other three cases, the group $H$ is a cyclic group of order 2, $p$, or $2p$. Thus, the required statement follows from the Schur–Wielandt theory [14, Corollary 3.2, Theorem 3.4]. □

In [10, p. 423], the automorphism groups of S-rings over a cyclic group $C_{pq}$, with primes $q \neq p$, were completely classified. This classification shows that for every Cayley scheme $\mathcal{Y}$ over the group $M = C_{2p}$ and for every group $N \leq \mathfrak{S}(\mathcal{Y})$, we have

$$\mathrm{Aut}(\mathcal{Y})^N = \mathrm{Aut}(\mathcal{Y}_N) \quad \text{and} \quad \mathrm{Aut}(\mathcal{Y})^{M/N} = \mathrm{Aut}(\mathcal{Y}_{M/N}). \tag{8}$$

Note that relations in (8) obviously hold also if $M = E_4$, or if $M = C_p$.

**Lemma 2** *Let $\mathcal{X}$ be a Cayley scheme over the group $G = E_4 \times C_p$. Suppose that the groups $U, L \in \mathfrak{S}$ are such that either $\mathcal{X} = \mathcal{X}_U \otimes \mathcal{X}_L$, or $\mathcal{X}$ is the $U/L$-wreath product. Then given $H \in \mathfrak{S}$,*

$$\mathrm{Aut}(\mathcal{X})^H = \mathrm{Aut}(\mathcal{X}_H) \quad \textit{for all } H \leq L, \tag{9}$$

*and*

$$\mathrm{Aut}(\mathcal{X})^{G/H} = \mathrm{Aut}(\mathcal{X}_{G/H}) \quad \textit{for all } H \geq U. \tag{10}$$

***Proof*** Let us prove formula (9); formula (10) can be proved in a similar way. Assume first that $\mathcal{X} = \mathcal{X}_U \otimes \mathcal{X}_L$. Then by the first part of formula (8) for $\mathcal{Y} = \mathcal{X}_L$, $M = L$, and $N = H$, we have

$$\mathrm{Aut}(\mathcal{X})^H = (\mathrm{Aut}(\mathcal{X}_U) \otimes \mathrm{Aut}(\mathcal{X}_L))^H = \mathrm{Aut}(\mathcal{X}_L)^H = \mathrm{Aut}(\mathcal{X}_H),$$

as is required. Now let $\mathcal{X}$ be the $U/L$-wreath product. Then again by the first part of (8) for $\mathcal{Y} = \mathcal{X}_{G/L}$, $M = G/L$, and $N = U/L$, we have

$$\mathrm{Aut}(\mathcal{X}_{G/L})^{U/L} = \mathrm{Aut}((\mathcal{X}_{G/L})_{U/L}) = \mathrm{Aut}(\mathcal{X}_{U/L}).$$

Similarly, by the second part of (8) for $\mathcal{Y} = \mathcal{X}_U$, $M = U$, and $N = L$, we have

$$\mathrm{Aut}(\mathcal{X}_U)^{U/L} = \mathrm{Aut}((\mathcal{X}_U)_{U/L}) = \mathrm{Aut}(\mathcal{X}_{U/L}).$$

The two above equalities show that the hypothesis of Theorem 4 is satisfied and hence $\mathrm{Aut}(\mathcal{X})^U = \mathrm{Aut}(\mathcal{X}_U)$. Applying the first of relations (8) again for $\mathcal{Y} = \mathcal{X}_U$, $M = U$, and $N = H$, we get

$$\mathrm{Aut}(\mathcal{X})^H = (\mathrm{Aut}(\mathcal{X})^U)^H = (\mathrm{Aut}(\mathcal{X}_U))^H = \mathrm{Aut}(\mathcal{X}_H),$$

as is required. □

## 5 Extension of Coherent Configuration and WL-Algorithm

### *5.1 Partial Order*

There is a natural partial order $\leq$ on the set of all coherent configurations on the same set $\Omega$. Namely, given two coherent configurations $\mathcal{X} = (\Omega, S)$ and $\mathcal{X}' = (\Omega, S')$, we set

$$\mathcal{X} \leq \mathcal{X}' \Leftrightarrow S^{\cup} \subseteq (S')^{\cup}.$$

The minimal and maximal elements with respect to this order are, respectively, the trivial and discrete coherent configurations.

## 5.2 *WL-Algorithm*

One of the most important properties of the partial ordering of coherent configurations comes from the fact that given a set $T \subseteq 2^{\Omega\times\Omega}$, there exists a unique minimal coherent configuration $\mathcal{X} = (\Omega, S)$, for which $T \subseteq S^{\cup}$ (in particular, the set of all coherent configurations on the same set form a join-semilattice). This coherent configuration is called the *coherent closure* of $T$ and can be constructed by the well-known Weisfeiler–Leman algorithm (WL-algorithm) [17, Sect. B] in time polynomial in sizes of $T$ and $\Omega$. To stress this fact, the coherent closure of $T$ is denoted by $\mathrm{WL}(T)$.

The *extension* of a coherent configuration $\mathcal{X}$ with respect to the set $T$ is defined to be the coherent closure of $S \cup T$ and is denoted by $\mathrm{WL}(\mathcal{X}, T)$. The following statement is a straightforward consequence of [17, Theorem 8.2].

**Theorem 6** *Let* $\mathcal{X} = (\Omega, S)$ *be a coherent configuration,* $T \subseteq 2^{\Omega\times\Omega}$*, and* $\mathcal{Y} = \mathrm{WL}(\mathcal{X}, T)$*. Then*

$$\mathrm{Aut}(\mathcal{Y}) = \{f \in \mathrm{Aut}(\mathcal{X}) : s^f = s \text{ for all } s \in T\}.$$

From Theorem 6, it follows that if $X$ is a graph with vertex set $\Omega$ and arc set $R$, then $\mathrm{Aut}(X)$ equals the automorphism group of the coherent closure $\mathrm{WL}(\{R\})$.

## 5.3 *Examples of Extensions*

For a coherent configuration $\mathcal{X}$ and an equivalence relation $E \in \mathcal{E}$, we denote by $\mathcal{X}_E$ the extension of $\mathcal{X}$ with respect to the set

$$T = \{1_\Delta : \Delta \in \Omega/E\}.$$

From Theorem 6, it follows that $\mathrm{Aut}(\mathcal{X}_E)$ consists of the automorphisms of $\mathcal{X}$ leaving each class of $E$ fixed. On the other hand, each automorphism of $\mathcal{X}$ permutes the classes of $E$. Thus, $\mathrm{Aut}(\mathcal{X}_E)$ equals the kernel of the natural epimorphism from $\mathrm{Aut}(\mathcal{X})$ to $\mathrm{Aut}(\mathcal{X})^{\Omega/E}$.

In the above notation, let $c \in \mathrm{Sym}(\Omega/E)$. Denote by $\mathcal{X}_c$ the extension of $\mathcal{X}$ with respect to the singleton $\{s\}$, where

$$s = s(E, c) = \bigcup_{\Delta\in\Omega/E} \Delta \times \Delta^c \tag{11}$$

(note that if $E = 1_\Omega$, then the relation $s$ coincides with the graph of the permutation $c$). Now by Theorem 6, the group $\mathrm{Aut}(\mathcal{X}_c)$ consists of all $f \in \mathrm{Aut}(\mathcal{X})$ such that $f^{\Omega/E}$ commute with $c$.

## 5.4 Extension of Cayley Schemes

Let $\mathcal{X}$ be a Cayley scheme over the group $G$. Fix a group $H \in \mathfrak{G}$ and consider the coherent configuration $\mathcal{Y} = \mathcal{X}_E$, where $E = E_H$. Then the set $\Phi = \Phi(\mathcal{Y})$ consists of the orbits of $H$ acting on $G$ by right multiplications. Besides, the equivalence relation $E$ is invariant with respect to the group $G_{right}$. Therefore, this group acts as an isomorphism group of the coherent configuration $\mathcal{Y}$. This enables us to define the coherent configuration

$$\mathcal{Y}^G = (G, \{s^G : s \in S_\mathcal{Y}\}), \tag{12}$$

where $S_\mathcal{Y}$ is the set of basis relations of $\mathcal{Y}$ and $s^G$ is the union of the relations $s^g$, $g \in G_{right}$. Obviously,

$$\mathcal{Y} \geq \mathcal{Y}^G \geq \mathcal{X}.$$

To find the group $\mathrm{Aut}(\mathcal{Y}^G)$, we note that if $\rho : G_{right} \to \mathrm{Sym}(\Phi)$ is the induced homomorphism, then

$$\mathrm{im}(\rho) = (G/H)_{right} \quad \text{and} \quad \ker(\rho) \leq \mathrm{Aut}(\mathcal{Y}).$$

In this situation, we can apply a result in [5, Theorem 2.2] saying that

$$\mathrm{Aut}(\mathcal{Y}^G) = G\,\mathrm{Aut}(\mathcal{Y}). \tag{13}$$

**Lemma 3** *In the above notation, suppose that the group $G$ is abelian. Then $\mathcal{Z} = \mathcal{Y}^G$ is a Cayley scheme over $G$. Moreover,*

*(1)* $\mathrm{Aut}(\mathcal{Y})^H = \mathrm{Aut}(\mathcal{Z})^H$ *and* $\mathcal{Y}_H = \mathcal{Z}_H$ *and*
*(2)* $\mathcal{Z}_{G/H}$ *is the trivial cyclotomic scheme over* $G/H$.

***Proof*** The fact that $\mathcal{Z}$ is a Cayley scheme over $G$ immediately follows from formula (13). This also implies statement (2), because the coherent configuration $\mathcal{Y}^{\Omega/E}$ is discrete. The aforementioned formula shows that the setwise stabilizer of the set $H$ in the group $\mathrm{Aut}(\mathcal{Z})$ coincides with $\mathrm{Aut}(\mathcal{Y})$. This proves the first equality of statement (1). The second one follows from statement (2) and [8, Theorem 2.1]. □

# 6 Structure of Cayley Schemes over $E_4 \times C_p$

## 6.1 Principal Equivalence Relation

In this section we are interested in the equivalence relations $E$ belonging to set $\mathcal{E} = \mathcal{E}(\mathcal{X})$, where $\mathcal{X} = (\Omega, S)$ is a homogeneous coherent configuration on $n = 4p$ points. The homogeneity of $\mathcal{X}$ implies that the valency $n_E$ of $E$ divides $n$. In what follows, we assume that $p \geq 5$ is a prime.

**Definition 1** We say that $E \in \mathcal{E}$ is a *principal* equivalence relation of $\mathcal{X}$ if one of the following statements holds

(E1) $n_E \geq p$ and $E$ is minimal (with respect to inclusion) and
(E2) $n_E \leq 4$ and $E$ has a complement with respect to tensor or generalized wreath product.

In case (E2), it is assumed that $\mathcal{E}$ contains no $E$ satisfying (E1).

It follows from Definition 1 that every principal equivalence relation of the coherent configuration $\mathcal{X}$ equals the equivalence closure of the union of at most two basis relations of $\mathcal{X}$. Since $|S| \leq n$, we immediately obtain the following statement.

**Lemma 4** *Given a homogeneous coherent configuration $\mathcal{X}$ of degree $n = 4p$, one can test in time* $\mathrm{poly}(n)$ *whether there exists a principal equivalence relation of $\mathcal{X}$, and (if so) find it within the same time.*

Clearly, for the trivial scheme of degree $n$, the equivalence relation $\Omega \times \Omega$ is principal. Let now $\mathcal{X}$ be a Cayley scheme over a group $G$ (of order $n$), and let $E$ be a principal equivalence relation of $\mathcal{X}$. If $\mathcal{X}$ is a cyclotomic scheme, then the Sylow $p$-subgroup $P$ of $G$ is characteristic in $G$, and hence $P \in \mathfrak{S}$. Since also $|P| = p$, the equivalence relation $E_P$ satisfies the condition (E1) and hence is a unique principal equivalence of $\mathcal{X}$.

**Lemma 5** *Every Cayley scheme over the group $G$ has a principal equivalence relation.*

***Proof*** Let $\mathcal{X}$ be a Cayley scheme over $G$. By the above remarks, we may assume that the scheme $\mathcal{X}$ is neither cyclotomic nor trivial. Therefore by Theorem 5, the set $\mathfrak{S}$ contains two proper subgroups $U$ and $L$ of the group $G$ such that

$$\mathcal{X} = \mathcal{X}_U \otimes \mathcal{X}_L \quad \text{or} \quad \mathcal{X} \text{ is the } U/L\text{-wreath product.} \tag{14}$$

Suppose that $\mathfrak{S}$ does not contain a principal equivalence relation satisfying condition (E1). Then every minimal subgroup of $\mathfrak{S}$ is of order at most 4. It follows that if $|L| \leq 4$, then $E_L$ is a principal equivalence satisfying condition (E2), and we are done. Thus, we may assume that

$$L \text{ is not minimal and } |L| > 4.$$

Then it is easily seen that $|L| = 2p$. Now if the first equality in formula (14) holds, then $E_U$ is a principal equivalence satisfying condition (E2). In the remaining case, $L$ contains a proper subgroup, say $H$, of order other than $p$. Therefore, $|H| = 2$. Besides, $\mathcal{X}$ being the $U/L$-wreath product is also the $U/H$-wreath product. Thus, $E_H$ is a principal equivalence satisfying condition (E2). □

## 6.2 A Principal Equivalence Relation of Large Valency

Let $\mathcal{X}$ be a Cayley scheme over the group $G$, $E$ a principal equivalence relation of $\mathcal{X}$, and let $\mathcal{Y} = \mathcal{X}_E$ be the coherent configuration defined in Sect. 5.3.

**Lemma 6** *In the above notation, assume that $n_E \geq p$. Then for each set $\Delta \in G/E$, either $\mathrm{Aut}(\mathcal{Y})^{\Delta} = \mathrm{Sym}(\Delta)$ or $\mathcal{Y}_{\Delta}$ is isomorphic to a proper cyclotomic scheme over $C_p$.*

***Proof*** Let us verify that $E$ is minimal in $\mathcal{E}(\mathcal{Z})$, where the Cayley scheme $\mathcal{Z} := \mathcal{Y}^G$ is defined by formula (12). First we note that

$$E \in \mathcal{E}(\mathcal{X}) \subseteq \mathcal{E}(\mathcal{Z}),$$

because $\mathcal{Z}$ is larger than $\mathcal{X}$. The minimality of $E$ in $\mathcal{E}(\mathcal{Z})$ is obvious if $n_E = p$. In the other two cases, we have $n_E = 2p$ or $4p$. This implies that $\mathcal{Z} = \mathcal{X}$ and the claim immediately follows from the definition of principal equivalence.

Set $H = H_E$. We claim that to prove the lemma, it suffices to verify the validity of the equality

$$\mathrm{Aut}(\mathcal{Z})^H = \mathrm{Aut}(\mathcal{Z}_H). \tag{15}$$

Indeed, by Lemma 3 we have $\mathrm{Aut}(\mathcal{Y})^H = \mathrm{Aut}(\mathcal{Z})^H$ and $\mathcal{Y}_H = \mathcal{Z}_H$. This implies that assuming (15),

$$\mathrm{Aut}(\mathcal{Y})^H = \mathrm{Aut}(\mathcal{Z})^H = \mathrm{Aut}(\mathcal{Z}_H) = \mathrm{Aut}(\mathcal{Y}_H).$$

By Corollary 1 applied to $\mathcal{X} = \mathcal{Z}$ and $G = H$, this implies that the scheme $\mathcal{Y}_H$ is either trivial and then $\mathrm{Aut}(\mathcal{Y}_H) = \mathrm{Sym}(H)$, or is proper cyclotomic. This proves our claim, because the scheme $\mathcal{Y}_{\Delta}$ with $\Delta \in G/E$, is isomorphic to the scheme $\mathcal{Y}_H$ (the isomorphism is induced by any permutation of the group $G_{right}$ that takes $H$ to $\Delta$).

Let us prove equality (15). If the scheme $\mathcal{X}$ is trivial, then we have $\mathcal{Y} = \mathcal{X}$, $H = G$, and the equality is obvious. So we may assume that

$$|H| \in \{p, 2p\}. \tag{16}$$

Then the scheme $\mathcal{Z} \geq \mathcal{X}$ is not trivial. If it is cyclotomic, then $|H| = p$ and the set $\mathfrak{S}(\mathcal{Z})$ contains the group $P \cong E_4$. Moreover, from statement (2) of Lemma 3 it follows that $\mathcal{Z}_P \cong \mathcal{Z}_{G/H}$ is the trivial cyclotomic scheme over $G/H \cong P$. According to [3, Lemma 2.3] this implies that $\mathcal{Z} \cong \mathcal{Z}_H \otimes \mathcal{Z}_P$. Thus, by Theorem 5 the scheme $\mathcal{Z}$ is proper tensor or generalized wreath product. Let us consider these two cases separately.

Let $\mathcal{Z} = \mathcal{Z}_U \otimes \mathcal{Z}_L$, where $U$ and $L$ are proper subgroups of $G$ that belong to $\mathfrak{S}(\mathcal{Z})$. Without loss of generality, we may assume that $L$ is of order $p$ or $2p$. Then in view of (16),

$$(|H|, |L|) \in \{(p, p), (2p, p), (p, 2p), (2p, 2p)\}.$$

By the minimality of $H$ the case $(2p, p)$ is impossible. By the same reason, in the case $(2p, 2p)$ we have $H = L$, for otherwise $H \cap L$ is the proper subgroup of $H$. Thus, in any case $H \leq L$. Therefore equality (15) immediately follows from Lemma 2.

Let now $\mathcal{Z}$ be the $U/L$-wreath product, where $U, L \in \mathfrak{S}(\mathcal{Z})$ are such that $1 < L \leq U < G$. By Lemma 2, to prove equality (15) it suffices to verify that

$$H \leq L. \tag{17}$$

To this end, suppose first that $H \not\leq U$. Then $E = E_H \not\subseteq E_U$ and hence the scheme $\mathcal{Z}$ has a basis relation $s \subseteq E \setminus E_U$. By the definition of generalized wreath product, we have $E_L \cdot s = s$. It follows that $E_L$ is contained in the minimal equivalence relation $F \in \mathcal{E}(\mathcal{Z})$ containing $s$. Therefore $L \leq H_F$. On the other hand, $H_F \leq H$ because $s \subset E$. Thus, $L \leq H_F \leq H$. By the minimality of $H$, this implies that $L = H$ which proves inclusion (17).

It remains to show that the statement

$$H \leq U \quad \text{and} \quad H \not\leq L \tag{18}$$

does not hold. Indeed, otherwise the minimality of $H$ implies that $H \cap L = 1$ and hence $|L| = 2$. Consequently, $\mathcal{Z}' = \mathcal{Z}_{G/H}$ is a Cayley scheme over the group $G' = G/H$ isomorphic to $E_4$. This scheme is the $U'/L'$-wreath product with $U' = UH/H$ and $L' = LH/H$, because $\mathcal{Z}$ is the $U/L$-wreath product. In view of assumption (18),

$$1 < L' \leq U' < G'.$$

Therefore the rank of $\mathcal{Z}'$ is less than 4. However, by statement (2) of Lemma 3, the rank of $\mathcal{Z}'$ equals 4, a contradiction. □

### 6.3 Summary

The following theorem summarizes what we proved in this section and shows a way how we are going to use the principal equivalence relation.

**Theorem 7** *Let $\mathcal{X} = (\Omega, S)$ be a Cayley scheme over $E_4 \times C_p$ $(p \geq 5)$, $E$ a principal equivalence relation of $\mathcal{X}$, and $\mathcal{Y} = \mathcal{X}_E$. Then*

*(1) if $n_E \leq 4$, then $\mathrm{Aut}(\mathcal{X})^{\Omega/E} = \mathrm{Aut}(\mathcal{X}_{\Omega/E})$,*
*(2) if $n_E \geq p$, then one of the following statements holds*

*(a) $\mathcal{Y}$ is quasi-trivial,*
*(b) for each $\Delta \in \Phi(\mathcal{Y})$, $\mathcal{Y}_\Delta$ is isomorphic to a proper cyclotomic scheme over $C_p$.*

***Proof*** If $n_E \leq 4$, then the required statement immediately follows from the definition of the principal equivalence relation and Lemma 2. Otherwise, we are done by Lemma 6 and the definition of a quasi-trivial coherent configuration. □

## 7 Finding a Representative Set of Semiregular $C_p$-Subgroups

In this section, we describe an efficient algorithm finding a *representative* set $B_p$ of the automorphism group $K$ of a coherent configuration of degree $4p$ (Sect. 7.1). By definition, $B_p$ consists of semiregular $C_p$-subgroups of $K$ such that every group in $\mathrm{Reg}(K, C_p)$ is $K$-conjugate to one of them. Then we estimate the running time of the algorithm and explain the implementation details (Sect. 7.2). In fact, the algorithm correctly finds $B_p$ if the input coherent configuration is isomorphic to a Cayley scheme over $G = E_4 \times C_p$; we prove this in Sect. 7.3.

### 7.1 The Main Algorithm

In the algorithm below, we make use of the algorithm from [6] that constructs a *cycle base* of a coherent configuration $\mathcal{X}$; the cycle base is defined to be a maximal set of pairwise non-conjugated full cycles of the group $K = \mathrm{Aut}(\mathcal{X})$. We always assume that $p \geq 5$ is a prime.

**Main Subroutine (MS)**

**Input**: A homogeneous coherent configuration $\mathcal{X}$ on $4p$ points.
**Output**: A set $B_p \subseteq \mathrm{Reg}(K, C_p)$, which is either empty or is a representative set of $K = \mathrm{Aut}(\mathcal{X})$.

**Step 1**. Find a principal equivalence relation $E$ of $\mathcal{X}$; if there is no such $E$, then output $B_p := \varnothing$.
**Step 2**. If $n_E \le 4$, then

Step 2.1. find a cycle base $C$ of the coherent configuration $\mathcal{X}_{\Omega/E}$.
Step 2.2. for each $c \in C$, find successively the relation $s = s(E, c)$ defined in (11) and the coherent configuration $\mathcal{X}_c = \mathrm{WL}(\mathcal{X}, s)$.
Step 2.3. output $B_p = \{P_c \in \mathrm{Reg}(K, C_p) : c \in C\}$, where $P_c$ is a Sylow $p$-subgroup of the group $K_c = \mathrm{Aut}(\mathcal{X}_c)$.

**Step 3**. (Here $n_E \ge p$.) If the coherent configuration $\mathcal{Y} = \mathcal{X}_E$ has a fiber $\Delta$ such that $|\Delta| \ne n_E$ or the homogeneous component $\mathcal{Y}_\Delta$ is not circulant, then output $B_p = \varnothing$.

**Step 4**. If the coherent configuration $\mathcal{Y}$ is quasi-trivial, then output $B_p = \{P\}$ with arbitrary $P \in \mathrm{Reg}(\mathrm{Aut}(\mathcal{Y}), C_p)$.

**Step 5**. Output $B_p = \mathrm{Reg}(K', C_p)$, where $K' \le \mathrm{Sym}(\Omega)$ is the direct product of arbitrarily chosen regular cyclic subgroups $P_\Delta \le \mathrm{Aut}(\mathcal{Y}_\Delta)$, $\Delta \in \Phi(\mathcal{Y})$.

## 7.2 *Analysis of the Running Time*

In the proof of the theorem below, we present detailed explanations of the steps of the algorithm MS and analyze the running time of each of them.

**Theorem 8** *The algorithm MS terminates in time* poly$(n)$. *In particular, the size of its output is polynomially bounded.*

***Proof*** Let us verify successively that each step of the algorithm MS runs in polynomial time.
**Step 1**. Here the required statement follows from Lemma 4.

**Step 2**. At Step 2.1, we apply the Main Algorithm from [6] that finds a cycle base of a coherent configuration in polynomial time. In our case, the input for this algorithm is the coherent configuration $\mathcal{X}_{\Omega/E}$ of degree $p$ or $2p$. By the well-known upper bound for the size of a cyclic base (see [13]), we have

$$|C| \le |\Omega/E| \le 2p. \tag{19}$$

At Step 2.2, the coherent configuration $\mathcal{X}_c$ found by the Weisfeiler–Leman algorithm is isomorphic to a coherent configuration extending the wreath product $\mathcal{Y} = \mathcal{X}_1 \wr \mathcal{X}_2$, where $\mathcal{X}_1$ is a trivial coherent configuration of degree $n_E = 2$ or 4, and $\mathcal{X}_2$ is the trivial cyclotomic scheme over the group $\langle c\rangle$ of order $2p$ or $p$, respectively. Therefore, the group $K_c$ defined at Step 2.3 is isomorphic to a subgroup of the group

$$\mathrm{Aut}(\mathcal{Y}) = \mathrm{Aut}(\mathcal{X}_1) \wr \mathrm{Aut}(\mathcal{X}_2) \cong \mathrm{Sym}(n_E) \wr C_{n/n_E}, \tag{20}$$

which is solvable and can be constructed efficiently. This enables to construct in polynomial time the group $K_c$ by the Babai–Luks algorithm [2] and then its Sylow $p$-subgroup $P_c$ by the Kantor algorithm (see [16]). Thus, in view of inequality (19), Step 2 terminates in time $\mathrm{poly}(n)$.

**Step 3**. This step involves the WL-algorithm to construct the coherent configuration $\mathcal{Y}$ and the Main Algorithm from [6] to find a cycle base $C_\Delta$ of the group $\mathrm{Aut}(\mathcal{Y}_\Delta)$: the coherent configuration $\mathcal{Y}_\Delta$ is circulant if and only if the set $C_\Delta$ is not empty. Since both the algorithms run in time $\mathrm{poly}(n)$, we are done.

**Step 4**. To verify whether the coherent configuration $\mathcal{Y}$ is quasi-trivial, it suffices to make use of Lemma 1. Next, assume that $\mathcal{Y}$ is quasi-trivial. Then one can efficiently find the decomposition (5) of the coherent configuration $\mathcal{Y}$ into the direct sum and enumerate the fibers of $\mathcal{Y}$ so that

$$\Phi(\mathcal{Y}) = \{\Delta_{ij} : i = 1, \ldots, m, \ j = 1, \ldots, m_i\}.$$

In our case, $1 \le m \le 4$ and $\sum_i m_i = n/n_E$. For each $i$, one can efficiently find a permutation $g_i \in \mathrm{Sym}(\Delta_{i1})$, which is the product of $n_E/p$ disjoint $p$-cycles. Then applying the second part of Theorem 3 for $\mathcal{X} = \mathcal{Y}$, we see that the permutation $g$ defined by formula (6) is an automorphism of the coherent configuration $\mathcal{Y}$. In particular, the group $P = \langle g \rangle$ is contained in $\mathrm{Reg}(\mathrm{Aut}(\mathcal{Y}), C_p)$.

**Step 5**. Here, we define $P_\Delta$ as the group generated by an arbitrary full cycle belonging to the set $C_\Delta$ found at Step 3. Then

$$|K'| = \prod_{\Delta \in \Omega/E} |P_\Delta| = |\Delta|^{|\Omega/E|} \le n^4.$$

Thus, all the elements of order $p$ in $K'$ and hence the set $B_p$ can be found in time $\mathrm{poly}(n)$. □

## 7.3 *The Correctness of the MS*

We keep the notation of Sect. 7.1 and denote by $\mathcal{K}_p$ the class of coherent configurations isomorphic to a Cayley scheme over a group $G = E_4 \times C_p$. Set

$$\mathrm{Reg}_p(K) = \{H_p : \ H \in \mathrm{Reg}(K, G)\},$$

where $H_p$ is the Sylow $p$-subgroup of the group $H$.

**Theorem 9** *Let $\mathcal{X}$ be a coherent configuration on $4p$ points, and let the group $K$ and set $B_p$ be as in the Main Subroutine. Then if $\mathcal{X} \in \mathcal{K}_p$, then $B_p$ is not empty and every group in $\mathrm{Reg}_p(K)$ has a $K$-conjugate in $B_p$. In particular, the set $B_p$ is representative.*

**Corollary 2** *If $B_p = \varnothing$, then $\mathcal{X} \notin \mathcal{K}_p$.*

**Proof of Theorem** 9. Without loss of generality, we may assume that $\mathcal{X}$ is a Cayley scheme over $G = E_4 \times C_p$. Then by Lemma 5, the set $\mathcal{E} = \mathcal{E}(X)$ contains a principal equivalence relation $E$. Therefore, at Step 1 the algorithm MS does not terminate. Let us verify that every group

$$sQ \in \mathrm{Reg}_p(K)$$

has a $K$-conjugate in $B_p$. To this end, let $H \in \mathrm{Reg}(K, G)$ be such that $Q = H_p$.

Assume first that $n_E \leq 4$. Then $n_E = 2$ or 4. Therefore, $H^{\Omega/E}$ is a regular cyclic group of order $n/n_E = rp$ with $r = 1$ or 2. Consequently, there exists a permutation $h \in H$ of order $rp$ such that $H^{\Omega/E}$ is generated by $h^{\Omega/E}$. In particular,

$$Q = H_p = \langle h \rangle_p.$$

On the other hand, $K^{\Omega/E} = \mathrm{Aut}(\mathcal{X}_{\Omega/E})$ by statement (1) of Theorem 7. Now, if $C$ is the cycle base found at Step 2.1, then there exists a permutation $k \in K$ and a full cycle $c \in C$ such that

$$(h^k)^{\Omega/E} = (h^{\Omega/E})^{k^{\Omega/E}} = c.$$

By the definition of the relation $s$ at Step 2.2, this immediately implies that $s^{h'} = s$, where $h' = h^k$. Since also $h' \in K$, we conclude that $h'$ belongs to the group $K_c$ constructed at Step 2.3. However, the order of $h'$ equals $rp$ (the order of $h$) and the order of $P_c$ is equal to $p$ (see formula (20)). Thus, there exists $k' \in K_c$ such that $(h'_p)^{k'} \in P_c$, where $h'_p$ is the $r$th power of $h'$. Now,

$$(Q)^{kk'} = (\langle h \rangle_p)^{kk'} = \langle (h')^{k'} \rangle_p = \langle (h'_p)^{k'} \rangle = P_c.$$

Thus, the set $B_p$ constructed at Step 2.3 contains the $K$-conjugate $P_c$ of the group $Q$, as required.

Assume that $n_E \geq p$. Then $n/n_E$ is less or equal to 4. Since $p \geq 5$, we conclude that $Q$ acts trivially on $\Omega/E$. It follows that

$$Q \leq \mathrm{Aut}(\mathcal{Y}), \tag{21}$$

where $\mathcal{Y}$ is the coherent configuration found at Step 3. Since $\mathcal{X}$ is a Cayley scheme over $G$, every fiber $\Delta \in \Phi(\mathcal{Y})$ is a class of the equivalence relation $E$, and hence $|\Delta| = n_E$. The group $\mathrm{Aut}(\mathcal{Y}_\Delta)$ contains a regular subgroup $G^\Delta$. This group is cyclic if $n_E \neq 4p$ and hence $\mathcal{Y}_\Delta$ is a circulant scheme in this case. Finally, $n_E = 4p$, then $\mathcal{Y}_\Delta$ is trivial and hence circulant. Thus, the algorithm MS does not terminate at Step 3.

Suppose first that the coherent configuration $\mathcal{Y}$ is quasi-trivial. Then in the notation of Theorem 3, we have

$$Q^{\Delta_{i1}} \leq \mathrm{Aut}(\mathcal{Y})^{\Delta_{i1}} = \mathrm{Sym}(\Delta_{i1}), \quad i = 1, \ldots, m.$$

However, $Q^{\Delta_{i1}}$ and $P^{\Delta_{i1}}$ are semiregular cyclic subgroups of order $p$ in the group $\mathrm{Sym}(\Delta_{i1})$, where $P$ is the group defined at Step 4. Therefore, there exists a permutation $g_i \in \mathrm{Sym}(\Delta_{i1})$ such that

$$(Q^{\Delta_{i1}})^{g_i} = P^{\Delta_{i1}}, \quad i = 1, \ldots, m.$$

Then $Q^g = P$, where the permutation $g \in K$ is defined by formula (6). By Theorem 3, we have $g \in \mathrm{Aut}(\mathcal{Y})$. Since the latter group is contained in $K$, the set $B_p$ constructed at Step 4 consists of the $K$-conjugate $P$ of the group $Q$, as required.

To complete the proof, we may assume that $\mathcal{Y}$ is not quasi-trivial. Then by statement (2) of Theorem 7, for each $\Delta \in \Phi(\mathcal{Y})$ the coherent configuration $\mathcal{Y}_\Delta$ is isomorphic to a proper cyclotomic scheme over $C_p$. In particular, we come to Step 5 with $|\Delta| = p$. Moreover,

$$\mathrm{Reg}(\mathrm{Aut}(\mathcal{Y}_\Delta), C_p) = \{P_\Delta\},$$

where $P_\Delta$ is the group found at Step 5. By inclusion (21), this implies that

$$\mathrm{Orb}(Q, \Omega) = \Phi(\mathcal{Y}).$$

Therefore, $Q^\Delta \in \mathrm{Reg}(\mathrm{Aut}(\mathcal{Y}_\Delta), C_p)$ and hence $Q^\Delta = P_\Delta$ for all $\Delta$. This implies that

$$Q \le \prod_{\Delta\in\Omega/E} Q^\Delta = \prod_{\Delta\in\Omega/E} P_\Delta = K',$$

where $K'$ is the group defined at Step 5. Thus, $Q$ is contained in the set $B_p$ found at this step and we are done. □

## 8 Proof of Theorem 2

Let $\mathcal{X}$ be a coherent configuration constructed from the graph $X$ by the WL-algorithm. Then $X$ is isomorphic to a Cayley graph over $G$ if and only if $\mathcal{X}$ is isomorphic to a Cayley scheme over $G$. Since also $\mathrm{Aut}(X) = \mathrm{Aut}(\mathcal{X}) := K$, it suffices to check in time $\mathrm{poly}(n)$ whether the set $\mathrm{Reg}(K, G)$ is not empty, and (if so) find an element of this set within the same time.

Without loss of generality, we assume $p \ge 5$. Let $B_p \subseteq \mathrm{Reg}(K, C_p)$ be the set constructed by the algorithm MS applied to the coherent configuration $\mathcal{X}$. By Theorem 8, this can be done in time $\mathrm{poly}(n)$. If the set $B_p$ is empty, then $\mathcal{X} \notin \mathcal{K}_p$ by Corollary 2 and hence the set $\mathrm{Reg}(K, G)$ is also empty. Thus, we may assume that $\mathcal{X} \in \mathcal{K}_p$ and hence $B_p \ne \varnothing$. For each group $P \in B_p$, we define a set

$$R(P) = \{H \in \mathrm{Reg}(C_P \cap K, G) : \ P \le H\}$$

of regular subgroups of $\mathrm{Sym}(\Omega)$, where $C_P$ is the centralizer of $P$ in $\mathrm{Sym}(\Omega)$. Note that the group $C_P$ is permutation isomorphic to the wreath product $C_p \wr \mathrm{Sym}(4)$. Therefore,

$$|C_P| = |C_p \wr \mathrm{Sym}(4)| = 24p^4 \leq n^4, \tag{22}$$

and the group $C_P \cap K$ can be found in time $\mathrm{poly}(n)$ by testing each permutation of $C_P$ for membership to the group $K$. Next, every group $H \in R(P)$ is generated by $P$ and two involutions $x, y \in C_P \cap K$. Since the number of such pairs $(x, y)$ does not exceed $|C_P|^2$, the following statement is a consequence of inequality (22).

**Lemma 7** *Given a group $P \in B_p$, the set $R(P)$ can be found in time* $\mathrm{poly}(n)$. *In particular, $|R(P)| \leq n^c$ for a constant $c > 0$.*

This lemma shows that to complete the proof, we need to verify the implication

$$\mathrm{Reg}(K, G) \neq \varnothing \quad \Rightarrow \quad \bigcup_{P \in B_p} R(P) \neq \varnothing.$$

To this end, it suffices to prove that every group $V \in \mathrm{Reg}(K, G)$ is $K$-conjugate to a group belonging to $R(P)$ for some $P \in B_p$. However by the above assumption, $\mathcal{X} \in \mathcal{K}_p$ and $V_p \in \mathrm{Reg}_p(K)$, where $V_p$ denotes the Sylow $p$-group of $V$. Therefore, by Theorem 9, there exists $P \in B_p$ and $k \in K$ such that $(V_p)^k = P$. It follows that

$$V^k \leq (C_K(V_p))^k = C_K((V_p)^k) = C_K(P) \leq C_P,$$

where $C_K(V_p)$ denotes the centralizer of $V_p$ in $K$. Since also $V^k \leq K$ and $V^k \cong G$, we conclude that $V^k \in R(P)$, as required.

**Acknowledgements** The paper was finished during the research stay of the second author at the Faculty of Applied Sciences of the University of West Bohemia in October, 2016. The first author is supported by the grants APVV-15-0220, VEGA 1/0487/17, Project LO1506 of the Czech Ministry of Education, Youth, and Sports.

## References

1. L. Babai, Isomorphism problem for a class of point symmetric structures. Acta Math. Acad. Sci. Hung. **29**(3), 329–336 (1977). https://doi.org/10.1007/BF01895854
2. L. Babai, E.M. Luks, Canonical labeling of graphs, in *Proceedings 15th ACM STOC* (1983), pp. 171–183. https://doi.org/10.1145/800061.808746
3. S. Evdokimov, I. Kovács, I. Ponomarenko, On schurity of finite abelian groups. Commun. Algebra **44**(1), 101–117 (2016), https://doi.org/10.1.1.755.7799
4. S. Evdokimov, I. Ponomarenko, A new look at the Burnside-Schur theorem. Bull. Lond. Math. Soc. **37**, 535–546 (2005). https://doi.org/10.1112/S0024609305004340
5. S. Evdokimov, I. Ponomarenko, Characterization of cyclotomic schemes and normal Schur rings over a cyclic group. St. Petersburg Math. J. **14**(2), 189–221 (2003), https://doi.org/10.1.1.22.8137

6. S. Evdokimov, I. Ponomarenko, Circulant graphs: recognizing and isomorphism testing in polynomial time. St. Petersburg Math. J. **15**, 813–835 (2004). https://doi.org/10.1016/j.endm.2005.06.002
7. S. Evdokimov, I. Ponomarenko, Permutation group approach to association schemes. Eur. J. Combin. **30**(6), 1456–1476 (2009). https://doi.org/10.1023/A:1018672019177
8. S. Evdokimov, I. Ponomarenko, Schemes of a finite projective plane and their extensions. St. Petersburg Math. J. **21**(1), 65–93 (2010). https://doi.org/10.1090/S1061-0022-09-01086-3
9. S. Evdokimov, I. Ponomarenko, Schurity of S-rings over a cyclic group and generalized wreath product of permutation groups. St. Petersburg Math. J. **24**(3), 431–460 (2013). https://doi.org/10.1090/S1061-0022-2013-01246-5
10. M. Klin, R. Pöschel, The König problem, the isomorphism problem for cyclic graphs and the method of Schur rings, in *Algebraic Methods in Graph Theory, Szeged*. Colloq. Math. Soc. János Bolyai, Vol. 25 (North-Holland, Amsterdam, 1978), pp. 405–434. https://doi.org/10.1007/s10468-019-09859-7
11. I. Kovács, M. Muzychuk, The group $C_p^2 \times C_q$ is a CI-group. Commun. Algebra **37**(10), 3500–3515 (2009). https://doi.org/10.1080/00927870802504957
12. A. Lang, *A Classification of the Supercharacter Theories of* $C_p \times C_2 \times C_2$ *for Prime* $p$, pp. 1–28 (2016), arXiv:1609.07182 [math.RT]
13. M. Muzychuk, On the isomorphism problem for cyclic combinatorial objects. Discrete Math. **197**, 589–606 (1999). https://doi.org/10.1016/S0012-365X(99)90119-X
14. M. Muzychuk, I. Ponomarenko, Schur rings. Eur. J. Combin. **30**(6), 1526–1539 (2009). https://doi.org/10.1016/j.ejc.2008.11.006
15. G. Ryabov, *Separability of Schur Rings over Abelian p-Groups*. Algebra and Logic **57**(1), 49–68 (2018)
16. A. Seress, *Permutation Group Algorithms*, vol. 152. Cambridge Tracts Mathematics (Cambridge University Press, Cambridge, 2003). https://doi.org/10.1017/CBO9780511546549
17. B. Weisfeiler (ed.), *On Construction and Identification of Graphs*, vol. 558. Lecture Notes Mathematics (1976). https://doi.org/10.1.1.135.8587
18. H. Wielandt, *Finite Permutation Groups* (Academic Press, New York, 1964)

# Tatra Schemes and Their Mergings

**Sven Reichard**

**Abstract** Starting from an imprimitive action of the group $\mathrm{PSL}(2, q)$ we construct a series of the association schemes of rank $2r$ on $(q+1)r$ points, where $r$ divides $q-1$. By merging classes in these schemes we obtain several series of interesting schemes, among them $P$-polynomial schemes related to distance-regular graphs first described by Mathon, and non-commutative schemes of rank 6.

**Keywords** Association schemes · Distance-regular graphs · Projective groups

## 1 Introduction

This work originated from questions by Nevo and Thurston [12] related to their investigation of certain simplicial complexes. These structures are invariant under certain transitive actions of simple projective groups.

A description of these objects in terms of permutation groups and association schemes led to the investigation of such actions in general. Small cases were checked empirically using computer algebra systems such as COCO [5] and GAP [13].

A first success was obtained for $\mathrm{PSL}(2, 29)$ acting on the 210 cosets of $D_{29}$. The action admits seven isomorphic distance-regular graphs of valency 29, which were described by Mathon [10, 11]. It also admits 3-class association schemes with valencies $(1, 6, 3 \cdot 29, 4 \cdot 29)$, which appeared to be new.

Further examples were found, and the following pattern emerged:

**Theorem 1** (see Sect. 5) *Let $q$ be an odd prime power, $r > 1$ an odd divisor of $q-1$, and $d$ the size of a difference set in the cyclic group of order $r$.*

*Then there exists a 3-class association scheme of order $r(q+1)$ and valencies $(1, r-1, dq, (r-d)q)$.*

Such difference sets exist for example whenever $q$ is an odd prime and $d = (q-1)/2$. This leads to a general construction for rank 4 association schemes. It is given below.

S. Reichard (✉)
Institut für Algebra, TU Dresden, Dresden, Germany
e-mail: sven.reichard@freenet.de



G. A. Jones et al. (eds.), *Isomorphisms, Symmetry and Computations in Algebraic Graph Theory*, Springer Proceedings in Mathematics & Statistics 305,
https://doi.org/10.1007/978-3-030-32808-5_8

The proofs are computer-free. Due to the elaboration of this construction during a trip through the mountains of Slovakia, we call these objects *Tatra schemes*.

A further class of association schemes is obtained by merging and consists of non-commutative schemes of rank 6. Such schemes were studied by Hanaki and Zieschang [7]. In addition to three families of "classical" examples, there is only one known construction by Drabkin and French [4] which for each Mersenne prime $p$ gives a scheme of order $p(p+1)$. Our construction leads to different orders:

**Theorem 2** (see Sect. 7) *Let* $r \equiv 3 \mod 4$ *be a prime, let* $q = 1 + mr$ *be a prime power. Then there exists a non-commutative scheme of rank 6 and of order* $r(q+1)$.

After recalling preliminaries and fixing notation in Sect. 2 we describe an action of the group $\mathrm{PSL}(2, q)$ and the corresponding schurian association scheme in Sect. 3. Section 4 is devoted to a distance-regular basis graph of this scheme that was mentioned above. In Sect. 5 we construct new schemes by merging classes using cyclic difference sets.

In Sect. 6 we give a general construction of mergings based on S-rings over cyclic groups leading to non-commutative schemes of rank 6 which are described in Sect. 7. We finish with some discussion of the origin of these results.

# 2 Preliminaries

## 2.1 Association Schemes

A *d-class association scheme* $W$ is a finite set $\Omega$ together with a set $\mathcal{R} = \{R_0, \dots, R_d\}$ of binary relations such that

1. the $R_i$ form a partition of $\Omega^2$;
2. $R_0$ is the diagonal of $\Omega^2$;
3. with each relation $R_i$, its inverse $R_{i'}$ is also in $\mathcal{R}$;
4. there are integers $p_{ij}^k$ for $0 \le i, j, k \le d$ such that for any $(x, y) \in R_k$ there are exactly $p_{ij}^k$ points $z \in \Omega$ with $(x, z) \in R_i$ and $(z, y) \in R_j$.

The set $\Omega$ together with the relations $R_i$ can be considered as an edge coloring of the complete graph on $\Omega$. Therefore, the basic relations are also referred to as the *colors* of $W$.

We can represent binary relations by matrices. We index the rows and columns by elements of $\Omega$. The adjacency matrix $A_i$ of $R_i$ has the entry 1 in row $x$ and column $y$ if $(x, y) \in R_i$, and 0 otherwise. We usually consider these matrices in characteristic 0, for convenience over $\mathbb{C}$.

Then the axioms of an association scheme can be reformulated in terms of those matrices:

- $\sum A_i = J$, where $J$ is the all one matrix;
- $A_0 = I$, the identity matrix;

- the $A_i$ form a linear basis for a matrix algebra which is closed under transposition; the $p_{ij}^k$ are the structure constants of that algebra with respect to this basis. In other words,
$$A_i A_j = \sum p_{ij}^k A_k \qquad \text{for all } 0 \leq i,\ j, k \leq d.$$

This algebra is the adjacency algebra $\mathrm{A}(W)$ of $W$ (in the commutative case also called the Bose-Mesner algebra); the basis $\{A_i\}$ is its standard basis.

In relation to an association scheme $W$ we can consider several types of automorphisms:

- A (proper) automorphism of $W$ is a permutation of the points which leaves each relation $R_i$ invariant. The set of all automorphisms forms the automorphism group $\mathrm{Aut}(W)$.
- A color automorphism of $W$ is a permutation of the points which leaves the set $\mathcal{R}$ of relations invariant. It induces a permutation on $\mathcal{R}$. The set of all color automorphism forms the color automorphism group $\mathrm{CAut}(W)$.
- A permutation of $\mathcal{R}$ is an algebraic automorphism of $W$ if it preserves the structure constants. The set of these automorphisms forms the algebraic automorphism group $\mathrm{AAut}(W)$.

We can see that $\mathrm{Aut}(W)$ is a normal subgroup of $\mathrm{CAut}(W)$, and that the action of the latter on $\mathcal{R}$ is a subgroup of $\mathrm{AAut}(W)$.

A scheme $U$ is a merging of a scheme $W$ if $\mathrm{A}(U) \leq \mathrm{A}(W)$. We write $U \leq W$. In this case, the basic relations of $U$ are unions of basic relations of $W$.

Given a group $G$ acting transitively on a finite set $\Omega$ we call the orbits of $G$ on $\Omega^2$ the 2-orbits of $G$ on $\Omega$.

The following facts are easy to check:

**Lemma 1** – *The 2-orbits of a finite transitive permutation group $G$ form an association scheme $V(G)$.*
- *For any association scheme, $W \leq V(\mathrm{Aut}(W))$.*

In the case of equality in the second statement, $W$ is called *schurian.* There is a one-to-one correspondence between schurian schemes and 2-closed permutation groups, that is, automorphism groups of systems of binary relations. (In fact, schurian schemes and 2-closed permutation groups are the closed objects of a Galois correspondence between binary relations and permutations, see [5] for details.) Non-schurian schemes are purely combinatorial objects.

**Lemma 2** ([5]) *Let $\mathcal{T} = \{T_i\}$ be a partition of the colors of an association scheme $W = (\Omega, \{R_j\})$. Then it defines a merging of $W$ if and only if*

- *$\{0\} \in \mathcal{T}$;*
- *for $T_a, T_b, T_c \in \mathcal{T}$ and $k, l \in T_c$,*

$$\sum_{i \in T_a} \sum_{j \in T_b} p_{ij}^k = \sum_{i \in T_a} \sum_{j \in T_b} p_{ij}^l.$$

## 2.2 Difference Sets

Let $H$ be a finite group, written additively. A difference set [3] in $H$ is a subset $S \subseteq H$ with the property that each non-identity element of $H$ can be expressed in the same number $\lambda$ of ways as a difference of two elements in $S$. We set $n = |H|$, $d = |S|$, and speak of an $(n, d, \lambda)$ difference set.

We have the obvious arithmetical requirement that

$$\lambda(n-1) = k(k-1).$$

If $H$ is a cyclic group we speak of a cyclic difference set. There are two simple constructions for cyclic difference sets (see [1], Chap. 5):

Singer : Consider a (desarguesian) $n$-dimensional projective space over a finite field of order $q$. Then any hyperplane is a difference set in the Singer cycle in its action on projective points. The parameters are $\left(\frac{q^{n+1}-1}{q-1}, \frac{q^n-1}{q-1}, \frac{q^{n-1}-1}{q-1}\right)$;

cyclotomic : Let $q = 4n-1$ be a prime power. The non-zero squares in the field of order $q$ form a difference set in its additive group. The parameters are $(4n-1, 2n-1, n-1)$.

# 3 The Main Construction

Let $q$ be an odd prime power, and let $F$ be a field of order $q$. Let $F^*$ be the multiplicative group of $F$, $K$ a subgroup of $F^*$, and $m = |K|$, where $q = 1 + mr$ and $r$ is odd. Let $C = F^*/K$, thus $|C| = r$.

Let $V = F^2$ be a two-dimensional vector space over $F$, and $\Omega$ the equivalence classes of non-zero vectors modulo $K$, i.e., the non-zero $K$-orbits. So

$$\Omega = (V \setminus \{0\})/K = \{Kv \mid v \in V \setminus \{0\}\}$$

and

$$|\Omega| = (q^2-1)/m = r(q+1).$$

Note that for $r = 1$, $\Omega$ is the projective line over $F$. We will call the elements of $\Omega$ *quasi-projective points* or simply points.

Define the form

$$\begin{aligned} \langle \cdot, \cdot \rangle : \Omega \times \Omega &\to C \\ (Ku, Kv) &\mapsto K \det(u, v). \end{aligned}$$

Here, for the vectors $u = (u_1, u_2)^T$ and $v = (v_1, v_2)^T$,

$$\det(u, v) = \det \begin{pmatrix} u_1 & v_1 \\ u_2 & v_2 \end{pmatrix} = u_1 v_2 - u_2 v_1.$$

**Lemma 3** *The form $\langle \cdot, \cdot \rangle$ is well-defined and symmetric.*

***Proof*** Let $k_1, k_2 \in K$. Let $u, v \in V \setminus \{0\}$. Then

$$K \det(k_1 u, k_2 v) = K k_1 k_2 \det(u, v) = K \det(u, v).$$

It is symmetric since $|F^*/K|$ is odd and therefore $(-1) \in K$.

**Lemma 4** *1. If $u, v \in V$ are linearly independent, then $w \in V$ is uniquely determined by $\det(u, w)$ and $\det(v, w)$.*
*2. For $\alpha = Ku \in \Omega$ and $x \in C$ the equation $\langle \alpha, \beta \rangle = x$ has exactly $q$ solutions.*

***Proof*** If for fixed vectors $u, v \in V$ and an unknown vector $w \in V$ we know that $\det(u, w) = x$, $\det(v, w) = y$, for some $x, y \in F$, we have

$$\begin{aligned} u_1 w_2 + u_2(-w_1) &= x \\ v_1 w_2 + v_2(-w_1) &= y \end{aligned}$$

If the vectors $u, v$ are linearly independent then this system of linear equations has rank 2 and hence a unique solution; therefore, $w$ is uniquely determined.

For (2) pick a vector $w$ which is linearly independent of $u$. For any $k \in K$ and any $x \in F$ there is a unique $v$ with $\det(u, v) = kx$ and $\det(w, v) = x$. So there are $mq$ vectors $v$ with $\langle Ku, Kv \rangle = x$; these correspond to $q$ points in $\Omega$.

If $\langle Ku, Kv \rangle = 0$, then $Kv = xKu$ for some $x \in C$. This allows us to define two binary relations on $\Omega$ for each such $x$:

- $(\alpha, \beta) \in R_x$ if and only if $\langle \alpha, \beta \rangle = 0$ and $\beta = x\alpha$;
- $(\alpha, \beta) \in S_x$ if and only if $\langle \alpha, \beta \rangle = x$.

In what follows we will denote by $A_x = A(R_x)$ and $B_x = A(S_x)$ the adjacency matrices of those relations.

**Theorem 3** *The set $\Omega$ together with the relations $R_x$ and $S_x$ for $x \in C$ form an association scheme $W$, with valencies $1^r, q^r$.*

*In particular, for $x, y \in C$:*

$$\begin{aligned} A_1 &= I_\Omega \\ \sum_{x \in C} (A_x + B_x) &= J_\Omega \\ A_x A_y &= A_{xy} \\ A_x B_y &= B_{xy} \end{aligned}$$

$$B_x A_y = B_{y/x}$$
$$B_x B_y = q A_{y/x} + m \sum_{z \in C} B_z$$

***Proof*** The first two equations are obvious. The statement on the valencies follows from Lemma 4. In what follows, let $\alpha = Ku$, $\beta = Kv$, and $\gamma = Kw$ be three points in $\Omega$, and let $x, y, z \in C$.

Assume that $(\alpha, \beta) \in R_x$, $(\beta, \gamma) \in R_y$, $(\alpha, \gamma) \in R_z$. Then $\gamma = z\alpha = y\beta = xy\alpha$, hence $z = xy$, and $\beta$ is uniquely determined by $\alpha, \gamma, x, y$. Therefore,

$$A_x A_y = A_{xy}.$$

Let $(\alpha, \gamma) \in S_z$. Then $\langle \alpha, \gamma \rangle = z$. If $(\alpha, \beta) \in S_x$ and $(\beta, \gamma) \in R_y$, then

$$\langle \alpha, \beta \rangle = x$$
$$\gamma = y\beta.$$

This determines $\beta$ uniquely, and $z = \langle \alpha, y\beta \rangle = xy$. Hence,

$$B_x A_y = B_{xy}.$$

Similarly, let $\alpha, \gamma \in S_z$. If $(\alpha, \beta) \in R_x$ and $(\beta, \gamma) \in S_y$, then $\beta = x\alpha$, and $y = \langle x\alpha, \gamma \rangle = xz$. So $z = y/x$, and thus

$$A_x B_y = B_{x/y}.$$

Let $(\alpha, \gamma) \in R_z$, or $\gamma = z\alpha$. We want to count the choices of $\beta$ which are linearly independent from $\alpha$, and thus from $\gamma$. So $(\alpha, \beta) \in S_x$, $(\beta, \gamma) \in S_y$. We have

$$\begin{aligned} x &= \langle \alpha, \beta \rangle \\ y &= \langle \beta, \gamma \rangle \\ &= \langle \beta, z\alpha \rangle \\ &= z \langle \alpha, \beta \rangle \\ &= xz. \end{aligned}$$

Hence $z = y/x$. By Lemma 4 there are exactly $q$ choices for $\beta$.

Finally, let $\langle \alpha, \gamma \rangle = z$. We want to count the number of choices for $\beta$ with

$$\begin{aligned} \langle \alpha, \beta \rangle &= x \\ \langle \beta, \gamma \rangle &= y. \\ \det(u, v) &\in Kx \\ \det(v, w) &\in Ky. \end{aligned}$$

Since $|K| = m$ we have $m^2$ choices for $v$, leading to exactly $m$ different choices for $\beta = Kv$.

Therefore,

$$B_x B_y = q A_{y/x} + m \sum_{z \in C} B_z.$$

To check that the scheme is closed under transposition we note that the matrices $B_x$ are symmetrical, and that

$$A_x^T = A_{x^{-1}}.$$

Hence, $W$ is an association scheme of rank $2r$.

## 4 Distance-Regular Basis Graphs of $W$

For the theory of distance-regular graphs, we refer to [2].

The basic relations $S_x$ for $x \in C$ are symmetric and antireflexive. In fact, we have the following:

**Theorem 4** *For $x \in C$ the graph $(\Omega, S_x)$ is a distance-regular, antipodal cover of the complete graph. Its intersection array is*

$$(q, (r-1)m, 1; 1, m, q)$$

***Proof*** The graphs $(\Omega, S_x)$ are all isomorphic, so let $x = 1$. Let

$$\begin{aligned} D_0 &= A_1 \\ D_1 &= B_1 \\ D_2 &= \sum_{x \neq 1} B_x \\ D_3 &= \sum_{x \neq 1} A_x \end{aligned}$$

Then Theorem 3 gives us

$$D_1^2 = B_1^2 = q A_1 + m \sum_x B_x = q D_0 + m D_1 + m D_2$$

$$D_1 D_2 = \sum_{x \neq 1} B_1 B_x = \sum_{x \neq 1} \left( q A_x + m \sum_y B_y \right) = q D_3 + (r-1)m(D_1 + D_2)$$

$$D_1 D_3 = \sum_{x \neq 1} B_1 A_x = \sum_{x \neq 1} B_x = D_2.$$

So $(\Omega, B_1)$ is distance-regular, and the intersection array can be read off the identities.

The matrices $D_i$ from the proof of Theorem 4 correspond to the binary relations of points at distance $i$ in the graph. Being at distance 3 is an equivalence relation; its equivalence classes are called *antipodal classes* of the graph.

Let $\mathcal{A}$ be the rank 4 scheme generated by $B_1$.

**Proposition 1** *The eigenvalue matrix of $\mathcal{A}$ is given by*

$$P = \begin{pmatrix} 1 & q & (r-1)q & r-1 \\ 1 & -1 & 1-r & r-1 \\ 1 & \sqrt{q} & -\sqrt{q} & -1 \\ 1 & -\sqrt{q} & \sqrt{q} & -1 \end{pmatrix}.$$

(We remark that the multiplicities of the eigenvalues can be computed as $m_0 = 1$, $m_1 = q$, $m_2 = m_3 = (r-1)(q+1)/2$.)

***Proof*** The intersection algebra of $\mathcal{A}$ is generated by the matrices $M_j$ which are formed by the intersection numbers:

$$(M_j)_{ik} = p_{ij}^k, \qquad 0 \le i, j, k \le 3.$$

According to Theorem 1.1.1 in Faradžev et al. [6], the map $D_j \mapsto M_j$ extends to an isomorphism of the generated algebras. In particular, the eigenvalues of $M_j$ and $D_j$ are the same. In our case, we have

$$M_0 = \begin{pmatrix} 1 & 0 & 0 & 0 \\ 0 & 1 & 0 & 0 \\ 0 & 0 & 1 & 0 \\ 0 & 0 & 0 & 1 \end{pmatrix}$$

$$M_1 = \begin{pmatrix} 0 & q & 0 & 0 \\ 1 & m & (r-1)m & 0 \\ 0 & m & (r-1)m & 1 \\ 0 & 0 & q & 0 \end{pmatrix}$$

$$M_2 = \begin{pmatrix} 0 & 0 & (r-1)q & 0 \\ 0 & (r-1)m & (r-1)^2 m & r-1 \\ 1 & (r-1)m & (r-1)^2 m & r-2 \\ 0 & q & (r-2)q & 0 \end{pmatrix}$$

$$M_3 = \begin{pmatrix} 0 & 0 & 0 & r-1 \\ 0 & 0 & r-1 & 0 \\ 0 & 1 & r-2 & 0 \\ 1 & 0 & 0 & r-2 \end{pmatrix}$$

These are the common eigenvectors of those matrices:

$$v_0 = (1, 1, 1, 1)^T$$
$$v_1 = (q, -1, -1, q)^T$$

$$v_2 = ((r-1)\sqrt{q}, r-1, -1, -\sqrt{q})^T$$
$$v_3 = ((r-1)\sqrt{q}, 1-r, 1, -\sqrt{q})^T.$$

The eigenvalues are given by $M_j \cdot v_i = p_j(i)v_i = a_{ij}v_i$, where

$$a_{ij} = \begin{pmatrix} 1 & q & (r-1)q & r-1 \\ 1 & -1 & 1-r & r-1 \\ 1 & \sqrt{q} & -\sqrt{q} & -1 \\ 1 & -\sqrt{q} & \sqrt{q} & -1 \end{pmatrix}.$$

## 5 A Rank 4 Merging Based on Difference Sets

The group $C = F^*/K$ is cyclic. Let us define a merging in $W$ by considering a difference set $D$ in this group.

**Theorem 5** *Let $D$ be an $(r, d, \lambda)$-difference set in $C$. Let*

$$E_0 = A_1$$
$$E_1 = \sum_{x \neq 1} A_x$$
$$E_2 = \sum_{x \in D} B_x$$
$$E_3 = \sum_{x \notin D} B_x.$$

*Then $E_0, E_1, E_2, E_3$ form the Bose-Mesner algebra of an association scheme with valencies $1, r-1, qd, q(r-d)$.*

***Proof*** This is a merging of the original scheme, so we have that $E_0 = I$ and $\sum E_i = J$. Moreover, all basic matrices are symmetric, and hence it is sufficient to compute products $E_iE_j$ with $1 \leq i \leq j$. The identities in Theorem 3 give us

$$\begin{aligned} E_1 \cdot E_1 &= \sum_{x \neq 1} \sum_{y \neq 1} A_x \cdot A_y \\ &= \sum_{x \neq 1} \sum_{y \neq 1} A_{xy} \\ &= (r-1)A_1 + (r-2)\sum_{x \neq 1} A_x \\ &= (r-1)E_0 + (r-2)E_1 \end{aligned}$$

$$\begin{aligned}
E_1 \cdot E_2 &= \sum_{x \neq 1} \sum_{y \in D} A_x B_y \\
&= \sum_{x \neq 1} \sum_{y \in D} B_{xy} \\
&= \sum_{x \in C} \sum_{y \in D} B_{xy} - \sum_{y \in D} B_y \\
&= d \sum_{x \in C} B_x - \sum_{x \in D} B_x \\
&= (d-1)E_2 + dE_3.
\end{aligned}$$

$$\begin{aligned}
E_1 \cdot E_3 &= E_1 \cdot (E_2 + E_3) - E_1 \cdot E_2 \\
&= \sum_{x \neq 1} \sum_{y \in C} A_x B_y - ((d-1)E_2 + dE_3) \\
&= (r-1) \sum_{x \in C} B_x - (d-1)E_2 - dE_3 \\
&= (r-d)E_2 + (r-d-1)E_3.
\end{aligned}$$

$$\begin{aligned}
E_2 \cdot E_2 &= \sum_{x \in D} \sum_{y \in D} B_x B_y \\
&= \sum_{x \in D} \sum_{y \in D} \left( q A_{y/x} + m \sum_{z \in C} B_z \right) \\
&= dqA_1 + \lambda q \sum_{x \neq 1} A_x + d^2 m \sum_{z \in C} B_z \\
&= dqE_0 + \lambda q E_1 + d^2 m (E_2 + E_3).
\end{aligned}$$

Similarly, we compute

$$\begin{aligned}
E_2 \cdot E_3 &= (d-\lambda)qE_1 + dm(r-d)(E_2+E_3) \\
E_3 \cdot E_3 &= (r-d)qE_0 + (r-2d+\lambda)qE_1 + (r-d)(q-1-dm)(E_2+E_3).
\end{aligned}$$

Hence the algebra generated by $E_0$, $E_1$, $E_2$, $E_3$ is closed under multiplication, and we get an association scheme.

In the small cases that were investigated by computer, the automorphism group of the resulting scheme is equal to the automorphism group of the original scheme $W$, hence the 3-class scheme is non-schurian. Our conjecture is that this holds in general.

## 6 Mergings Based on S-Rings

Let $G$ be a finite group. Recall that the group ring $\mathbb{Z}[G]$ consists of all formal sums $\sum_{g\in G}\alpha_g \cdot g$, where $\alpha_g \in \mathbb{Z}$. For a set $T \subseteq G$ we let $\underline{T} = \sum_{g\in T} 1 \cdot g$, and we set $\underline{g} := \underline{\{g\}}$.

We identify $g$ and $\underline{g}$ which allows us to consider each element of the group ring as a linear combination of group elements. This gives us a multiplication on $\mathbb{Z}[G]$: Set $\underline{g} \cdot \underline{h} = \underline{g \cdot h}$, and extend this to all ring elements using distributivity. This, in fact, yields a ring that is commutative if and only if $G$ is abelian.

As was mentioned before, $C = F^*/K$ is a cyclic group of order $r$. An S-ring $\mathcal{A}$ over a group $C$ is a partition $C = T_0 \cup \cdots \cup T_s$ such that in the group ring $\mathbb{Z}[G]$ we have $\underline{T_0} = 1$,

$$\underline{T_i} \cdot \underline{T_j} = \sum_k p_{ij}^k \underline{T_k}$$

for all $i$, $j$ between 0 and $s$, and $\sum_{g\in T_i} g^{-1} = \underline{T_{i'}}$ for some $i'$. $\mathcal{A}$ is commutative.

Given an S-ring $\mathcal{A}$ over $C$ we define $W_{\mathcal{A}}$ as follows: For $0 \le i \le s$ we set

$$C_i = \sum_{x\in T_i} A_x$$

$$D_i = \sum_{x\in T_i} B_x$$

Here, the matrices $A_x$ and $B_x$ are the basis matrices of the adjacency algebra of the scheme $W$ defined in Theorem 3.

**Theorem 6** *The matrices $C_i$ and $D_i$, $0 \le i \le s$, form the standard basis of the adjacency algebra of a scheme $W_{\mathcal{A}}$ of rank $2(s+1)$.*

*The valencies of $W$ are $|T_i|$ and $q|T_i|$, for $0 \le i \le s$.*

Thus, $W_{\mathcal{A}}$ is a merging of the scheme $W$ from Sect. 3.

***Proof*** Since the relations corresponding to $A_i$ and $B_i$ have degrees 1 and $q$, respectively, we obtain the stated valencies for $W$ from the definition of $C_i$ and $D_i$.

It is clear that $I = C_0$ and $J = \sum_i (C_i + D_i)$.

In Theorem 3 we have seen that $A_x A_y = A_{xy}$ for $x, y \in C$. Therefore, $x \mapsto A_x$ is a group homomorphism from $C$ to the group $\langle A_i \rangle$. This homomorphism can be extended linearly to a ring homomorphism $\varphi : \mathbb{Z}[C] \to W$.

We have

$$\varphi(T_i) = C_i.$$

Next, we recall that for all $x$,

$$B_x = A_x B_1 = B_1 A_{1/x}$$

and hence,

$$D_i = C_i B_1 = B_1 C_{i'}.$$

We compute the pairwise products of the given matrices. Let $i$, $j$ be arbitrary. Then:

$$\begin{aligned}
C_i \cdot C_j &= \left(\sum_{x\in T_i} A_x\right) \cdot \left(\sum_{y\in T_j} A_y\right)\\
&= \sum_{x\in T_i}\sum_{y\in T_j} A_x A_y\\
&= \sum_{x\in T_i}\sum_{y\in T_j} A_{xy}\\
&= \sum_{x\in T_i}\sum_{y\in T_j} \varphi(xy)\\
&= \varphi\left(\sum_{x\in T_i}\sum_{y\in T_j} xy\right)\\
&= \varphi\left(T_i \cdot T_j\right)\\
&= \varphi\left(\sum_k p_{ij}^k T_k\right)\\
&= \sum_k p_{ij}^k \varphi(T_k)\\
&= \sum_k p_{ij}^k C_k,
\end{aligned}$$

where the $p_{ij}^k$ are the structure constants of the S-ring, and also $\mathcal{A}$.

$$C_i \cdot D_j = C_i C_j B_1 = \sum_k p_{ij}^k C_k B_1 = \sum_k p_{ij}^k D_k.$$

$$\begin{aligned}
D_i \cdot C_j = B_1 C_{i'} C_j &= B_1 \sum_k p_{i'j}^k C_k = \sum_k p_{i'j}^k B_1 C_k\\
&= \sum_k p_{i'j}^k D_{k'} = \sum_k p_{i'j}^{k'} D_k.
\end{aligned}$$

$$\begin{aligned}
D_i D_j &= B_1 C_{i'} C_j B_1 \\
&= B_1 \left( \sum_k p^k_{i'j} C_k \right) B_1 \\
&= B_1^2 \sum_k p^{k'}_{i'j} C_k \\
&= \left( q A_1 + m \sum_x B_x \right) \left( \sum_k p^{k'}_{i'j} C_k \right) \\
&= \left( q A_1 + m \sum_x B_1 A_{1/x} \right) \left( \sum_k p^{k'}_{i'j} C_k \right) \\
&= \left( q A_1 + m B_1 \sum_x A_x \right) \left( \sum_k p^{k'}_{i'j} C_k \right) \\
&= \sum_k q p^{k'}_{i'j} C_k + m B_1 \sum_k p^{k'}_{i'j} \sum_x A_x C_k \\
&= \sum_k q p^{k'}_{i'j} C_k + m B_1 \sum_k p^{k'}_{i'j} \sum_l C_l C_k \\
&= \sum_k q p^{k'}_{i'j} C_k + m B_1 \sum_k p^{k'}_{i'j} |T_k| \sum_l C_l \\
&= \sum_k q p^{k'}_{i'j} C_k + \sum_l \left( m \sum_k p^{k'}_{i'j} |T_k| \right) B_1 C_l \\
&= \sum_k q p^{k'}_{i'j} C_k + \sum_l \left( m \sum_k p^{k'}_{i'j} |T_k| \right) D_{l'} \\
&= \sum_k q p^{k'}_{i'j} C_k + \sum_l \left( m \sum_k p^{k'}_{i'j} |T_k| \right) D_l
\end{aligned}$$

So the linear span of the $C_i$ and the $D_i$ is closed under multiplication and we get a merging of the scheme $W$.

**Corollary 1** *If the merging $W_{\mathcal{A}}$ is commutative then $p^k_{ij} = p^{k'}_{ij}$ for all $i$, $j$, $k$.*

***Proof*** We have seen above that

$$\begin{aligned}
D_i D_j &= \sum_k q p^{k'}_{i'j} C_k + \sum_l \left( m \sum_k p^{k'}_{i'j} |T_k| \right) D_l \\
D_j D_i &= \sum_k q p^{k'}_{j'i} C_k + \sum_l \left( m \sum_k p^{k'}_{j'i} |T_k| \right) D_l
\end{aligned}$$

Since the matrices $C_k$ and $D_k, 0 \leq k \leq s$ are linearly independent we get that $D_i D_j = D_j D_i$ if and only if $p^k_{i'j} = p^k_{j'i}$ for all $k$.

Replacing $i'$ and $k'$ by $i$ and $k$ we get

$$p^k_{ij} = p^k_{j'i'} = p^{k'}_{ij}.$$

# 7 Non-commutative Schemes of Rank 6

It is a well-known elementary fact that any group of the order less than 6 is abelian, and that there is up to isomorphism exactly one non-abelian group of order 6. As schemes are generalizations of groups it is not surprising that any scheme of rank at most 5 is commutative ([9], Sect. 4.1). However, there are infinitely many non-commutative schemes of rank 6, and their classification is a natural and interesting problem that is still open.

These schemes have been studied systematically by Hanaki and Zieschang [7]. Herman et al [8] recently investigated feasible parameter sets. A class of non-standard examples was given by Drabkin and French [4]. The result of the previous section allows us to construct another class of examples.

Let $p$ be a prime, $p \equiv 3 \mod 4$. Consider the additive cyclic group $G$ of order $p$. We define three subsets of $G$: $T_0 = \{0\}$, $T_1 = \{x^2 \mid x \neq 0\}$ and $T_2 = G \setminus (T_0 \cup T_1)$. This gives us a non-symmetric scheme of rank 3, also known as a doubly regular tournament.

**Theorem 7** ([2]) *The basic quantities $T_0, T_1, T_2$ form an S-ring over $G$ with the following structure constants:*

| | 0 | 1 | 2 | Valency |
|---|---|---|---|---|
| | 1 | 0 | 0 | |
| 0 | 0 | 1 | 0 | 1 |
| | 0 | 0 | 1 | |
| | 0 | 0 | 2n + 1 | |
| 1 | 1 | n | n | 2n +1 |
| | 0 | n + 1 | n | |
| | 0 | 2n + 1 | 0 | |
| 2 | 0 | n | n + 1 | 2n + 1 |
| | 1 | n | n | |
| Valency | 1 | 2n + 1 | 2n + 1 | 4n + 3 |

*where $p = 4n + 3$.*

The non-trivial basis graphs of these S-rings are also known as Paley tournaments.

**Corollary 2** *Let $r \equiv 3 \mod 4$ be a prime, let $q = 1 + mr$ be a prime power. Then there exists a non-commutative scheme of rank 6 and of order $r(q + 1)$.*

***Proof*** Applying the construction from Sect. 6 to Paley tournaments gives us a scheme $W_{\mathcal{A}}$ of rank 6. For $i = j = k = 1$ we have $k' = 2$, and $p_{ij}^{k} = n \neq n + 1 = p_{ij}^{k'}$. Hence $W_{\mathcal{A}}$ is non-commutative by Corollary 1.

If $r = 4n + 3$, then Theorem 6 gives us the subdegrees $1, 2n + 1, 2n + 1, q$, $q(2n + 1), q(2n + 1)$. The relation of valency $q$ corresponds to the distance-regular graph $\Gamma$ considered earlier. In particular, the automorphism group of the rank 6 merging is a subgroup of the automorphism group of the distance-regular graph $\Gamma$.

It appears that the schemes constructed in Corollary 2 are non-schurian. More specifically, the automorphism group of the distance-regular basis graph of valency $q$ has rank at least $r$, and it has rank $2r$ if $q$ is a prime. This has been verified for all small examples with less than 1000 points. Examining the results leads us to the following:

*Conjecture 1* – The action of $Aut(\Gamma)$ on the antipodal classes is faithful.
– If $q$ is a prime then any automorphism of $\Gamma$ fixing three classes are trivial.

In fact, we can prove the first part. It would follow from the second part — which holds for all small examples — that $Aut(\Gamma)$ contains $PSL(2, q)$ with index at most 2. This would imply that its rank is at least $r$, and this gives non-Schurity.

## 8 Discussion

This project has had quite a long history. E. Nevo and W. Thurston studied certain simplicial complexes invariant under an action of projective groups. After Thurston's death, Nevo approached M. Klin, who looked at it under the aspect of association schemes.

Klin's student D. Kalmanovich studied these schemes and performed numerous experiments. At some point, the author was involved. However, a theoretical explanation of the abundance of mergings did not appear.

The numerical evidence led Klin to suggest a connection to cyclic difference sets, which led to the construction in Sect. 5. It reached its final form on a bus trip from Banská Bystrica to Poprad in Slovakia which led from the Low Tatra to the High Tatra mountain ranges. For this reason, we like to refer to these schemes as 'Tatra schemes'.

Non-commutative schemes of rank 6 are studied by Hanaki and Zieschang [7]. There are three classes of 'classical' examples which are are well-understood:

1. Coxeter schemes corresponding to spherical buildings of rank 2;
2. Semidirect products with kernels of rank 3;
3. Schemes with a thin normal closed subset of order 3.

Drabkin and French [4] have given an additional construction of schemes of order $p(p + 2)$, where $p$ is a Mersenne prime. It is interesting to note that this is similar to the order of the schemes constructed here: In fact, let $q = p + 1, m = 1, r = p$, then

$q$ is a power of 2, $q = 1 + rm$, and $p(p + 2) = r(q + 1)$, corresponding to the order of the Tatra schemes. However, since we require $q$ to be an odd prime power, this is a new construction. It remains to be investigated if there are additional connections and whether a unified construction can be given.

The author wishes to thank M. Klin for the initial problem as well as motivation and input along the way. He also wants to thank D. Kalmanovich, S. Gyürki, I. Ponomarenko, M. Muzychuk and P.-H. Zieschang for helpful discussion. Ilia Ponomarenko in particular helped a lot with the presentation of the material.

Parts of this project have been funded under the project: Mobility—enhancing research, science, and education at the Matej Bel University, ITMS code 26110230082.

## References

1. L.D. Baumert, *Cyclic difference sets*, vol. 182. Lecture Notes in Mathematics (Springer, Berlin, 1971)
2. A.E. Brouwer, A.M. Cohen, A. Neumaier, *Distance-Regular Graphs* (Springer, Berlin, 1989)
3. R.H. Bruck, Difference sets in a finite group. Trans. Am. Math. Soc. **78**, 464–481 (1955)
4. B. Drabkin, C. French, On a class of non-commutative imprimitive association schemes of rank 6. Commun. Algebra **43**(9), 4008–4041 (2015)
5. I. Faradzev, M. Klin, Computer package for computations with coherent configurations, in *Proceedings of ISSAC '91* ed. by S.M. Watt (1991), pp. 219–223
6. I.A. Faradžev, M.H. Klin, M.E. Muzichuk, Cellular rings and groups of automorphisms of graphs, in *Investigations in Algebraic Theory of Combinatorial Objects*, vol. 84. Mathematics Application (Soviet Ser.) (Kluwer Academic Publishers, Dordrecht, 1994), pp. 1–152
7. A. Hanaki, P.-H. Zieschang, On imprimitive noncommutative association schemes of order 6. Commun. Algebra **42**(3), 1151–1199 (2014)
8. A. Herman, M. Muzychuk, B. Xu, Noncommutative reality-based algebras of rank 6 (2016), https://arxiv.org/abs/1608.08463
9. D.G. Higman, Coherent configurations I. Ordinary representation theory. Geometriae Dedicata **4**(1), 1–32 (1975)
10. R. Mathon, 3-class association schemes, in *Proceedings of the Conference on Algebraic Aspects of Combinatorics (Univ. Toronto, Toronto, Ont., 1975)* (Congressus Numerantium, No. XIII. Utilitas Math., Winnipeg, Man., 1975), pp. 123–155
11. R. Mathon, Lower bounds for Ramsey numbers and association schemes. J. Comb. Theory B **42**, 122–127 (1987)
12. E. Nevo, Polyhedrons and PBIBDs from hyperbolic manifolds (2013), arXiv:1308.3651v1
13. The GAP Group. *GAP – Groups, Algorithms, and Programming, Version 4.7.8* (2015), http://www.gap-system.org

Printed by Printforce, the Netherlands